Environmental Microbiology of Aquatic and Waste Systems

Nduka Okafor

Environmental Microbiology of Aquatic and Waste Systems

 Springer

Nduka Okafor
Department of Biological Sciences
Clemson University
Clemson, South Carolina
USA
nokafor@g.clemson.edu

The Author acknowledges that there are instances where he was unable to trace or contact the copyright holder for permission to reproduce selected materials in this volume. The Author has included complete source references for all such materials and takes full responsibility for these matters. If notified, the publisher will be pleased to rectify any errors or omissions at the earliest opportunity.

ISBN 978-94-007-1459-5   e-ISBN 978-94-007-1460-1
DOI 10.1007/978-94-007-1460-1
Springer Dordrecht Heidelberg London New York

Library of Congress Control Number: 2011930674

Printed on acid-free paper

Springer is part of Springer Science+Business Media (www.springer.com)

This book is dedicated to
The Okafor-Ozowalu family of Nri, Anaocha Local Government Area,
in the Anambra State of Nigeria, especially grandchildren Chinelo,
Ndiya, Kenechukwu, Victor, Toochukwu and Chinedum
and
Very Rev Fr. Prof Emmanuel M P Edeh, CSSp, OFR,
for his contribution in furthering access to education
in Nigeria by founding Madonna and Caritas Universities
as well several other educational establishments

Preface

This book is a result of the course on environmental microbiology that I taught at the University of Northern Iowa, Cedar Falls, Iowa, USA. The book seeks to place the main actors in matters of environment microbiology, namely, the microorganisms, on center stage. Many first degree courses in microbiology do not cover the detailed taxonomy of microorganisms, which facilitates their being pigeonholed into specific taxonomic niches in the environment. Because of this, the book looks, as much as possible, at the biology, especially the taxonomy, of the microorganisms.

To this end, the taxonomy of not only bacteria and fungi, but also protozoa, algae, viruses and even the smaller macroorganisms, such as nematodes and rotifers, which are usually associated with microorganisms in aquatic and waste environments, are also looked at.

Viruses are given special attention because, until recently, it was thought they were not important in aquatic environments; the modern understanding is that they are not only abundant, but play a crucial role in the sustenance of the various components of the biological ecosystem in aquatic environments.

The book will be useful to a wide range of undergraduates and beginning graduate students in microbiology, general biology, aquatic science, public health and civil (sanitary) engineering. Practitioners in sanitary engineering and public health will also find it of interest.

I thank my family for their encouragement.

Mableton, GA, USA   Nduka Okafor
October 1, 2010

Contents

Part I

Introduction

Nature, Properties, and Distribution of Water

1

Abstract

The water molecule is composed of two hydrogen atoms attached to an oxygen atom with a V-shape formation and an angle of 105° between the hydrogen atoms. It has a slightly negative charge at the oxygen end and a slight positive charge at the hydrogen end. This makes the water molecule polar, i.e., having poles like a magnet. This property makes water molecules attract each other, thus giving it unique properties such as rising in capillary tubes, plant roots, and blood vessels against gravity. Water has a high latent heat enabling it to absorb large quantities of heat before a rise in temperature; this enables the waters of the oceans to affect the earth's temperature in a gradual manner. Water is an important solvent. Water forms about 71% of the earth's surface, but most of it is saline. Freshwater which is required for domestic and industrial use, and for agriculture forms only about 2.5%. Water is distributed in the atmosphere as clouds; on the earth's surface as oceans, seas, rivers, and lakes; and underground.

Keywords

Water properties • Water influence on environment • Hydrologic cycle • Global water distribution

Liquid water, H_2O (synonyms: aqua, dihydrogen monoxide, hydrogen hydroxide), is transparent, odorless, tasteless, and ubiquitous. It is colorless in small amounts but exhibits a bluish tinge in large quantities (Chaplin 2009a).

As will be seen below, it has unique properties because of the structure of the water molecule. Water has influence on human affairs far beyond its apparent ordinariness. The influence of water on humans and their existence on planet earth, on account of its peculiar properties will be discussed below.

1.1 Physical and Chemical Properties of Water and Their Consequences

1.1.1 Molecular Structure of Water and Its Strong Surface Tension

Water has a very simple V-shaped atomic structure. This is because the atoms in the water molecule are arranged with the two H–O bonds at an angle of about 105° rather than on directly opposite sides of the oxygen atom. The two hydrogen atoms are bonded to

N. Okafor, *Environmental Microbiology of Aquatic and Waste Systems*,
DOI 10.1007/978-94-007-1460-1_1,

one side of an oxygen atom, resulting in the V-shaped structure mentioned above. On account of this, a water molecule has a positive charge on the side where the hydrogen atoms are, and a negative charge on the other side, where the oxygen atom is (see Fig. 1.1). Oxygen has a higher electronegativity than hydrogen. Because oxygen has a slight negative charge while the hydrogen has a slight positive charge, water is a polar molecule, since there is a net negative charge toward the oxygen end (the apex) of the V-shaped molecule and a net positive charge at the hydrogen end. Therefore, water molecules tend to attract each other, because opposite electrical charges attract. As will be seen below, this molecular polarity causes water to be a powerful solvent and is responsible for its strong surface tension. It is also responsible for the high specific heat of water (Pidwirny 2006; Chaplin 2009a; Anonymous 2009a).

The elements surrounding oxygen in the periodic table, namely, nitrogen, fluorine, phosphorus, sulfur, and chlorine, all combine with hydrogen to produce gases under standard conditions. The reason that oxygen hydride (water) forms a liquid is that it is more electronegative than all of these elements (other than fluorine).

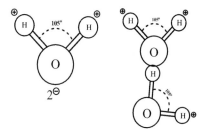

Fig. 1.1 The molecular structure of water (Modified from Chaplin 2009a). *Left*: A water molecule showing the charges. *Right*: Two water molecules attracting each other

1.1.2 The High Surface Tension of Water and Capillarity in Plants

Water molecules as a whole have no net charge, but the oxygen end has a slight negative charge (since the electrons tend to stay on the side of the large oxygen nucleus), and the hydrogen end has a slight positive charge. For this reason, water is referred to as a polar molecule because it has a positive end and a negative end, analogous to the north and south poles of a bar magnet. On account of this, the negative (oxygen) end of one water molecule forms a slight bond with the positive (hydrogen) end of another water molecule.

Because of the above, water has a high surface tension. In other words, water is adhesive and elastic, and tends to aggregate in drops rather than spread out over a surface as a thin film (see Fig. 1.2). This is a result of water molecules attracting each other because of the positive charge at one end and the negative charge at the other. The strong surface tension also causes water to stick to the sides of vertical structures despite gravity's downward pull. Water's high surface tension allows for the formation of water droplets and waves, plants' capillary of water (and dissolved nutrients) from their roots to their leaves, and the movement of blood through tiny vessels in the bodies of some animals (Sharp 2002; Pidwirny 2006; Chaplin 2009a).

1.1.3 The Three Physical States of Water, and the Floatation of Ice

Water is unique in that it is the only natural substance on earth that is found in all three physical states – liquid, solid (ice), and gas (steam) – at the temperatures normally found on earth. When the water molecule

Fig. 1.2 Attraction of water molecules to create high surface tension (Copyright Michael Pidwirny; www.our-planet-earth.net. Reproduced with permission from Pidwirny 2006)
Note: In this illustration the water film is two layers thick

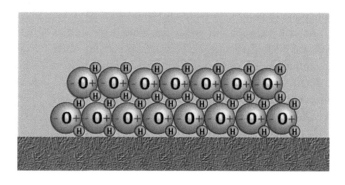

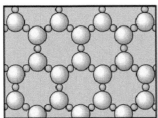

Ordered Molecular Structure of Frozen Water

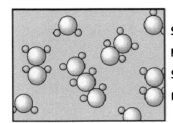

Semi–Ordered Molecular Structure of Liquid Water

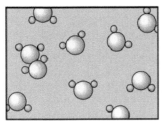

Random Molecular Structure of Vaporized Water

Fig. 1.3 Arrangement of molecules in steam, liquid water, and ice (Copyright Michael Pidwirny; www.our-planet-earth.net. Reproduced with permission from Pidwirny 2006). Note the highly organized molecules in ice, the semi-ordered arrangement in liquid water, and the random molecular structure in water vapor

makes a physical phase change, its molecules arrange themselves in distinctly different patterns. In ice, the molecules are highly organized; in liquid water, they are semi-ordered, while the molecules are randomly arranged in water vapor (see Fig. 1.3).

The molecular arrangement taken by ice (the solid form of the water molecule) leads to an increase in volume and a decrease in density. Expansion of the water molecule at freezing allows ice to float on top of liquid water.

Water molecules exist in liquid form over an important range of temperature from 0°C to 100°C. This range allows water molecules to exist as liquid in most places on earth (Anonymous 2009b).

1.1.4   The Thermal Properties of Water and Their Effect on Climate

Because of hydrogen bonding between water molecules, the latent heats of fusion and of evaporation and the heat capacity of water are all unusually high. For these reasons, water serves both as a heat-transfer medium (e.g., ice for cooling and steam for heating) and as a temperature regulator (the water in lakes and oceans helps regulate the climate). Latent heat, also called heat of transformation, is the heat given up or absorbed by a unit mass of a substance as it changes from a solid to a liquid, from a liquid to a gas, or the reverse of either of these changes. Incorporated in the changes of state are massive amounts of heat exchange. This feature plays an important role in the redistribution of heat energy in the earth's atmosphere. In terms of heat being transferred into the atmosphere, approximately 75% of this process is accomplished by the evaporation and condensation of water (see The Hydrological Cycle below: Fig. 1.4).

Water has the second highest specific heat capacity of any known chemical compound, after ammonia. Specific heat is the amount of energy required to change the temperature of a substance. Because water has a high specific heat, it can absorb large amounts of heat energy before it begins to get hot. It also means that water releases heat energy slowly when situations cause it to cool. Water's high specific heat allows for the moderation of the Earth's climate, through the heat in oceans, and helps organisms regulate their body temperature more effectively. This explains why water is used for cooling, say in automobile radiators, and why the temperature change between seasons is gradual rather than sudden, especially near the oceans.

Finally, water conducts heat more easily than any liquid except mercury. This fact causes large bodies of liquid water like lakes and oceans to have essentially a uniform vertical temperature profile (Table 1.1).

1.1.5   The Change in the Nature of Water at Different Temperatures and Lake Temperatures

An unusual property of water is that the solid state, ice, is not as dense as liquid water because of the geometry of the hydrogen bonds (see above) which are formed only at lower temperatures. For almost all other substances the solid form has a greater density than the liquid form. On account of this unusual property the solid form, ice, floats on its liquid form. Freshwater at standard atmospheric pressure is most dense at 3.98°C, and will sink by convection as it cools to that

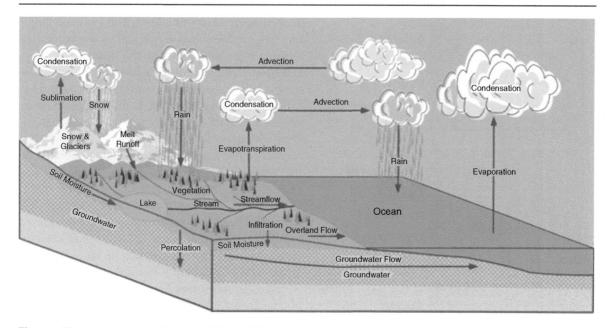

Fig. 1.4 The hydrologic cycle (Copyright Michael Pidwirny; www.our-planet-earth.net. Reproduced with permission (Pidwirny 2006))

Table 1.1 Some properties of liquid water and their consequences on the environment (Modified from Sigee 2005)

S/no	Property	Comparison with other substances	Importance to environment
1	Density	Maximum density at 4°C, not at freezing; expands at freezing (both are unusual properties)	In lakes prevents freezing and causes seasonal stratification
2	Melting and boiling points	Abnormally high	Permits water to exist as liquid at earth's surface; moderates temperatures by preventing extremes
3	Heat capacity	Highest of any liquid, except ammonia	Permits only gradual changes in climatic temperatures
4	Heat of vaporization	One of the highest known	Important in heat transfer atmosphere and in oceans; and moderates temperatures by preventing extremes
5	Surface tension	Very high	Regulates drop formation in clouds and rain
6	Absorption of radiation	Large in infra-red and ultra-violet regions; little in visible light area	Important in control of photosynthesis in water bodies, and in the control of atmospheric temperature
7	Solvent properties	Excellent solvent for ionic salts and polar molecules because of bipolar nature	Important in transfer of dissolved materials in hydrological systems and biological entities

temperature, and if it becomes colder it will rise instead. This reversal causes deep water to remain warmer than shallower freezing water. Because of this, ice in a body of water will form first at the surface and progress downward, while the majority of the water underneath will hold a constant 4°C. This effectively insulates a lake floor from the cold in temperate climates (see Table 1.2).

1.1.6   The Low Electrical Conductivity of Pure Water

Pure water has a low electrical conductivity, but this increases significantly upon solvation of a small amount of ionic material water such as hydrogen chloride. Thus the risks of electrocution are much greater in water with impurities than in waters in which

Table 1.2 Density of water molecules at various temperatures (Modified from SIMetric.co.uk; http://www.simetric.co.uk/si_water.htm; Anonymous 2009b)

Temperature (°C)	Density (gm/cm³)
0 (solid)	0.9150
0 (liquid)	0.9999
4	1.0000
20	0.9982
40	0.9922
60	0.9832
80	0.9718
100 (gas)	0.0006

impurities are absent. Any electrical properties observable in water are from the ions of mineral salts and carbon dioxide dissolved in it. Water does self-ionize where two water molecules become one hydroxide anion and one hydronium cation, but not enough to carry sufficient electric current to do any

$$2H_2O \rightleftarrows H_3O^+ + OH^-$$

work or harm for most operations. In pure water, sensitive equipment can detect a very slight electrical conductivity of 0.055 μS/cm at 25°C. Water can also be electrolyzed into oxygen and hydrogen gases but in the absence of dissolved ions, this is a very slow process since very little current is conducted.

1.1.7   The High Chemical Reactiveness of Water

Water is very chemically reactive, reacting with certain metals and metal oxides to form bases, and with certain oxides of nonmetals to form acids. It reacts with certain organic compounds to form a variety of products, e.g., alcohols from alkenes. Although, as seen above, completely pure water is a poor conductor of electricity, it is a much better conductor than most other pure liquids because of its self-ionization, i.e., the ability of two water molecules to react to form a hydroxide ion, OH^-, and a hydronium ion, H_3O^+. Its polarity and ionization are both due to the high dielectric constant of water.

1.1.8   The pH of Water

Water in a pure state has a neutral pH. As a result, pure water is neither acidic nor basic. Water changes its pH when substances are dissolved in it. Rain has a naturally acidic pH of about 5.6 because it contains dissolved carbon dioxide and sulfur dioxide.

1.1.9   The High Solvent Power of Water

Because water is a polar compound, it is a good solvent. Water is a very strong solvent; it is referred to as the universal solvent because it dissolves many types of substances. The substances that will mix well and dissolve in water (e.g., salts) are known as "hydrophilic" (water-loving) substances, and those that do not mix well with water (e.g., fats and oils) are known as "hydrophobic" (water-fearing) substances. The ability of a substance to dissolve in water is determined by whether or not the substance can match or better the strong attractive forces that water molecules generate between other water molecules. If a substance has properties that do not allow it to overcome these strong intermolecular forces, the molecules are "pushed out" from among the water and do not dissolve. Water is able to dissolve a large number of different chemical compounds and this feature enables it to carry dissolved nutrients in runoff, infiltration, groundwater flow, and living organisms.

1.2   Importance and Uses of Water

1.2.1   Composition of Biological Objects

Most living things consist of a large percentage of water: In humans, for example, it constitutes about 92% of blood plasma, about 80% of muscle tissue, and about 60% of red blood cells. This is because water is required for the biochemical reactions taking place in them.

1.2.2   Drinking by Man and Animals and Domestic Use

All living things need water for their metabolism. Water is absolutely essential for survival. A person may survive for a month without food, but only about a week without water. In order to be clean enough for human consumption, water usually has to be treated in some way to remove harmful microorganisms and chemicals. In the home, water is used for cooking, cleaning, and disposal of wastes as sewage.

1.2.3   Irrigation and Aquaculture

Throughout the world, irrigation for growing crops is probably the most important use of water (except for drinking). Almost 60% of all the world's freshwater withdrawals goes towards irrigation uses. Large-scale farming cannot provide food for the world's large populations without the irrigation of crop fields by water from rivers, lakes, reservoirs, and wells. Without irrigation, crops could never be grown in the deserts of countries such as California and Israel.

Another agricultural use to which water is put is in aquaculture for growing fish.

1.2.4   Power Generation

Power generation with water is in two forms: Hydroelectric and thermoelectric.

1. *Hydroelectric* power contributes about 12% of total power output in the USA. This kind of power generation is restricted to regions where water can fall from a high altitude, and in the process cause the rotation of huge turbines which leads to the generation of electric power.
2. *Thermoelectric* production of electrical power results in one of the largest uses of water in the United States and worldwide. In thermoelectric power generation, water is heated either by petroleum oil or gas or by heat generated in a nuclear reaction. The steam produced is used to drive turbines which generate the electricity. In the USA, in the year 2000, about 195,000 million gallons of water each day were used to produce electricity (excluding hydroelectric power). This represented about 52% of fresh surface-water withdrawals.

1.2.5   Transportation

Water is a primary medium for transporting heavy goods. Huge ships ply the oceans and large lakes of the world, carrying raw materials and manufactured products. Water transportation, although relatively slow is one of the cheapest means of transporting heavy goods.

1.2.6   Recreation

Water is used for many recreational purposes, as well as for exercising and for sports. Some of these include swimming, waterskiing, boating, fishing, and diving. In addition, some sports, like ice hockey and ice skating, take place on ice.

1.2.7   Human Affairs

Water has played an important part in human affairs for centuries, and still does (Keeley 2005).

1. *Religion*: Water is considered a purifier in most religions. Major faiths that incorporate ritual washing (ablution) include Hinduism, Christianity, Islam, Judaism, and Shinto. Water baptism is a central sacrament of Christianity; it is also a part of the practice of other religions, including Judaism and Sikhism. In addition, a ritual bath in pure water is performed for the dead in many religions including Judaism and Islam.

   Many religions also consider particular bodies of water to be sacred or at least auspicious, for example, the River Ganges in Hinduism.

   Greek philosophers believed that water was one of the four classical elements along with fire, earth, and air. Water was also one of the five elements in traditional Chinese philosophy, along with earth, fire, wood, and metal. Among the Igbos in Nigeria, water and rainfall are considered blessings; rainfall during a celebration is considered a good omen.
2. *Politics*: Water has caused frictions and even wars between nations: for example, the Ganges is disputed between India and Bangladesh, while the Golan Heights which provides 770 million cubic meters of water per year is disputed between Israel and Syria.

1.3   The Hydrologic Cycle

The continuous circulation of water from the earth to the atmosphere and from the atmosphere back to the earth is known as the hydrologic cycle. The hydrologic cycle is powered by the energy of the sun. The cycle operates through evaporation, transpiration in plants, condensation in clouds, and precipitation.

Although the hydrologic cycle balances what goes up with what comes down, one phase of the cycle is "frozen" in the colder regions of the world during the winter season. During the winter in the colder parts of the world, most of the precipitation is simply stored as snow or ice on the ground. Later, during the spring snow and ice melt, releasing huge quantities of water in relatively short periods; this leads to heavy spring runoff and, sometimes, flooding.

The heating of the ocean water by the sun is the key process that keeps the hydrologic cycle in motion. Water evaporates when heated by the sun, then falls as precipitation in the form of rain, hail, snow, sleet, drizzle, or fog. On its way to earth, some precipitation may evaporate or, when it falls over land, be intercepted by vegetation before reaching the ground. The cycle proceeds by the following processes (see Fig 1.4):

1. *Evaporation*
   As water is heated by the sun, the surface molecules become sufficiently energized to break free of the attractive force binding them together, and then *evaporate* and rise as invisible vapor in the atmosphere. As much as 40% of precipitation can be lost by evaporation.

2. *Transpiration*
   Water vapor is also emitted from plant leaves by the process of transpiration. Every day an actively growing plant *transpires* five to ten times as much water as it can hold at any one time.

3. *Condensation*
   As water vapor rises, it cools as it encounters the lower temperatures of the upper atmosphere and eventually *condenses*, usually on tiny particles of dust in the air. When it condenses it becomes a liquid again or turns directly into a solid (ice, hail, or snow). These water particles then collect and form clouds. Clouds are formed when air containing water vapor is cooled below a critical temperature called the dew point and the resulting moisture condenses into droplets on microscopic dust particles (condensation nuclei) in the atmosphere.

4. *Precipitation*
   Precipitation in the form of rain, snow, and hail comes from clouds. Clouds move around the world, propelled by air currents. When they rise over mountain ranges, they cool, becoming so saturated with water that water begins to *fall* as rain, snow, or hail, depending on the temperature of the surrounding air.

5. *Runoff*
   Runoff is the visible *flow* of water from land into rivers, creeks and lakes, and the oceans following the precipitation of atmospheric water as snow or rain. Runoff water runs overland into nearby streams and lakes; the steeper the land and the less porous the soil, the greater the runoff. Overland flow is particularly visible in urban areas. Rivers join each other and eventually form one major river that carries all of the sub-basins' runoff into the ocean.

6. *Percolation*
   Water moves downward through cracks and pores in soil and rocks to the water table. Water can move up by capillary action; it can move vertically or horizontally under the earth's surface until it re-enters a surface water system. Some of the precipitation and melted snow moves downwards, *percolates* or infiltrates through cracks, joints, and pores in soil, and rocks until it reaches the water table where it becomes groundwater.

7. *Groundwater*
   Subterranean water is held in cracks and pore spaces. Depending on the geology of the area, the groundwater can flow to support streams. It can also be tapped by wells. Some groundwater is very old and may have been there for thousands of years.

1.4 Classification of Waters

Water may be classified into *natural* and *artificial* waters.

1.4.1 Natural Waters

Natural waters may be grouped thus:
1. Atmospheric waters:
   (a) Rain
   (b) Hail
   (c) Snow
2. Surface waters:
   (a) Streams
   (b) Ponds
   (c) Lakes
   (d) Rivers and estuaries
   (e) Oceans

3. Ground (or underground) waters:
  (a) Springs
  (b) Wells
  (c) Underground streams

1.4.2   Artificial Waters

Artificial waters, all of which are surface waters, are man-made, and include:
1. Reservoirs
2. Dams
3. Oxidation ponds
4. Man-made lakes
5. Canals

1.5   Global Distribution of Water on Earth and Its Study

Much of the world's surface (71%) is covered by the oceans, which have a mean depth of 3.8 km and whose water is highly saline. The oceans and seas hold 97.5% of all the world's waters, the remaining 2.5% being freshwater. Of this freshwater, 68.9% is locked in glaciers and ice-sheets, while most of the rest (29.9%) is *groundwater*. Much of the world's freshwater is held in rivers and lakes and constitutes 0.3%. Figure 1.5 and Table 1.3 show the global distribution of water.

The seas and oceans of the world form a continuous body of water which are linked to each other. They are,

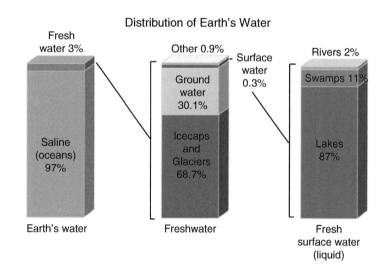

Fig. 1.5 Global distribution of water (From http://ga. water.usgs.gov/edu/ waterdistribution.html. With permission) Credit: United States Geological Services (USGS))

Table 1.3 Volumes of water in global water bodies (Modified from Gleick 1996)

Water source	Water volume (cubic kilometers)	Percent of freshwater	Percent of total water
Oceans, seas, and bays	1,338,000,000	–	96.5
Ice caps, glaciers, and permanent snow	24,064,000	68.7	1.74
Groundwater	23,400,000	–	1.7
Fresh	10,530,000	30.1	0.76
Saline	12,870,000	–	0.94
Soil moisture	16,500	0.05	0.001
Ground ice and permafrost	300,000	0.86	0.022
Lakes	176,400	–	0.013
Fresh	91,000	0.26	0.007
Saline	85,400	–	0.006
Atmosphere	12,900	0.04	0.001
Swamp water	11,470	0.03	0.0008
Rivers	2,120	0.006	0.0002
Biological water	1,120	0.003	0.0001
Total	1,386,000,000	–	100

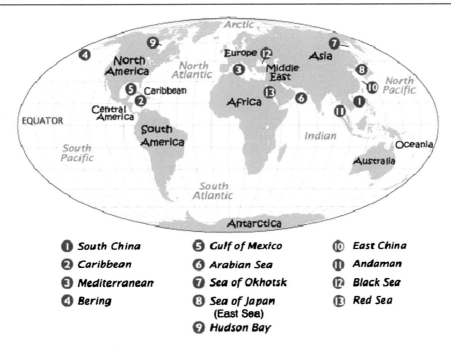

Fig. 1.6 Seas and oceans of the world (Copyright Compare Infobase Limited; www.mapsofworld.com. Reproduced with permission from Anonymous 2009d)
Note: The names of the oceans are in written in full blue and those of the seas are numbered in red

Table 1.4 Area and depth of the seas and oceans of the world (Modified from www.infoplease.com; Anonymous 2009c. With permission)

Name	Area Sq km	Average depth m	Greatest known depth m	Place of greatest known depth
Pacific Ocean	155,557,000	4,028	11,033	Mariana Trench
Atlantic Ocean	76,762,000	3,926	9,219	Puerto Rico Trench
Indian Ocean	68,556,000	3,963	7,455	Sunda Trench
Southern Ocean	20,327,000	4,000–5,000	7,235	South Sandwich Trench
Arctic Ocean	14,056,000	1,205	5,625	77°45′N; 175°W
Mediterranean Sea	2,965,800	1,429	4,632	Off Cape Matapan, Greece
Caribbean Sea	2,718,200	2,647	6,946	Off Cayman Islands
South China Sea	2,319,000	1,652	5,016	West of Luzon
Bering Sea	2,291,900	1,547	4,773	Off Buldir Island
Gulf of Mexico	1,592,800	1,486	3,787	Sigsbee Deep
Okhotsk Sea	1,589,700	838	3,658	146°10′E; 46°50′N
East China Sea	1,249,200	188	2,782	25°16′N; 125°E
Hudson Bay	1,232,300	128	183	Near entrance
Japan Sea	1,007,800	1,350	3,742	Central Basin
Andaman Sea	797,700	870	3,777	Off Car Nicobar Island
North Sea	575,200	94	660	Skagerrak
Red Sea	438,000	491	2,211	Off Port Sudan
Baltic Sea	422,200	55	421	Off Gotland

however, divided into about six oceans and about 13 seas (see Fig. 1.6). They vary greatly in size and depth, the largest and deepest being the Pacific Ocean (see Table 1.4).

Tables 1.5 and 1.6 give the world's longest rivers and largest lakes. At any given time rivers and lakes hold 0.33% of freshwater, while the atmosphere holds 0.035% (Anonymous 2010a, b).

Table 1.5 The top ten longest rivers of the world (From http://www.infoplease.com/toptens/worldrivers.html; Anonymous 2010a. With permission)

S/No	Name	Length (km)	Location	Destination
1	Nile	6,690	Egypt	Mediterranean Sea
2	Amazon	6,570	Amapa-Para, Brazil	Atlantic Ocean
3	Mississippi-Missouri	6,020	Louisiana, USA	Gulf of Mexico
4	Yangtze	5,980	Kiangsu, China	East China Sea
5	Yenisey	5,870	Russia	Yenisey Gulf in Kara Sea
6	Amur	5,780	China	Tatar Strait
7	Ob-Irtysh	5,410	Russia	Gulf of Ob in Kara Sea
8	Plata-Parana	4,880	Argentina-Uruguay	Atlantic Ocean
9	Yellow (Hwang)	4,840	Shantung, China	Yellow Sea
10	Congo (Zaire)	4,630	Angola-Zaire	Atlantic Ocean

Table 1.6 The top ten largest lakes of the world (From http://www.infoplease.com/toptens/largelakes.html; Anonymous 2010b. With permission)

Lake	Location	Volume (km³)	Area (km²)	Salinity
Caspian Sea	Azerbaijan, Iran, Kazakhstan, Russian Federation, Turkmenistan	78,200 km³	374,000	Mostly saline
Baikal (Ozero Baykal)	Russian Federation	23,600 km³	31,500	Fresh
Tanganyika	Burundi, Congo (Democratic Republic), Tanzania, Zambia	19,000 km³	32,900	Mostly fresh
Superior	Canada, United States of America	12,100 km³	82,100	Fresh
Malawi (Nyasa, Niassa)	Malawi, Mozambique, Tanzania	7,775 km³	–	Fresh
Michigan	United States of America	4,920 km³	57,750	Fresh
Huron	Canada, United States of America	3,540 km³	59,500	Fresh
Victoria	Kenya, Tanzania, Uganda	2,760 km³	68,460	Fresh
Great Bear Lake	Canada	2,292 km³	31,326	Fresh
Great Slave	Canada	2,088 km³	29,568	Fresh

The science which studies waters on earth is *hydrology*. The branch of hydrology which studies the oceans is *oceanography*; the branch of hydrology which studies inland waters running or standing, fresh or saline, is *limnology* (from the Greek: limne = lake, logos = study). Limnology includes the study of (natural and man-made) lakes and ponds, rivers and streams, wetlands, and groundwaters.

The water which man requires for his domestic and industrial purposes, however, is freshwater, which forms only a small proportion by volume of the earth's waters (Fig. 1.5 and Table 1.3). Although the abundant waters of the oceans and seas can be de-salinated to provide freshwater, the present methods of doing so are uneconomical.

Freshwater is thus a relatively scarce commodity. For this reason, it is purified and reused in some countries. Water which has been used and rain water draining into the ground, all enter, and form part of, the hydrologic cycle.

References

Anonymous (2009a). Water structure and science. http://www.btinternet.com/~martin.chaplin/index.html. Accessed 10 Sept 2009.

Anonymous (2009b). Table of density of pure and tap water and specific gravity. SIMertic.co.uk. http://www.simetric.co.uk/si_water.htm. Accessed on 12 June 2009.

Anonymous (2009c). Area and depth of the seas and oceans of the world. www.infoplease.com. Accessed 6 Dec 2009.

Anonymous (2009d). Seas and oceans. http://www.infoplease.com/ipa/A0001773.html. Accessed 10 Dec 2009.

Anonymous (2010a). The top ten longest rivers of the world. http://www.infoplease.com/toptens/worldrivers.html. Accessed 9 Dec 2009.

Anonymous (2010b). The top ten largest lakes of the world http://www.infoplease.com/toptens/largelakes.html. Accessed 9 Dec 2009.

Chaplin, M (2009a). Water structure and science. http://www. btinternet.com/~martin.chaplin/index.html. Accessed 10 Sept 2009.

Gleick, P.H. (1996). Basic water requirements for human activities: Meeting basic needs. *Water International, 21*, 83–92.

Keeley, J. (2005). Water in history. In J. H. Lehr (Ed.), *Water encyclopedia* (Vol. 4, pp. 726–732). Hoboken: Wiley.

Pidwirny, M. (2006). Physical properties of water. *Fundamentals of physical geography*, 2nd Edn. http://www. physicalgeography.net/fundamentals/8a.html. Accessed 10 Jan 2009.

Sharp, K. A. (2002). Water: Structure and properties. In *Encyclopedia of life sciences* (Vol. 19, pp. 512–519). London: Nature Publishing Group.

Sigee, D. (2005). *Freshwater microbiology: Biodiversity and dynamic interactions of microorganisms in the aquatic environment*. Chichester: Wiley.

Peculiarities of Water as an Environmental Habitat for Microorganisms

Abstract

As a habitat for the existence of microorganisms, water has properties not found in other natural microbial habitats such as soil, and plant and animal bodies; indigenous aquatic microorganisms are adapted to these conditions. Natural waters are generally low in nutrient content (i.e., they are oligotrophic); what nutrients there are, are homogeneously distributed in the water. The movement of water freely transports microorganisms; to counter this and offer themselves some protection, many aquatic organisms are either stalked or arranged in colonies immersed in gelatinous materials. To enable free movement in water, many aquatic microorganisms and/or their gametes have locomotory structures such as flagella. Microorganisms are often adapted to, and occupy particular habitats in the water body; some occupy the air–water interphase (neuston), while others live in the sediment of water bodies (benthic). The conditions which affect aquatic microorganisms are temperature, nutrient, light, salinity turbidity, water movement. The methods for the quantitative study of aquatic microorganisms are cultural methods (plate count and MPN), direct methods (microscopy and flow cytometry), and the determination of microbial mass. The microscopy methods are light (optical), epi-flourescence, confocal laser scanning microscopy, transmission electron microscopy, and scanning electron microscopy. Microbial mass may be direct (weight after oven-drying) or indirect (turbidity, CO_2 release, etc).

Keywords

Aquatic microorganisms • Oligotrophic • Aquatic habitats • Flagellar motility • Epi-flourescence microscopy • Flow cytometry

2.1 The Peculiar Nature of Water as an Environment for Microbial Habitation

Microorganisms exist in different natural environments such as the soil, the animal intestine, the air, and water. Each of these habitats has peculiar characteristics to which the organisms living therein must adapt.

Aquatic microorganisms generally experience highly fluctuating and highly varying environmental conditions. These conditions differ not only from one aquatic macro-environment to another, but also in different sub-locations in the same aquatic macro-environment.

N. Okafor, *Environmental Microbiology of Aquatic and Waste Systems*,
DOI 10.1007/978-94-007-1460-1_2, © Springer Science+Business Media B.V. 2011

This variability contributes to the well-recognized microbial biodiversity in aquatic environments. On account of this, because the freshwater macro-environment is clearly different from the marine macro-environment, the microorganisms attached to organic matter of the same composition would be expected to be different.

Various habitats can be recognized within an aquatic environment. Each habitat is characterized by one or more microbial communities.

This chapter will discuss the peculiarities of the aquatic environment as dwelling places for microorganisms as well as the methods used in quantitative study of microorganisms found in water.

The aquatic environment can itself be divided broadly into freshwater and saline. The microbiology of the freshwater environment will be discussed in Chap. 5 and that of the saline (marine) environment in Chap. 6.

Water differs from other natural microbial environments such as soil, plant, and animal bodies in a number of ways, namely (Sigee 2005):

1. *The low concentration of nutrients*
   Natural bodies of water are generally oligotrophic, i.e., low in nutrients. The concentration of nutrients available to a microbial cell in the environment of natural waters such as the open sea or rivers away from shorelines is usually low, when compared with the concentrations found in other microbial habitats such as soil crevices or plant and animal bodies. The result of this is that the truly indigenous aquatic microorganisms must be able to subsist under conditions of low nutrient availability, which may be unfavorable to their terrestrial counterparts of the same group. For example, *Escherichia coli* is known to die out quickly in distilled water, whereas aquatic indigenes such as *Pseudomomas* spp. and *Achromobacter* spp. do not.

2. *Relative homogeneity of the properties of water*
   Because of the vastness of the aquatic environment in comparison with the size of the individual microbial cell, the aquatic environment is fairly homogeneous in terms of nutrients, pH, etc. For example, metabolic products released by aquatic microorganisms are continuously diluted away in natural waters, say in a river. Such products, therefore, hardly accumulate in natural waters in the same way as they could theoretically do in the soil. Two samples of soils taken an inch apart from each other may have completely different properties in terms of pH, the population of microorganisms, etc. This is not the case with water which can be said to be more homogeneous both in space and time than other natural environments.

3. *The movement of water*
   Natural bodies of water generally flow. Even when there is no apparent gross movement of a body of water, minor movements induced by the wind take place regularly. Consequently, many truly aquatic microorganisms are attached to larger bodies which provide them with support, and stop them from drifting. In order to provide this attachment, some aquatic bacteria are stalked, for example, the aquatic bacterium *Caulobacter* spp. Others form filaments which enable attachment only at one end leaving the rest of the filament free for the absorption of nutrients. An example will be found among the adults of the sessile ciliated Protozoa such as those found among the *Suctoria* (see Chap. 4). Still others form themselves into gelatinous balls or masses which can offer slightly better protection against the hazards of moving waters than would be available to single individuals. Examples of organisms which form colonial units or are immersed in gelatinous masses are *Zooglea* spp. among the Bacteria, and *Pandorina* spp. among the Algae.

   With respect to the terminology for describing water movements or flow of freshwaters, those which are still or exhibit little flow, e.g., ponds or some lakes, are known as *limnetic* waters, while those in which the movement or flow is rapid, as in some rivers and streams, are known as *lotic*.

4. *Freedom of movement of microorganisms in aquatic environments*
   Because of the vastness of the water environment in comparison with the size of microorganisms, aquatic microorganisms are afforded movement without impediment over comparatively huge distances in a way which is not available to organisms in soil or other environments inhabited by microorganisms. Most truly aquatic microorganisms possess flagella or cilia which enable them or their reproductive cells to move about freely in aquatic environments. This is presumably to help them move around freely in search of food or around those areas of the water body such as decomposing animal or plant bodies which may have slightly higher concentrations of nutrients than the rest of the water. In illustration of this, the truly aquatic fungi, i.e., *Phycomycetes* are the only group among the fungi in which flagellated gametes are found. Similarly, among the algae which are recognized as mostly aquatic, flagellated cells or reproductive structures are found in all of the algae except the *Rhodophyceae* (Red Algae) and the *Cyanophyceae* or Blue green algae.

2.2 Ecological Habitats of Microorganisms in Aquatic Environments

In the ecological study of plants and animals, a *population* is a group of living things of the same kind living in the same place at the same time, while a *habitat* is the place where a population lives (Pugnaire and Valladares 2007). These concepts are applied here to the microorganisms living in water. In aquatic environments, various, sometimes overlapping ecological populations of microorganisms exist at various depths of the water column. The various populations inhabiting various aquatic habitats will be described below (Matthews 1972):

1. *Planktonic organisms*

   Planktonic organisms are populations of free floating organisms in water. They are the fodder (or food) upon which the smaller aquatic life, especially krills and small fish, subsist. Planktons include algae (Diatoms or *Bacillariophyceae*) and Dinoflagellates (*Dinophyceae*) as well as Protozoa.

2. *Tectonic organisms*

   In contrast to planktonic organisms, tectonic organisms are those which, while subject to aquatic flow, also have locomotory means of their own, and hence can also move apart from being carried with the flow of water, e.g., ciliated protozoa, flagellated bacteria, etc.

3. *"Aufwuchs" (periphyton)*

   This is a term derived from the German; it indicates that the aquatic microorganism in question is attached to something. On the basis of the nature of their support, "aufwuchs" are divided into the following categories:

   (a) Epiphytic – attached to the surfaces of plants
   (b) Epizootic – attached to the surface of animals
   (c) Epilithon – a community of microorganisms attached to rocks and stones of the aquatic environment
   (d) Epixylon – the microorganisms found on the fallen woods in water bodies
   (e) Episammon – the community of organisms attached to sand grains

4. *Benthic organisms, benthos*

   Organisms inhabiting the bottom sediment of aquatic environments (i.e., bottom or mud-dwelling organisms) constitute the benthic community or benthos.

5. *Neuston*

   The microorganisms which are found at the surface of an aquatic environment, exactly at the air–water interface, are referred to as neuston.

6. *Pleuston*

   The specific term "pleuston" may be used to denote the organisms occupying the air–water interface in a marine biota. These neuston and pleuston can be regarded as specialized communities as their air–water interface habitat is subjected to widely fluctuating environmental conditions.

7. *Seston epipelon*

   Sestons are the particulate matter suspended in bodies of water such as lakes and seas. Some authors apply it to all particulates, including plankton, organic detritus, and inorganic material. Seston produced by nekton, i.e., large swimming animals, may also act as a habitat along with other organic debris. Microorganisms inhabiting these detritus matters and other fine sediment surface are referred to as epipelon.

2.3 Foreign Versus Indigenous Aquatic Bacteria

Microorganisms are constantly being washed into surface waters from the soil; thus the microbial population of the near-side of inland waters is very similar to the surrounding soil especially after rains. Those which are not truly indigenous to water soon die. These are the foreign, migrant, or allochthonous organisms of water, while the indigenous organisms, known as authochthonous, survive in the water (Fig. 2.1).

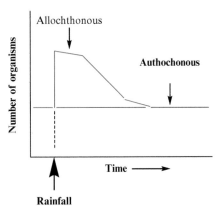

Fig. 2.1 Relative populations of allochthonous and authochthonous aquatic microorganisms after rainfall (Drawn by the author: see text)

2.4   Challenges of Aquatic Life: Factors Affecting the Microbial Population in Natural Waters

Various physical and chemical factors affect the microbial population in natural waters. The physical factors affecting the microbial populations in aquatic environments are floatation, temperature, nutrients, and light, while the chemical factors include nutrients, salinity, and pH (Sigee 2005).

1. *Floatation*

   Flotation or placement in the water column is a challenge faced by all aquatic organisms. For example, it is crucial to phytoplankton (microscopic algae) to stay in the photic zone, where there is access to sunlight. The small size of most phytoplankton, plus a special oily substance in the cytoplasm of cells, helps keep these organisms afloat.

   Zooplankton (microscopic protozoa) use a variety of techniques to stay close to the water surface. These include the secretion of oily or waxy substances, possession of air-filled sacs similar to the swim bladders of fish, and special appendages that assist in floating. Some zooplankton even tread water.

   Fishes have special swim bladders, which they fill with gas to lower their body density. By keeping their body at the same density as water, a state called "neutral buoyancy," fishes are able to move freely up and down.

2. *Temperature*

   Temperature is one of the main factors affecting the growth of microorganisms. Psychrophiles (low temperature loving) are microorganisms which have an optimum temperature of growth of 0–5°C, whereas thermophiles (high temperature loving) have optimum temperatures of 60°C and above. Mesophiles (middle temperature loving) have optimum temperatures of 20–40°C. The temperature of natural waters varies from about 0°C in polar regions to 75–80°C in hot springs. In tropical regions it is about 25–30°C, and in temperate regions about 15°C in the summer and lower in the colder months.

   The specific thermal capacity of water is very high, hence large water bodies are able to either absorb or lose great amount of heat energy without much change in their temperature. Because of this, many aquatic organisms including microbes experience very stable temperature conditions. For instance, about 90% of the marine habitats maintain a constant temperature of about 4°C which encourages the growth of psychrophilic microorganisms. A good example of a marine psychrophile is *Vibrio marinus*, while a good example of a thermophile is *Thermus aquaticus* (optimum temperature 70–72°C) which grows in hot springs and whose thermostable DNA polymerase is used in the very useful procedure used to amplify DNA, the Polymerase Chain Reaction (PCR) see Chap. 3.

   The temperature of non-marine water bodies like lakes, streams, and estuaries shows seasonal variations in temperate countries as indicated above, with corresponding changes in the microbial population. Such bodies in tropical countries are more stable in temperature and hence have a more constant group of organisms.

   Eurythermal species are those that can survive in a variety of temperatures. Eurythermality generally characterizes species that live near the water surface, where temperatures change depending on the seasons or the time of day. Species that occupy deeper waters generally experience more constant temperatures, are intolerant of temperature changes, and are described as stenothermal.

3. *Nutrients*

   The supply of nutrients is one of the major determinants of microbial density and variety in aquatic environments. Both organic and inorganic nutrients are important, and the nutrients may either be present dissolved or as particulate material. Aquatic environments with limited nutrient content are said to be *oligotrophic* and those with a high nutrient content are said to be *eutrophic*.

   The open (the pelagic zone of the) sea has a stable and very low nutrient load. But, nearshore water shows variations in nutrient load due to the addition of nutrients from domestic and industrial wastewaters. A shortage of inorganic nutrients, particularly, nitrogen and phosphorus, may limit algal growth. However, the presence of these nutrients in unusually large quantities often lead to excessive growth of algae and cyanobacteria, a condition known as *eutrophication*. Heavy metals like mercury have an inhibitory effect on microorganisms, but certain microorganisms have developed resistance toward these heavy metals.

4. *Light*

Almost all forms of life in aquatic habitats are either directly or indirectly influenced by light, because primary food production is mainly through photosynthesis. Since algae and photosynthetic bacteria are involved in photosynthesis, they are the most important forms with respect to light.

Light is important in the spatial distribution of microorganisms in an aquatic environment, especially the photosynthetic forms. Photosynthetic microorganisms are mainly restricted to the upper layers of aquatic systems, the photic zone, where effective penetration of light occurs. Although photosynthesis is confined to the upper 50–125 m of the water bodies, the depth of the photic zone may vary depending upon the latitude, season, and turbidity. Apart from the decrease in the quantity of light with depth, there is also a change in the color of the light able to penetrate water. Of the component rays of visible light, blue light is the most transmitted to the lower depths of water, while the red is the least.

Light may also be bactericidal. It has been suggested that the reason for the death of sewage bacteria in seawater is due to light. Sometimes, there may be a reduction in the bacterial activity without necessarily killing them. For example, reduction in the rate of oxidation has been shown to occur in *Nitrosomonas* spp., and *Nitrobacter* spp. due to high light intensities. Cell division may also be related to the day and night variation. The diatom *Nitzschia* spp. divides mostly in the light, while the dinoflagellate *Ceratium* spp. divides during darkness.

5. *Salinity*

Aquatic species also have to deal with salinity, the level of salt in the water. Some marine species, including sharks and most marine invertebrates, simply maintain the same salinity level in their tissues as is in the surrounding water. Some marine vertebrates, however, have lower salinity in their tissues than is in seawater. These species have a tendency to lose water to the environment. They make up for this by drinking seawater and excreting excess salt through their gills. Freshwater aquatic species have the opposite problem – a tendency to absorb too much water. These species must constantly expel water, which they do by excreting a dilute urine.

Species that occupy both freshwater and marine habitats at different stages of their life cycle must transition between two modes of maintaining water balance. Salmon hatch in freshwater, mature in the ocean, and return to freshwater habitats to spawn. Eels, on the other hand, hatch in salt water, migrate to freshwater environments where they mature, and return to the ocean to spawn.

Salinity is particularly variable in coastal waters, because oceans receive variable amounts of freshwater from rivers and other sources. Species in coastal habitats must be tolerant of salinity changes and are described as *euryhaline*. In the open ocean, salinity levels are generally constant, and species that live there cannot tolerate salinity changes. These organisms are described as *stenohaline*.

A wide range of salinity (i.e., the content of NaCl, or salt) occurs in natural waters. For example, salinity is near zero in freshwater and almost at saturation level in salt lakes, such as those found in the State of Utah in the USA. A clear distinction can be made between the flora and fauna of fresh and saltwater systems due to the variation in the salinity levels. The dissolved salt concentration of sea water varies from 33 to 37 g per kg of water. The maximum level of salinity has been noted in Red Sea where it is 44 parts per thousand. Thus sea water characteristically contains a high salt content. Even within the ocean some variations, though small, may occur in salinity. Concentrations of salts are normally low in shallow offshore regions and near river mouths. Some rise in salinity in a water body may be due to the evaporation of water or ice formation. Decrease in salinity is also possible by the inflow of rain water or snow precipitation.

6. *Turbidity*

Since turbidity of water in an aquatic environment inhibits the effective penetration of light, it is also considered an important factor affecting microbial life. Turbidity is caused by suspended materials, which include particles of mineral material originating from land, detritus, particulate organic material such as cellulose, hemicellulose, and chitin fragments and suspended microorganisms.

Particulate matter in a water body may provide a substratum to which various microorganisms adhere. Many marine microorganisms, including bacteria and protozoa are attached to solid substrata. The attached bacterial communities are called epibacteria or periphytes.

7. *Water movements (currents)*

Water currents are obvious in lentic or limnetic habitats like rivers. In aquatic environments, including rivers and oceans, currents help aerate the water and cause the mixing of nutrients. The movement of water in ocean geothermal vents also helps in mixing nutrients. In deep waters, the water currents may move in opposite direction to those on the surface and usually they are slower. The velocity of water within a water body, especially in rivers have been found to influence the nutrient uptake and metabolism. Studies on biofilms have also shown that, under phosphorus limiting conditions, a small increase in water velocity can increase the uptake of phosphorus.

2.5 Methods for the Enumeration of Microorganisms in the Aquatic Environment

It may sometimes be necessary, as part of understanding microbial ecology in aquatic systems, to determine the number of microorganisms in a body of water. This section looks at the various methods used.

The methods for enumerating organisms in the aquatic environment may be divided broadly into cultural methods, direct microscopic methods, and determination of microbial mass.

2.5.1 Cultural Methods

1. *The plate count*

   The plate count on agar is one of the oldest techniques in the science of bacteriology (Buck 1977; Anonymous 2006). It has a number of *advantages*, including the following:

   (a) It is easy to perform.

   (b) It is scientifically uncomplicated.

   (c) It is relatively inexpensive.

   (d) It allows cultivable organisms to develop.

   (e) Within the limits of its shortcomings given below, it permits the quantitative comparison of organisms from different habitats.

   (f) It also facilitates the study of the different substrates on which organisms grow, and hence their physiology, through the incorporation of the substrates in the agar medium.

   Useful as it is, it has a number of *disadvantages* which constitute major handicaps in some circumstances:

   (a) The method selects organisms which can grow in the particular chemical components of the medium and the temperature, and other environmental conditions for cultivating the organism.

   (b) Total counts are calculated from plate counts on the assumption that each colony counted has arisen from a single bacterial cell; this is not necessarily always the case, due to possible clumping of the bacterial cells.

   (c) The procedure selects organisms which are cultivable, and molecular methods show that many more organisms are in natural environments than are counted by the plate count method.

2. *The Most Probable Number (MPN) method*

   The most probable number (MPN) technique is an important technique in estimating microbial populations in soils, waters, and agricultural products. Soils and food, and even some bodies of water are heterogeneous; therefore, exact cell numbers of an individual organism can be impossible to determine. The MPN technique is used to estimate microbial population sizes in such situations. The technique does not rely on quantitative assessment of individual cells; instead, it relies on specific qualitative attributes of the microorganism being counted. The important aspect of MPN methodology is the ability to estimate a microbial population size based on a process-related attribute that is based on the physiological activity of the organism.

   The principle of methodology for the MPN technique is the dilution and incubation of replicated cultures across several serial dilution steps. This technique relies on the pattern of positive and negative test results following inoculation of a suitable test medium (usually with a pH sensitive indicator dye) with the test organism (Colwell 1977; Anonymous 2006).

   The assumptions underlying the technique are as follows:

   (a) The organisms are randomly and evenly distributed throughout the sample.

   (b) The organisms exist as single entities, not as chains, pairs or clusters and they do not repel one another.

   (c) The proper growth medium, temperature, and incubation conditions have been selected to allow even a single viable cell in an inoculum to produce detectable growth.

   (d) The population does not contain viable, sublethally injured organisms that are incapable of growth in the culture medium used.

   (e) The MPN techniques also assume that all test organisms occupy a similar volume.

Fig. 2.2 The basis of the MPN technique (Drawn by the author: see text)

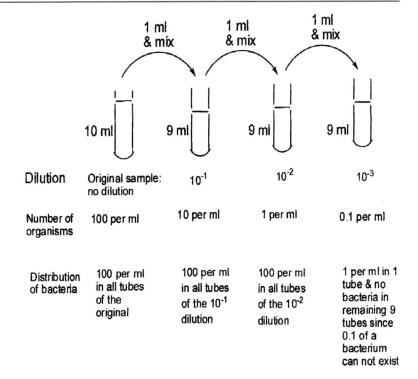

This procedure has advantages and disadvantages. Some of the *advantages* of the MPN methods are as follows:

(a) MPN methodology results in more uniform recovery of a microbial population across different portions of a body of water.

(b) Unlike direct quantitative procedures, it measures only live and active organisms; direct methods measure living and dead organisms.

The *disadvantages* are:

(a) The procedures of MPN are laborious.

(b) The MPN procedures also have a lower order of precision than direct counts.

The principle of the MPN method resides in the dilution procedure for counting bacteria. In dilution counting, a serial tenfold dilution of the water whose bacterial content is to be determined is made. A loopful of the dilution is streaked out on a suitable medium. For the dilution technique to be valid, enough dilutions must be made for growth *not* to occur at one of such dilutions; hence the phrase "dilution to extinction" is applied to the dilution method. The number of organisms in the original liquid is assumed to be the reciprocal of the dilution in which growth occurs, just before the dilution at which no growth occurs. Thus, if in series of tenfold dilutions, growth occurs at dilution 10^3 but not at dilution 10^4; the number of organisms in the original dilution is taken to be 10^3 or 100 (more accurately the number should be stated as more than 10^3 but less than 10^4. The rationale behind the MPN method is illustrated in Fig. 2.2).

The MPN is thus counting by the method of "dilution to extinction" made more accurate by the use of several tubes. For precise estimates, a very large number of tubes should be inoculated at each dilution. The confidence intervals can be narrowed by inoculating more tubes at each dilution or by using a smaller dilution ratio; a twofold dilution gives a greater precision than four-, five-, or tenfold dilution. If the microbial population size is not known, it is best to use a tenfold dilution with at least five (but preferably more) tubes at each dilution. If some idea of the microbial population size is known, it is best to use a dilution of 2, even if fewer tubes are used. The MPN is read from tables. Details of this method are available in *Standard Methods for the Determination of Water and Wastewater* (Anonymous 2006). The number of organisms is recorded in terms of Most Probable Numbers (MPN). Figures obtained in MPN determinations are indices of numbers which, more probably than any others, would give the results shown by laboratory examination. It should be emphasized that they are not actual numbers which, when determined with the plate count, are usually much lower. MPN index and 95% confidence limits for various combinations of positive results, when five tubes are used for dilution (10, 1, and 0.1 ml), are given in Table 2.1.

3. *The membrane filtration method*

The membrane filtration method consists in filtering a measured volume of the water sample through a membrane composed of cellulose esters and other materials. The bacteria present are retained on the surface of the membrane, and by incubating the membrane face upward on a suitable agar or liquid medium and temperature, colonies develop on the membrane which can then be counted. The volume of liquid chosen will depend on the expected bacterial density of the water and should be such that colonies developing on the membrane lie between 10 and 100.

The *advantages* of the filtration method over the multiple-tube method are:

(a) Rapidity: It is possible to obtain direct counts of coliforms and *E. coli* in 18 h without the use of probability tables.

(b) Labor, media, and glassware are also saved.

(c) For testing water for fecal contamination, neither spore-bearing anaerobes nor mixtures of organisms which may give false presumptive reactions in MacConkey broth cause false positive results on membranes.

(d) A sample may be filtered on the spot or in a nearby laboratory with limited facilities instead of taking the liquid to the proper laboratory.

Membranes have, however, the following *disadvantages*:

(a) They are unsuitable for waters of high turbidities, and in which the required organisms are low in number, since they can be blocked before enough organisms have been collected.

(b) For water testing, when non-coliforms predominate over coliforms, the former may overgrow the membrane and make counting of coliforms difficult.

(c) Similarly, in water testing, if non-gas producing lactose-fermenters predominate in the water, false results will be obtained.

Membrane filters are used to count various microorganisms including bacteriophages. *Standard Methods for the Determination of Water and Wastewater* (Anonymous 2006) contains detailed information on this.

4. *Dilution of water for viable counts (plate count or MPN)*

Depending on the load of bacteria expected in water, the water will need to be diluted before the determination of its bacterial content. Table 2.2 gives a possible range of dilutions of water and other consumable liquids, whose bacterial load may be determined by plate count or MPN determinations.

2.5.2 Direct Methods

Direct methods involve direct visualization of the microorganisms using a microscope. The impulses may be transmitted to a set up which may be read off as graphs or other visual end points. As will be seen below, some direct methods are often combined with some of the cultural methods described above.

2.5.2.1 Light Microscopy

The most common method of enumerating total microbial cells is the direct counting of cell suspension in a counting chamber of known volume using a microscope. One such counting chamber is the Neubauer counting chamber which has grids to facilitate counting, usually done under a light microscope. More sophisticated direct counting methods do not use the human eye, but instead use sensors which according to the desired program may detect size or different kinds of organisms depending on which dye is used to identify which organism. The various approaches are discussed under the title microscopy, below.

1. *Optical (light) microscopy*

The optical or light microscope uses visible light and a system of lenses to magnify images of small samples. The image can be detected directly by the eye, imaged on a photographic plate or captured digitally. In light microscopy, the wavelength of the light limits the resolution to around 0.2 μm. The shorter the wave length the greater the magnification of the image; hence, in order to gain higher resolution, the use of an electron beam with a far shorter wavelengths is used in electron microscopes. The light for the light microscope may be daylight.

Table 2.1 Most probable number index and 95% confidence limits for five tube, three dilution series (From Standard Methods for the Examination of Water and Wastewater; Anonymous 2006. With permission)

Number of tubes giving positive reaction out of			MPN index per	95% confidence limits:	
5 of 10 ml	5 of 1 ml	5 of 0.1 ml	100 ml	Lower	Upper
0	0	0	<2	–	–
0	0	1	2	<0.5	7
0	1	0	2	<0.5	7
0	2	0	4	<0.5	11
1	0	0	2	<0.5	7
1	0	1	4	<0.5	11
1	1	0	4	<0.5	11
1	1	1	6	<0.5	15
1	2	0	6	<0.5	15
2	0	0	5	<0.5	13
2	0	1	7	1	17
2	1	0	7	1	17
2	1	1	9	2	21
2	2	0	9	2	21
2	3	0	12	3	28
3	0	0	8	3	24
3	0	1	11	4	29
3	1	0	11	4	29
3	1	1	14	6	35
3	2	0	14	6	35
3	2	1	17	7	40
4	0	0	13	5	38
4	0	1	17	7	45
4	1	0	17	7	46
4	1	1	21	9	55
4	1	2	26	12	63
4	2	0	22	9	56
4	2	1	26	9	78
4	3	0	27	9	80
4	3	1	33	11	93
4	4	0	34	12	93
5	0	0	23	7	70
5	0	1	31	11	89
5	0	2	43	15	110
5	1	0	33	11	93
5	1	1	46	16	120
5	1	2	63	21	150
5	2	0	49	17	130
5	2	1	70	23	170
5	2	2	94	28	220
5	3	0	79	25	190
5	3	1	110	31	250
5	3	2	140	37	340
5	3	3	180	44	500
5	4	0	130	35	300
5	4	1	170	43	490
5	4	2	220	57	700

(continued)

Table 2.1 (continued)

Number of tubes giving positive reaction out of			MPN index per	95% confidence limits:	
5 of 10 ml	5 of 1 ml	5 of 0.1 ml	100 ml	Lower	Upper
5	4	3	280	90	850
5	4	4	350	120	1,000
5	5	0	240	68	750
5	5	1	350	120	1,000
5	5	2	540	180	1,400
5	5	3	920	300	3,200
5	5	4	1,600	640	5,800
5	5	5	$\geq 1,600$	–	–

Table 2.2 Possible dilution of water and some consumable fluids prior to viable counts determination (Modified from science. kennesaw.edu/~bensign/aqmeth/bacteriawaterquality.doc)

Item	Suggested dilutions								
	0	10^1	10^2	10^3	10^4	10^5	10^6	10^7	10^8
Iced tea	+	+	+						
Bottled water	+	+	+						
Pasteurized milk	+	+	+						
Drinking water	+	+	+						
Swimming pool		+	+	+					
Wells, springs		+	+	+					
Stagnant lakes, ponds			+	+	+				
River water				+	+	+			
Toilet water					+	+	+		
Raw sewage effluent						+	+	+	+

However, bright light similar to day light may be produced from mercury-vapor lamps or xenon-arc lamps. The Neubauer counter has wide applications including being used to count blood cells in animal and human pathology laboratories, apart from being used for bacterial counts.

2. *Fluorescence microscopy*

In fluorescence microscopy, the object being studied is labeled with a fluorescent dye which gives the object one color, say red, but emits another color, say green. Some materials, for example, chlorophyll are innately fluorescent and are said to be auto-fluorescent. Most current fluorescence microscopes are operated in the *epi*-illumination mode (illumination and detection from one side of the sample) and hence the system is known as *epifluorescence microscopy*. The excitation, illumination, or oncoming light strikes the object on one (usually, the top) side and, as will be seen below, this further decreases the amount of excitation light entering the detector. The major components of a fluorescence microscope are (Fig. 2.3):

(a) The light source (Xenon or Mercury arc-discharge lamp).

(b) The excitation (illumination) filter (which filters away light of other wavelengths leaving one of a specific wavelength or color).

(c) The dichroic mirror (or dichromatic beam splitter) which simultaneously reflects the illumination light to the object and allows the reflected fluorescent light (of a different wavelength or color from the illumination light) to pass through.

(d) The emission filter, which allows the much weaker fluorescence or reflected light to pass through, and filters away any illuminating light coming from the object (see Fig. 2.3). The filters and the dichroic mirror are selected in accordance with wavelengths of the illumination light and the reflected light peculiar to the fluorescent dye used.

The epifluorescence microscope is widely used in biology. It is used to count microorganisms directly in fresh and marine water as well as in milk, clinical specimens, and in environmental materials. The

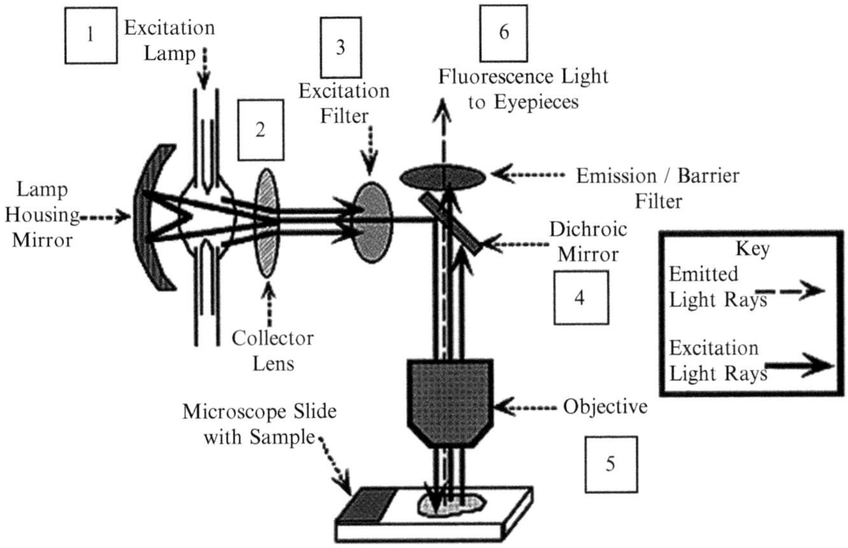

Fig. 2.3 Setup illustrating the principle of the epifluorescence microscope (From http://international.abbottmolecular.com/ DiagramoftheFluorescenceMicroscope_8843.aspx. With permission)

bacterial cells are captured on the surface of poly-carbonate membrane filters, stained with a fluorescent dye such as acridine orange, and visualized using epifluorescence microscopy. The fluoro-chrome (fluorescent dye) which is currently popular is 4′, 6′ diamidino-2-phenyl indole (DAPI). When using fluorescence techniques, the sample is either stained and then filtered onto membrane filters, or the cells are stained on the filters. Fluorochrome – stained – cells counts give an estimate of total living heterotrophic bacterial population in aquatic environments. However, a large proportion of the total cells are sometimes in a metabolically inactive state.

To get an estimate of metabolically active cells, a dehydrogenase stain assay can be employed. In this method, the active dehydrogenase enzyme reduces a synthetic water soluble, membrane permeable, straw colored tetrazolium salt converting it to a pink-red, water insoluble formozan. The proportion of cells that have accumulated pink formozan through enzymatic dehydrogenation of tetrazolium substrate represents metabolically active cells.

It has also been used to characterize planktonic procaryotic populations. An image analysis system may be used to digitize the video image of auto-fluorescing or fluorochrome-stained cells in the microscope field. The digitized image can then be stored, edited, and analyzed for total count or individual cell size and shape parameters, and results can be printed as raw data, statistical summaries, or histograms (Sieracki et al. 1985).

3. *The confocal laser scanning microscope*

The *confocal laser scanning microscope* (CLSM or LSCM) is now often used in biology (Prasad et al. 2007).

It is called "confocal" (meaning the same focus) because the final image has the same focus as the point of focus in the object. When an object is imaged in the fluorescence microscope, the signal produced is from the full thickness of the specimen; on account of this, most of the image is out of focus to the observer. The pinhole aperture of the confocal microscope blocks out this out-of-focus light (dotted lights in Fig. 2.4) and thus light from above and below the point of focus in the object. Filtering away some of the light reduces the amount of light and thus also reduces the "visibility" of the focused part of the specimen. To make up for this, laser beams are used which produce extremely bright light at a fixed wavelength. Highly sensitive photo-multiplier-detectors (PMTs) are used to pick up and multiply the reduced laser beam striking the image. The laser is focused on a fixed portion of the specimen (a square or rectangle) at a time using a

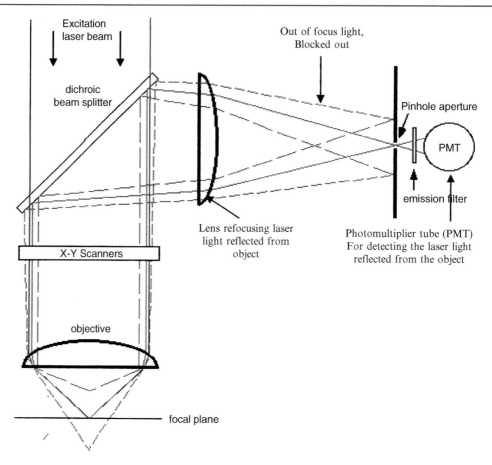

Fig. 2.4 Diagram illustrating the principle of the confocal microscope (From http:/www.gonda.ucla.edu/bri_core/confocal.htm; Schibler 2010. With permission)

motorized system of mirrors controlled by a computer. The computer also allows the system to scan sequential planes in the (opposite direction) Z-direction, and create overlays of all the in-focus Z section, and store them. The information can also be used to create three-dimensional images, or movie rotations of well-stained specimens (Schibler 2010).

The confocal microscope has several advantages over conventional optical microscopy including:

(a) Controllable depth of field through controllable three-dimensional (3D) reconstructions.

(b) The elimination of image degrading out-of-focus information, thus obtaining high resolution images.

(c) The ability to collect serial optical sections from thick specimens.

The key feature of confocal microscopy is its ability to produce sharp images of thick specimens at various depths. Images are taken point-by-point and

reconstructed with a computer, rather than projected through an eyepiece. First developed in 1953, it took another 30 years and the development of lasers for confocal microscopy to become a standard technique toward the end of the 1980s.

In this system, a laser beam passes through a light source aperture and then is focused by an objective lens into a small focal volume within a fluorescent specimen. A mixture of emitted fluorescent light as well as reflected laser light from the illuminated spot is then recollected by the objective lens. A beam splitter separates the light mixture by allowing only the laser light to pass through and reflecting the fluorescent light into the detection apparatus. After passing a pinhole, the fluorescent light is detected by a photo-detection device (photomultiplier tube [PMT]) transforming the light signal into an electrical one which is recorded by a computer.

The detector aperture obstructs the light that is not coming from the focal point, as shown by the dotted gray line in the image. The out-of-focus points are thus suppressed: most of their returning light is blocked by the pinhole. This results in sharper images compared to conventional fluorescence microscopy techniques and permits one to obtain images of various X-Y axis planes; for depth, the object is also scanned in the Z axis plane (Schibler 2010).

2.5.2.2 Electron Microscopy

Electron microscopes were developed due to the limitations of the light microscopes which are limited by the physics of light to 500× or 1000× magnification and a resolution of 0.2 µm. The desire in the early 1930s to study the fine details of the interior structures of organic cells such as the nucleus, mitochondria, etc., fueled this need. The transmission electron microscope (TEM) was the first type of electron microscope to be developed and is patterned exactly on the light transmission microscope except that a focused beam of electrons is used instead of light to "see" through the specimen. It was developed in Germany in 1931. The first of the other type of electron microscope, the scanning electron microscope (SEM) came out in 1942, but the commercial came out about 1965.

1. *The transmission electron microscope (TEM)*
   The ray of electrons is produced by a pin-shaped cathode heated up by electric current. The electrons are produced in a vacuum at high electric voltage. The higher the voltage, the shorter are the electron waves and the higher is the power of resolution. Modern powers of resolution range from 0.2 to 0.3 nm and magnification is around 300,000×. The electron microscope is like the light microscope; however, the "lens" is an electric coil generating an electromagnetic field. Specimens are thin, no more than 100 nm thick, and are "stained" with heavy metal salts to make them visible. The formed image is made visible on a fluorescent screen, or it is captured on photographic material. Photos taken with electron microscopes are always black and white. The degree of darkness corresponds to the electron density (i.e., differences in atom masses) of the candled preparation.

2. *The scanning electron microscope (SEM)*
   The path of the electron beam within the scanning electron microscope differs from that of the TEM and is based on television techniques. The method is suitable for the showing of preparations with electrically conductive surfaces. Biological objects have thus to be made conductive by coating with a thin layer of heavy metal, usually gold. The power of resolution is smaller than in transmission electron microscopes, but the depth of focus is much higher. Scanning electron microscopy is therefore also well suited for very low magnifications. The surface of the object is scanned with the electron beam point by point whereby secondary electrons are set free. The intensity of this secondary radiation is dependent on the angle of inclination of the object's surface. The secondary electrons are collected by a detector placed at an angle at the side above the object. The image appears a little later on a viewing screen.

A comparison of the light, transmission, and scanning microscopes is given in Table 2.3.

2.5.2.3 Flow Cytometry

Flow cytometry comes from "cyto" for cell, and "meter" for measure in the word cytometer. The technology involves the use of a beam of laser light projected through a liquid stream that contains cells, or other particles in the size range of 0.2–150 µm diameter, which when struck by the focused light give out signals which are picked up by detectors. These signals are then converted for computer storage and data analysis, and can provide information about various cellular properties.

Cells flow one at a time through a region of interaction at the rate of over 1,000 cells per second. The biophysical properties detected are then correlated with the biological and biochemical properties of interest. The high throughput of cells allows for rare cells, which may have inherent or inducible differences, to be easily detected and identified from the remainder of the cell population.

In order to make the measurement of biological/biochemical properties of interest easier, the cells are usually stained with fluorescent dyes which bind specifically to cellular constituents. The dyes are excited by the laser beam and emit light at longer wavelengths. This emitted light is picked up by detectors, and the signals are converted to digital so that they may be stored for later display and analysis.

Flow cytometers also have the ability to selectively deposit cells from particular populations into tubes, or other collection vessels. These selected cells can then

Table 2.3 Comparison of the properties of the light, and transmission and scanning electron microscopes (From http://universe-review.ca/R11-13-microscopes.htm#top; Anonymous 2010)

Characteristic	Compound microscope	Transmission electron microscope	Scanning electron microscope
Resolution (average)	500 nm	10 nm	2 nm
Resolution (special)	100 nm	0.5 nm	0.2 nm
Magnifying power	up to 1,500×	up to 5,000,000×	~ 100,000×
Depth of field	Poor	Moderate	High
Type of objects	Living or non-living	Non-living	Non-living
Preparation technique	Usually simple	Skilled	Easy
Preparation thickness	Rather thick	Very thin	Variable
Specimen mounting	Glass slides	Thin films on copper grids	Aluminum stubs
Field of view	Large enough	Limited	Large
Source of radiation	Visible light	Electrons	Electrons
Medium	Air	Vacuum	Vacuum
Nature of lenses	Glass	One electrostatic + a few em. lenses	One electrostatic + a few em. lenses
Focusing	Mechanical	Current in the objective lens coil	Current in the objective lens coil
Magnification adjustments	Changing objectives	Current in the projector lens coil	Current in the projector lens coil
Specimen contrast	By light absorption	By electron scattering	By electron scattering

be used for further experiments, cultured, or stained with another dye/antibody and reanalyzed. The data can be displayed in a number of different formats, each having advantages and disadvantages. The common methods are histogram, or dotplots. A flow cytometer typically consists of the following:

1. *Flow chamber*

   Cells flow through the flow chamber one at a time very quickly, about 10,000 cells in 20 s or more often 500 cells per second.

2. *Laser*

   A small laser beam of very bright light hits the cells as they pass through the flow chamber. The way the light bounces off each cell gives information about the cell's physical characteristics. Light bounced off at small angles is called *forward scatter*. Light bounced off in other directions is called *side scatter*.

3. *Light detector*

   The light detector processes the light signals and sends the information to the computer. Forward scatter tells you the size of the cell. Side scatter tells if the cell contains granules. Each type of cell in the immune system has a unique combination of forward and side scatter measurements, allowing you to count the number of each type of cell.

4. *Filters*

   The filters direct the light emitted by the fluorochromes (fluorescent dyes) to the color detectors.

5. *Color detectors*

   As the cells pass through the laser, the fluorochromes attached to the cells absorb light and then emit a specific color of light depending on the type of fluorochrome. The color detectors collect the different colors of light emitted by the fluorochromes and send them to a computer.

6. *Computer*

   The data from the light detector and the color detectors is sent to a computer and plotted on a histogram.

   Flow cytometers or flourescence activated cell sorters are instruments which analyze optical parameters of many individual cells in a short space of time. The sample flows through an orifice and across the path of a light source, usually a laser. Optical parameters like light scatter and fluorescence are measured on each cell by photodetectors. The signals received are processed by a computer. On the basis of predefined optical properties, cells can be sorted and collected automatically. Some of the advantages of flow cytometry are:

   (a) Rapid analysis of relatively large sample sizes, i.e., 10^4–10^6 cells per minute.

   (b) Modifications of the technique allow detection of smaller and less abundant cells in natural water samples. A well-known application of the flow cytometer is the analysis of natural phytoplankton by auto-fluorescence. By this technique, cyanobacterial phototrophs are enumerated and separated from other phytoplankton based on phycoerythrin and chlorophyll fluorescence. The former fluoresces orange and the latter red.

2.5.3 Determination of Bacterial Mass

Bacterial mass may be determined by direct weight measurement, or indirectly by the measurement of metabolic activities of the organism. For the determination of growth yields, wet or dry weight estimations are commonly used. For evaluating metabolic and enzymatic activities, protein or nitrogen content of a bacterial suspension is determined.

2.5.3.1 Direct Methods

After centrifuging the cells, wet weight can be determined. Dry weight is determined after drying the cells to constant weight.

Nitrogen content and total carbon content can also be determined using well-known laboratory procedures such as the Association of Official Agricultural Chemists (AOAC) procedures.

2.5.3.2 Indirect Methods

Bacterial mass may be determined by measuring the turbidity of cell suspensions; production of carbon dioxide, oxygen uptake, or production of acid can also be related to microbial mass after a calibration of the values against known quantities. These methods are particularly useful with very low concentration of cells. Chlorophyll determination is also a useful method, especially for algae; synthesis of ATP has also been used as a function of rate of microbial activity or total biomass (Hodson et al. 1976).

References

Anonymous. (2006). *Standard methods for the examination of water and wastewater.* Washington, DC: American Public Health Association, American Water Works Association, and Water Environmental Association.

Anonymous (2010). Microscopes. http://universe-review.ca/R11-13-microscopes.htm#top. Accessed 11 Aug 2010.

Buck, J. D. (1977). The plate count in aquatic microbiology. In J. W. Costerton & R. R. Colwell (Eds.), *Native aquatic bacteria: Enumeration, activity and ecology* (pp. 19–25). Philadelphia: American Society for Testing and Materials.

Colwell, R. R. (1977). Enumeration of specific populations by the most-probable number (MPN) method. In J. W. Costerton & R. R. Colwell (Eds.), *Native aquatic bacteria: Enumeration, activity and ecology* (pp. 56–64). Philadelphia: American Society for Testing and Materials.

Hodson, R. E., Holm-Hansen, O., & Azam, F. (1976). Improved methodology for ATP determination in marine environments. *Marine Biology, 34,* 143–149.

Matthews, J. E. (1972). *Glossary of aquatic ecological terms.* Environmental Protection Agency. Distributed By: National Technical Information Service, U.S. Department of Commerce.

Prasad, V., Semwogerere, V., & Weeks, D. (2007). Confocal microscopy of colloids. *Journal of Physics: Condensed Matter, 19,* 113102.

Pugnaire, F., & Valladares, F. (Eds.). (2007). *Functional plant ecology* (2nd ed.). Baton Rogue: CRC Press.

Schibler, M. (2010). http://www.gonda.ucla.edu/bri_core/confocal.htm. Private communication.

Sieracki, M. E., Johnson, P. W., & Sieburth, J. M. (1985). Detection, enumeration, and sizing of planktonic bacteria by image-analyzed epifluorescence microscopy. *Applied and Environmental Microbiology, 49,* 799–810.

Sigee, D. (2005). *Freshwater microbiology: Biodiversity and dynamic interactions of microorganisms in the aquatic environment.* Chichester: Wiley.

Part II

Biological Aspects of Microorganisms in Aquatic Environments

Aspects of the Molecular Biology of Microorganisms of Relevance to the Aquatic Environment

3

Abstract

Our increased knowledge of living things at the molecular level has greatly influenced the modern approach in biology. Thus, for example, the taxonomy of microorganisms is no longer based mostly on their morphology, but also on the sequence of bases in the genes of the 16S RNA of the small subunit of the ribosomes. This chapter looks at selected basic molecular biology topics of relevance to the environment. The processes of transcription and translation in protein synthesis are discussed. The principles behind some molecular procedures such as the polymerase chain reaction and micro-arrays are discussed. Many microorganisms in environments such as water and soil are not culturable and are studied only with molecular biology. Metagenomics, or the culture-independent genomic analysis of an assemblage of microorganisms, in environments such as marine or freshwater has potential to answer fundamental questions in microbial ecology.

Keywords

Protein synthesis • Polymerase chain reaction • Microarrays • Gene identification • Metagenomics

In recent times giant strides have been taken in harnessing our knowledge of the molecular basis of many biological phenomena. Many new techniques such as the polymerase chain reaction (PCR) and DNA sequencing have arrived on the scene. In addition, major projects involving many countries such as the human genome project have taken place. Coupled with all these exciting technological developments, new vocabulary such as genomics has arisen. These have transformed the approaches used in microbiology. For example, with molecular biology techniques, it is now possible to study the microbial populations of any environment by using culture-independent analyses of 16S rRNA gene sequences (see for example Elshahed et al. 2007; Bansal 2005).

This chapter will discuss only selected aspects of molecular biology in order to provide a background for understanding some of the newer directions of environmental microbiology. The discussion will be kept as simplified and as brief as possible, just enough in complexity and length to achieve the purpose of the chapter. The student is encouraged to look at many excellent publications in this field.

3.1 Protein Synthesis

Proteins are very important in the metabolism of living things. They are in hormones for transporting messages around the animal body; they are used as storage

N. Okafor, *Environmental Microbiology of Aquatic and Waste Systems*,
DOI 10.1007/978-94-007-1460-1_3, © Springer Science+Business Media B.V. 2011

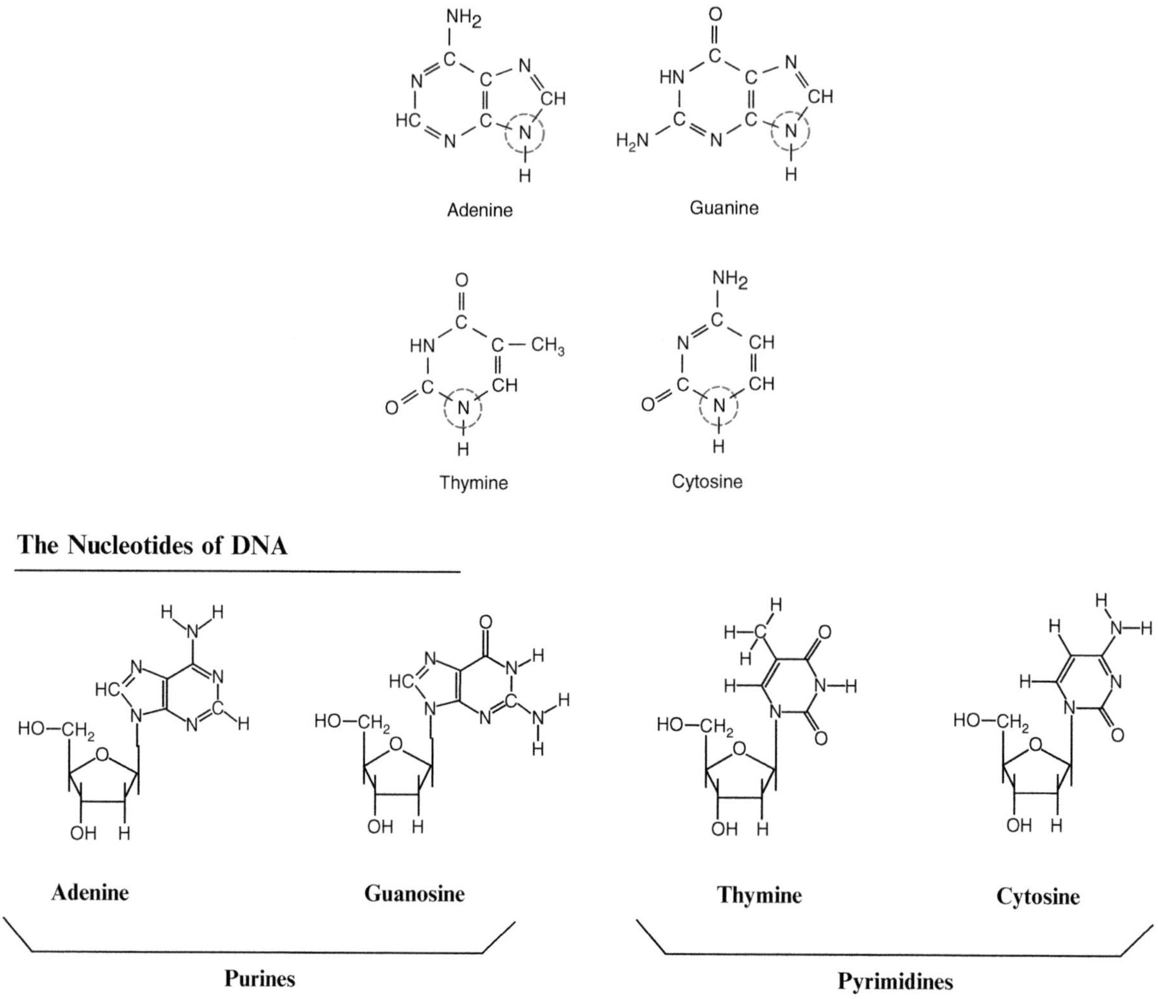

Adenine

Guanine

Thymine

Cytosine

The Nucleotides of DNA

Adenine

Guanosine

Thymine

Cytosine

Purines

Pyrimidines

Fig. 3.1 The nucleic acid bases

such as in the whites of eggs of birds and reptiles and in seeds; they transport oxygen in the form of hemoglobin; they are involved in contractile arrangements which enable movement of various animal body parts, through contractile proteins in muscles; they protect the animal body in the form of antibodies; they are in membranes where they act as receptors, participate in membrane transport and antigens, and they form toxins such as diphtheria and botulism. The most important function, if it can be so termed, is that they form the basis of enzymes which catalyze all the metabolic activities of living things; in short, proteins and the enzymes formed from them are the engines of life.

Notwithstanding the bewildering diversity of living things, varying from bacteria to protozoa to algae to maize to man, the same 20 amino acids are found in all

living things. On account of this, the principles affecting proteins and their structure are same in all living things.

The macromolecules linked to heredity are deoxyribonucleic acid (DNA) and ribonucleic acid (RNA). The genetic information which determines the potential properties of a living thing is carried in the DNA present in the nucleus, except in some viruses where it is carried in RNA. DNA is also present in the plant organelles mitochondria and chloroplasts.

DNA consists of four nucleotides, adenine, cytosine, guanine, and thymine. RNA is very similar except that uracil replaces thymine (see Fig. 3.1). RNA occurs in the nucleus and in the cytoplasm as well as in the ribosomes.

The processes of protein synthesis will be summarized briefly below. In protein synthesis, infor-

Fig. 3.2 Cloverleaf-
shaped transfer RNA
(From The Internet
Encyclopedia of Science;
http://www.daviddarling.
info/encyclopedia/T/tRNA.
html. Reproduced with
permission)

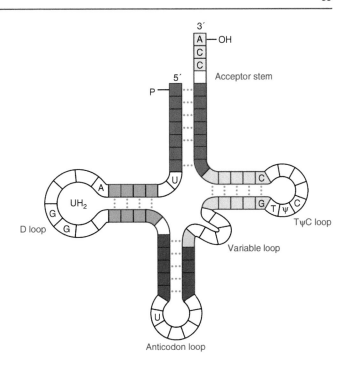

mation flow is from DNA to RNA via the process of transcription, and thence to protein via translation. Transcription is the making of an RNA molecule from a DNA template. Translation is the construction of a polypeptide from an amino acid sequence from an RNA molecule (see Fig. 3.2). The only exception to this is in retroviruses where reverse transcription occurs: a single-stranded DNA is transcribed from a single-stranded RNA (the reverse of transcription).

3.1.1 Transcription

An enzyme, RNA polymerase, opens the part of the DNA to be transcribed. Only one strand of DNA, the template or sense strand, is transcribed into RNA. The other strand, the anti-sense strand is not transcribed. The anti-sense strand is used in making ripe tomatoes to remain hard. The RNA transcribed from the DNA is the messenger or m-RNA (see Fig. 3.3). Because some students appear to be confused by the various types of RNA, it is important that we mention at this stage that there are two other types of RNA besides m-RNA. These are ribosomal or r-RNA and transfer or t-RNA.

During this step, mRNA goes through different types of maturation including one in which the non-coding sequences are eliminated, in a process known as splicing.

When m-RNA is formed, it leaves the nucleus in eukaryotes (there is no nucleus in prokaryotes!) and moves to the ribosomes.

3.1.2 Translation

In all cells, ribosomes are the organelles where proteins are synthesized. They consist of two-thirds of ribosomal RNA, r-RNA, and one-third protein. Ribosomes consist of two subunits, a smaller subunit and a larger subunit. In prokaryotes, typified by *E. coli*, the smaller unit is 30S and larger 50S. *S* is Svedberg units, the unit of weights recorded in an ultra centrifuge for the two parts. The length of r-RNA differs in each. The 30S unit has 16S r-RNA and 21 different proteins. The 50S subunit consists of 5S and 23S r-RNA and 34 different proteins. The smaller subunit has a binding site for the m-RNA. The larger subunit has two binding sites for t-RNA.

The *messenger RNA (mRNA)* is the "blueprint" for protein synthesis and is transcribed from one strand of the DNA of the gene; it is translated at the ribosome into a polypeptide sequence. Translation is the synthesis of protein from amino acids on a template of messenger RNA in association with a ribosome. The bases on m-RNA code for amino acids in triplets or codons, that is, three bases code an amino acid. Sometimes, different

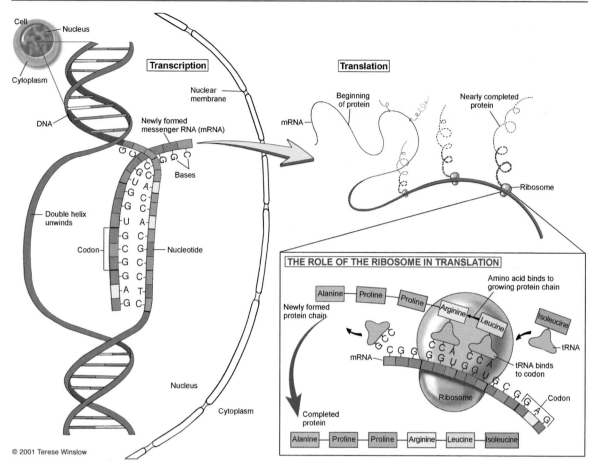

Fig. 3.3 Summary of protein synthesis activities (Copyright Terese Winslow; http://www.teresewinslow.com/. With permission) Note: There are two phases in protein synthesis: Transcription and Translation. In transcription, one strand of the DNA is used as a template for synthesizing messenger RNA (mRNA). In translation, the ribosome binds to this mRNA, and moves along it three nucleotides at a time, each triplet (known as a codon) being responsible for the synthesis of an amino acid. The polypeptide chain is increased one amino acid at a time, until a codon which does not code for any amino acid (a stop or non-sense codon) is reached.

triplet bases may code for the same amino acid. Thus the amino acid glycine is coded by four different codons: GGU, GGC, GGA, and GGG. However, a codon is always for one amino acid. There are 64 different codons; three of these UAA, UAG, and UGA are stop codons and stop the process of translation. The remaining 61 code for the amino acids in proteins (see Table 3.1). Translation of the message generally begins at AUG, which also codes for methionine. For AUG to act as a start codon, it must be preceded by a ribosome binding site. If that is not the case, it simply codes for methionine (Purvees et al. 1995).

Promoters are sequences of DNA that are the start signals for the transcription of mRNA. Terminators are the stop signals. mRNA molecules are long (500–10,000 nucleotides).

Ribosomes are the sites of translation. The ribosomes move along the mRNA and bring together the amino acids for joining into proteins by enzymes.

Transfer RNAs (tRNAs) carry amino acids to mRNA for linking and elongation into proteins. Transfer RNA is basically cloverleaf-shaped. tRNA carries the proper amino acid to the ribosome when the codons call for them. At the top of the large loop are three bases, the anticodon, which is the complement of the codon. There are 61 different tRNAs, each having a different binding site for the amino acid and a different anticodon. For the codon UUU, the complementary anticodon is AAA. Amino acid linkage to the proper tRNA is controlled by the aminoacyl-tRNA synthetases. Energy for binding the amino acid to tRNA comes from ATP conversion to adenosine monophosphate (AMP).

Table 3.1 The genetic code – codons

	U	C	A	G	
U	UUU = Phe	UCU = Ser	UAU = Tyr	UGU = Cys	U
	UUC = Phe	UCC = Ser	UAC = Tyr	UGC = Cys	C
	UUA = Leu	UCA = Ser	UAA = Stop	UGA = Stop	A
	UUG = Leu	UCG = Ser	UAG = Stop	UGG = Trp	G
C	CUU = Leu	CCU = Pro	CAU = His	CGU = Arg	U
	CUC = Leu	CCC = Pro	CAC = His	CGC = Arg	C
	CUA = Leu	CCA = Pro	CAA = Gln	CGA = Arg	A
	CUG = Leu	CCG = Pro	CAG = Gln	CGG = Arg	G
A	AUU = Ile	ACU = Thr	AAU = Asn	AGU = Ser	U
	AUC = Ile	ACC = Thr	AAC = Asn	AGC = Ser	C
	AUA = Ile	ACA = Thr	AAA = Lys	AGA = Arg	A
	AUG = Met	ACG = Thr	AAG = Lys	AGG = Arg	G
G	GUU = Val	GCU = Ala	GAU = Asp	GGU = Gly	U
	CUC = Val	GCC = Ala	GAC = Asp	GCG = Gly	C
	GUA = Val	GCA = Ala	GAA = Glu	GGA = Gly	A
	GUG = Val	GCG = Ala	GAG = Glu	GGG = Gly	G

Note: (i) Most codons for a given amino acid differ only in the last (third) base of the triplet (exceptions: Leu, Arg, Ser); (ii) One codon (AUG or Met) also signals the START of a polypeptide chain; (iii) Three codons (UAA, UAG, and UGA) are used to signal the END of a polypeptide chain (STOP codons)
AUG start codon; *UAA, UAG,* and *UGA* stop (nonsense) codons
Amino Acids: *Phe* phenylalanine, *Leu* leucine, *Ile* isoleucine, *Met* methionine, *Val* valine; *Ser* serine, *Pro* proline, *Thr* threonine, *Ala* alanine, *Tyr* tyrosine; *His* histidine, *Gln* glutamine, *Asn* asparagine, *Lys* lysine, *Asp* aspartic acid; *Glu* glutamic acid, *Cys* cysteine, *Trp* tryptophan, *Arg* arginine, *Gly* glycine

In summary, during translation, the ribosome binds to the mRNA at the start codon (AUG) that is recognized only by the initiator tRNA. During this stage, complexes, composed of an amino acid linked to tRNA, sequentially bind to the appropriate codon in mRNA by forming complementary base pairs with the tRNA anticodon. The ribosome moves from codon to codon along the mRNA. Amino acids are added one by one, translated into polypeptidic sequences dictated by DNA and represented by mRNA. At the end, a release factor binds to the stop codon, terminating translation and releasing the complete polypeptide from the ribosome.

Elongation terminates when the ribosome reaches a stop codon, which does not code for an amino acid and hence not recognized by tRNA.

After protein has been synthesized, the primary protein chain undergoes folding; secondary, tertiary, and quadruple folding occurs. The folding exposes chemical groups which confer their peculiar properties on the protein. Protein folding is the process by which a protein assumes its functional shape or conformation. All protein molecules are simple unbranched chains of amino acids, but it is by coiling into specific three-dimensional shapes that they are able to perform their biological functions.

3.2 The Polymerase Chain Reaction

The Polymerase Chain Reaction (PCR) is a technology used to amplify small amounts of DNA. The PCR technique was invented in 1985 by Kary B. Mullis while working as a chemist at the Cetus Corporation, a biotechnology firm in Emeryville, California. So useful is this technology that Muillis won the Nobel Prize for discovery in 1993, 8 years later. It has found use in a wide range of situations, from the medical diagnosis to microbial systematics and from courts of law to the study of animal behavior.

The requirements for PCR are:
(a) The DNA or RNA to be amplified
(b) Two primers
(c) The four nucleotides found in the nucleic acid
(d) A heat stable DNA derived from the thermophilic Archaebacterium, *Thermus aquaticus, Taq* polymerase

The Primer: A primer is a short segment of nucleotides which is complementary to a section of the DNA which is to be amplified in the PCR reaction.

Primers are annealed to the denatured DNA template to provide an initiation site for the elongation of the new DNA molecule. For PCR, primers must be duplicates of nucleotide sequences on either side of the piece of DNA of interest, which means that the exact order of the primers' nucleotides must already be known. These flanking sequences can be constructed in the lab, or purchased from commercial suppliers.

The Procedure: There are three major steps in a PCR, which are repeated for 30 or 40 cycles. This is done on an automated cycler (thermocycler), which can heat and cool the tubes with the reaction mixture in a very short time.

(a) *Denaturation* at 94°C

The unknown DNA is heated to about 94°C, which causes the DNA to denature and the paired strands to separate.

(b) *Annealing* at 54°C

A large excess of primers relative to the amount of DNA being amplified is added and the reaction mixture cooled to allow double-strands to anneal; because of the large excess of primers, the DNA single strands will bind more to the primers, instead of with each other.

(c) *Extension* at 72°C

This is the ideal working temperature for the polymerase. Primers that are on positions with no exact match, get loose again (because of the higher temperature) and do not give an extension of the fragment. The bases (complementary to the template) are coupled to the primer on the 3′ side (the polymerase adds dNTP's from 5′ to 3′, reading the template from 3′ to 5′ side, bases are added complementary to the template). The process of the amplification in the PCR process is shown in Fig. 3.4.

3.2.1 Some Applications of PCR in Environmental Biotechnology

PCR is extremely efficient and simple to perform. It is useful in biotechnology in the following areas:

(a) To generate large amounts of DNA for genetic engineering, or for sequencing, once the flanking sequences of the gene or DNA sequence of interest is known

(b) To determine with great certainty the identity of an organism to be used in an environmental biotechnology, such as those to be used in bioremediation

(c) To determine rapidly which organism is the cause of contamination in a production process so as to eliminate its cause, provided the primer appropriate to the contaminant is available

3.3 Microarrays

The availability of complete genomes from many organisms is a major achievement of biology. Aside from the human genome, the sequencing of the complete genomes of many microorganisms has been completed and the sequences are now available at the website of The Institute for Genomic Research (TIGR), a nonprofit organization located in Rockville, MD with its website at www.tigr.org. At the time of writing, TIGR had the complete genome of 294 microorganisms on its website (268 bacteria, 23 Archae, and 3 viruses). The major challenge is now to decipher the biological function and regulation of the sequenced genes. One technology important in studying functional microbial genomics is the use of DNA Microarrays (Hinds et al. 2002a).

Microarrays are microscopic arrays of large sets of DNA sequences that have been attached to a solid substrate using automated equipment. These arrays are also referred to as microchips, biochips, DNA chips, and gene chips. It is best to refer to them as microarrays so as to avoid confusing with computer chips.

DNA microarrays are small, solid supports onto which the sequences from thousands of different genes are immobilized at fixed locations. The supports themselves are usually glass microscope slides; silicon chips, or nylon membranes may be used. The DNA is printed, spotted, or, actually, directly synthesized onto the support mechanically at fixed locations or addresses. The spots themselves can be DNA, cDNA, or oligonucleotides (Fig. 3.5).

The process is based on hybridization probing. Single-stranded sequences on the microarray are labeled with a fluorescent tag or flourescein, and hybridized to, in fixed locations on the support. In microarray assays, an unknown sample is hybridized to an ordered array of immobilized DNA molecules of known sequence to produce a specific hybridization pattern that can be analyzed and compared to a given standard. The labeled DNA strand in solution is generally called the target, while the DNA immobilized on the microarray is the probe, a terminology opposite that is used in

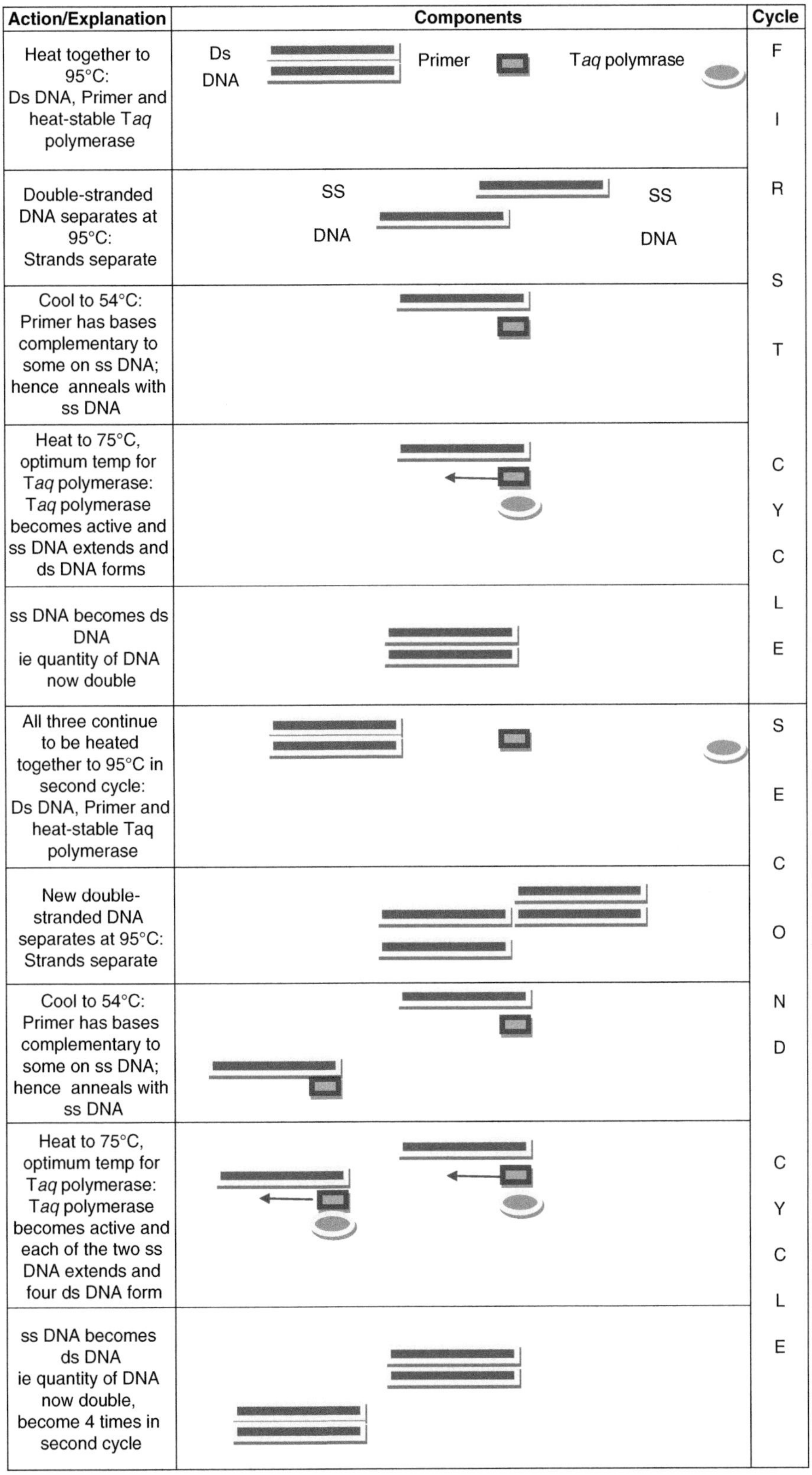

Action/Explanation	Components	Cycle
Heat together to 95°C: Ds DNA, Primer and heat-stable Taq polymerase	Ds DNA Primer Taq polymrase	F I
Double-stranded DNA separates at 95°C: Strands separate	SS DNA SS DNA	R S
Cool to 54°C: Primer has bases complementary to some on ss DNA; hence anneals with ss DNA		T
Heat to 75°C, optimum temp for Taq polymerase: Taq polymerase becomes active and ss DNA extends and ds DNA forms		C Y C
ss DNA becomes ds DNA ie quantity of DNA now double		L E
All three continue to be heated together to 95°C in second cycle: Ds DNA, Primer and heat-stable Taq polymerase		S E C
New double-stranded DNA separates at 95°C: Strands separate		O
Cool to 54°C: Primer has bases complementary to some on ss DNA; hence anneals with ss DNA		N D
Heat to 75°C, optimum temp for Taq polymerase: Taq polymerase becomes active and each of the two ss DNA extends and four ds DNA form		C Y C L
ss DNA becomes ds DNA ie quantity of DNA now double, become 4 times in second cycle		E

Fig. 3.4 Diagrammatic representation of PCR

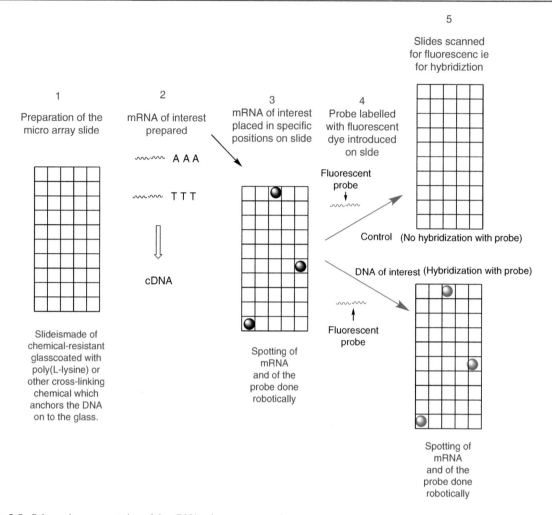

Fig. 3.5 Schematic representation of the cDNA microarray procedure

Southern blot. Microarrays have the following advantages over other nucleic acid–based approaches:

(a) High throughput: Thousands of array elements can be deposited on a very small surface area enabling gene expression to be monitored at the genomic level. Also, many components of a microbial community can be monitored simultaneously in a single experiment.

(b) High sensitivity: Small amounts of the target and probe are restricted to a small area ensuring high concentrations and very rapid reactions.

(c) Differential display: Different target samples can be labeled with different fluorescent tags and then hybridized to the same microarray, allowing the simultaneous analysis of two or more biological samples.

(d) Low background interference: Non-specific binding to the solid surface is very low resulting in easy removal of organic and fluorescent compounds that attach to microarrays during fabrication.

(e) Automation: Microarray technology is amenable to automation making it ultimately cost-effective when compared with other nucleic acid technologies.

3.3.1 Applications of Microarray Technology

Microarray technology is still young but yet it has found use in a few areas which have importance in microbiology in general as well as in.

(a) Disease diagnosis

(b) Drug discovery

(c) Toxicological research

(d) Environmental microbiology (Zhou 2002)

Microarrays are particularly useful in studying gene function. A microarray works by exploiting the ability of a given mRNA molecule to bind specifically to, or hybridize to, the DNA template from which it originated. By using an array containing many DNA samples, it is possible to determine, in a single experiment, the expression levels of hundreds or thousands of genes within a cell by measuring the amount of mRNA bound to each site on the array. With the aid of a computer, the amount of mRNA bound to the spots on the microarray is precisely measured, generating a profile of gene expression in the cell. It is thus possible to determine the bioactive potential of a particular microbial metabolite as a beneficial material in the form of a drug or its deleterious effect (Madigan and Martinko 2006).

When a diseased condition is identified through microarray studies, experiments can be designed which may be able to identify compounds, from microbial metabolites or other sources, which may improve or reverse the diseased condition.

In the environment, it can be used to determine the nature of the microbial population without necessarily isolating the organisms (Hinds et al. 2002b).

3.4 Sequencing of DNA

3.4.1 Sequencing of Short DNA Fragments

DNA sequencing is the determination of the precise sequence of nucleotides in a sample of DNA. Two methods developed in the mid-1970s are available: The Maxim and Gilbert method and the Sanger method. Both methods produce DNA fragments which are studied with gel electrophoresis. The Sanger method is more commonly used and will be discussed here. The Sanger method is also called the dideoxy method, the enzymic method. The dideoxy method gets its name from the critical role played by synthetic analogues of nucleotides that lack the -OH at the 3' carbon atom (star position): dideoxynucleotide triphosphates (ddNTP) (see Fig. 3.6). When (normal) deoxynucleotide triphosphates (dNTP) are used, the DNA strand continues to grow, but when the dideoxy analogue is incorporated, chain elongation stops because there is no 3' − ' -OH for the next nucleotide to be attached to.

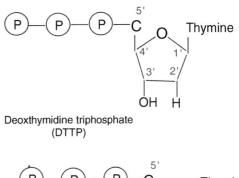

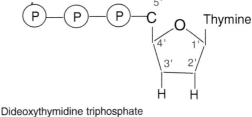

Fig. 3.6 Normal and dedeoxy nucleotides

For this reason, the dideoxy method is also called the chain termination method.

For Sanger sequencing, a single strand of the DNA to be sequenced is mixed with a primer, DNA polymerase I, an excess of normal nucleotide triphosphates and a limiting (about 5%) of the dideoxynucleotides labeled with a fluorescent dye, each ddNTP being labeled with a different fluorescent dye color. This primer will determine the starting point of the sequence being read, and the direction of the sequencing reaction. DNA synthesis begins with the primer and terminates in a DNA chain when ddNTP is incorporated in place of normal dNTP. Because all four normal nucleotides are present, chain elongation proceeds normally until, by chance, DNA polymerase inserts a dideoxy nucleotide instead of the normal deoxynucleotide. The result is a series of fragments of varying lengths. Each of the four nucleotides is run separately with the appropriate ddNTP. The mix with the ddCTP produces fragments with C (cytosine); that with ddTTP (thymine) produces fragments with T terminals, etc. The fluorescent strands are separated from the DNA template and electrophoresed on a polyacrilamide gel to separate them according to their lengths. If the gel is read manually, four lanes are prepared, one for each of the four reaction mixes. The reading is from the bottom of the gel up, because the smaller the DNA fragment the faster it is on the gel. A picture of the sequence of the nucleotides can be read from the gel (see Fig. 3.7). If the system is automated, all four are

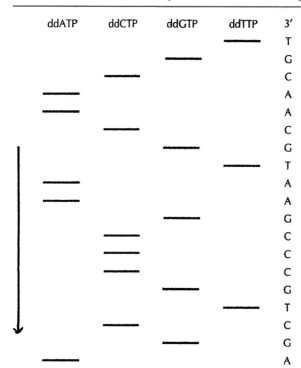

Fig. 3.7 Diagram illustrating autoradiograph of a sequencing gel of the chain terminating DNA sequencing method (*Arrow* shows direction of the electrophoresis. By convention the autoradiograph is read from *bottom* to the *top*)

mixed and electrophoresed together. Because the ddNTPs are of different colors, a scanner can scan the gel and record each color (nucleotide) separately. This can be used for relatively short fragments of DNA, 700–800 nucleotides.

3.4.2 Sequencing of Genomes or Large DNA Fragments

The best example of the sequencing of a genome is perhaps the human genome, which was completed a few years ago. During the sequencing of the human genome, two approaches were followed: The use of bacterial artificial chromosomes (BACs) and the short gun approach.

3.4.2.1 Use of Bacterial Artificial Chromosomes

The Human Genome Project was publicly funded, and the National Institutes of Health and the National Science Foundation have funded the creation of "libraries" of BAC clones. Each BAC carries a large piece of human genomic DNA in the order of 100–300 kb. All of these BACs overlap randomly, so that any one gene is probably on several different overlapping BACs. Those BACs can be replicated as many times as necessary, so there is a virtually endless supply of the large human DNA fragment. In the publically funded project, the BACs are subjected to shotgun sequencing (see below) to figure out their sequence. By sequencing all the BACs, we know enough of the sequence in overlapping segments to reconstruct how the original chromosome sequence looks.

3.4.2.2 Use of the Shotgun Approach

An innovative approach to sequencing the human genome was pioneered by a privately funded sequencing project: Celera Genomics. The founders of this company realized that it might be possible to skip the entire step of making libraries of BAC clones. Instead, they blast apart the entire human genome into fragments of 2–10 kb and sequence those. Now, the challenge is to assemble those fragments of sequence into the whole genome sequence. It was like having hundreds of 500-piece puzzles, each being assembled by a team of puzzle experts using puzzle-solving computers. Those puzzles are like BACs – smaller puzzles that make a big genome manageable. Celera threw all those puzzles together into one room and scrambled the pieces. They, however, have scanners that scan all the puzzle pieces and huge powerful computers to fit the pieces together.

3.5 The Open Reading Frame and the Identification of Genes

Regions of DNA that encode proteins are first transcribed into messenger RNA and then translated into protein. By examining the DNA sequence alone, we can determine the putative sequence of amino acids that will appear in the final protein. In translation, codons of three nucleotides determine which amino acid will be added next in the growing protein chain. The start codon is usually AUG, while the stop codons are UAA, UAG, and UGA. The Open Reading Frame (ORF) is that portion of a DNA segment which will putatively code for a protein; it begins with a start codon and ends with a stop codon.

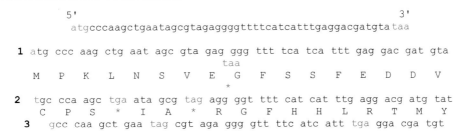

Fig. 3.8 Sequence from a hypothetical DNA fragment (From Cooper (2008), on behalf of Board of Regents, University of Wisconsin; http://bioweb.uwlax.edu/genweb/molecular/seq_anal/translation/translation.html. Reproduced with permission)

Table 3.2 Some Internet tools for the gene discovery in DNA sequence bases (Modified from Fickett 1996)

Category	Services	Organism(s)	Web address
Database search	BLAST; search sequence bases	Any	blast@ncbi.nlm.nih.gov
	FASTA; search sequence bases	Any	fasta@ebi.ac.uk
	BLOCKS; search for functional motifs	Any	blocks@howard.flicr.org
	Profilescan	Any	http://ulrec3.unil.ch.
	MotifFinder	Any	motif@genome.ad.jp
Gene identification	FGENEH; integrated gene identification	Human	service@theory.bchs.uh.edu
	GeneID; integrated gene identification	Vetebrate	geneid@bircebd.uwf.edu
	GRAIL; integrated gene identification	Human	grail@ornl.gov
	EcoParse; integrated gene identification	*Escherichia coli*	

Once a gene has been sequenced, it is important to determine the correct open reading frame. Every region of DNA has six possible reading frames, three in each direction because a codon consists of three nucleotides. The reading frame that is used determines which amino acids will be encoded by a gene. Typically, only one reading frame is used in translating a gene (in eukaryotes), and this is often the longest open reading frame. Once the open reading frame is known, the DNA sequence can be translated into its corresponding amino acid sequence.

For example, the sequence of DNA in Fig. 3.8 can be read in six reading frames: three in the forward and three in the reverse direction. The three reading frames in the forward direction are shown with the translated amino acids below each DNA sequence. Frame 1 starts with the "a," Frame 2 with the "t," and Frame 3 with the "g." Stop codons are indicated by an "*" in the protein sequence. The longest ORF is in Frame 1.

Genes can be identified in a number of ways, which are discussed below:

1. *Using computer programs*

    As was shown above, the open reading frame (ORF) is deduced from the start and stop codons. In prokaryotic cells which do not have many extrons (intervening non-coding regions of the chromosome), the ORF will in most cases indicate a gene. However, it is tedious to manually determine ORF and many computer programs now exist which will scan the base sequences of a genome and identify putative genes. Some of the programs are given in Table 3.2. In scanning a genome or DNA sequence for genes (i.e., in searching for functional ORFs), the following are taken into account in the computer programs:

    (a) Usually, functional ORFs are fairly long and do not usually contain less than 100 amino acids (i.e., 300 codons).

    (b) If the types of codons found in the ORF being studied are also found in known functional ORFs, then the ORF being studied is likely to be functional.

    (c) The ORF is also likely to be functional if its sequences are similar to functional sequences in genomes of other organisms.

    (d) In prokaryotes, the ribosomal translation does not start at the first possible (earliest 5′) codon. Instead, it starts at the codon immediately downstream of the Shine–Dalgardo binding site sequences. The Shine–Dalgardo sequence is a short sequence of nucleotides upstream of the translational start site that binds to ribosomal RNA and thereby brings the ribosome to the initiation codon on the mRNA. The computer program searches for a Shine–Dalgardo sequence and finding it helps to indicate not only which start codon is used, but also that the ORF is likely to be functional.

(e) If the ORF is preceded by a typical promoter (if consensus promoter sequences for the given organism are known, check for the presence of a similar upstream region).

(f) If the ORF has a typical GC content, codon frequency, or oligonucleotide composition of known protein-coding genes from the same organism, then it is likely to be a functional ORF.

2. *Comparison with Existing Genes*
Sometimes, it may be possible to deduce not only the functionality or not of a gene (i.e., a functional ORF), but also the function of a gene. This can be done by comparing an unknown sequence with the sequence of a known gene available in databases such as The Institute for Genomic Research (TIGR) in Maryland (Rogic et al. 2001).

3.6   Metagenomics

Metagenomics is the genomic analysis of the collective genome of an assemblage of organisms or "metagenome." Metagenomics describes the functional and sequence-based analysis of the collective microbial genomes contained in an environmental sample (Fig. 3.9). Other terms have been used to describe the same method, including environmental DNA libraries, zoolibraries, soil DNA libraries, eDNA libraries, recombinant environmental libraries, whole genome treasures, community genome, and whole genome shotgun sequencing. The definition applied here excludes studies that use PCR to amplify gene cassettes or random PCR primers to access genes of interest since these methods do not provide genomic information beyond the genes that are amplified.

Many environments have been the focus of metagenomics, including soil, the oral cavity, feces, and aquatic habitats, as well as the "hospital metagenome" a term intended to encompass the genetic potential of

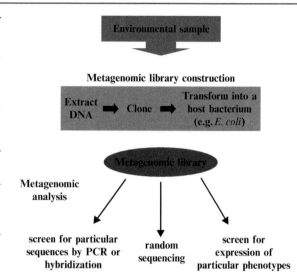

Fig. 3.9 Schematic procedure for metagenomic analysis (Modified from Riesenfeld et al. 2004)

organisms in hospitals that contribute to public health concerns such as antibiotic resistance and nosocomial infections.

Uncultured microorganisms comprise the majority of the planet's biological diversity. In many environments, as many as 99% of the microorganisms cannot be cultured by standard techniques, and the uncultured fraction includes diverse organisms that are only distantly related to the cultured ones. Therefore, culture-independent methods are essential to understand the genetic diversity, population structure, and ecological roles of the majority of microorganisms in a given environmental situation. Metagenomics, or the culture-independent genomic analysis of an assemblage of microorganisms, has the potential to answer fundamental questions in microbial ecology. Several markers have been used in metagenomics, including 16S mRNA, and the genes encoding DNA polymerases, because these are highly conserved (i.e., because they remain relatively unchanged in many groups). The marker most commonly used, however, is the sequence of 16S mRNA. The procedure in metagenomics is described in Fig. 3.9. Its potential application in biotechnology and environmental microbiology is that it can facilitate the identification of uncultured organisms whose role in a multi-organism environment such as sewage or the degradation of a recalcitrant chemical soil may be hampered because of the inability to culture the organism.

References

Bansal, K. A. (2005). Bioinformatics in microbial biotechnology – a mini review. *Microbial Cell Factories, 2005*(4), 19–30.

Cooper, S. (2008). Translation and open reading frames. http://bioweb.uwlax.edu/genweb/molecular/seq_anal/translation/translation.html. Accessed 30 Aug 2010.

Fickett, J. W. (1996). The gene identification problem: An overview for developers. *Computers & Chemistry, 20*, 103–118.

Elshahed, M. S., Youssef, N. H., Luo, Q., Najar, F. Z., Roe, B. A., Sisk, T. M., Bühring, S. I., Hinrichs, K. U., & Krumholz, L. R. (2007). Phylogenetic and metabolic diversity of Planctomycetes from anaerobic, sulfide- and sulfur-rich Zodletone Spring Oklahoma. *Applied and Environmental Microbiology, 73*, 4707–16.

Hinds, J., Liang, K. G., Mangan, J. A., & Butecer, P. D. (2002a). Glass slide microarrays for bacterial genomes. In *Methods in microbiology* (Vol. 33, pp. 83–99). Amsterdam: Academic.

Hinds, J., Witney, A. A., & Vaas, J. K. (2002b). Microarray design for bacterial genomes. In *Methods in microbiology* (Vol. 33, pp. 67–82). Amsterdam: Academic.

Madigan, M., & Martinko, J. M. (2006). *Brock biology of microorganisms* (11th ed.). Upper Saddle Rive: Pearson Prentice Hall.

Purvees, W. K., Orians, Helter (1995) Life: The Science of Biology (4th ed.). Sinaurs Associates. Sunderland, MA.

Riesenfeld, C. S., Schloss, P. D., & Handelsman. (2004). Metagenomics: Genomic analysis of microbial communities. *Annual Review of Genetics, 38*, 525–52.

Rogic, S., Mackworth, A. K., & Ouellette, F. B. F. (2001). Evaluation of gene-finding programs on mammalian sequences. *Genome Research, 11*, 817–832.

Zhou, J. (2002). Microarrays: Applications in environmental microbiology. In *Encyclopedia of environmental microbiology* (Vol. 4, pp. 1968–1979). New York: Wiley Interscience.

Taxonomy, Physiology, and Ecology of Aquatic Microorganisms

4

Abstract

The principles behind the taxonomy of the microorganisms, especially the molecular approach (using the sequence of the 16S RNA in the small subunit of the ribosome) in the identification of bacteria, are discussed. The detailed taxonomy of bacteria, fungi, algae, protozoa, and viruses (including bacteriophages) is discussed, and emphasis is laid on those microorganisms which are aquatic. The chapter includes information on some of the smaller macroorganisms found in water such as nematodes and rotifers. The activities of aquatic microorganisms in photosynthesis, and the global cycling of nitrogen and sulfur is discussed.

Keywords

Taxonomy of microbial groups • Photosynthesis in aquatic microorganisms • Aquatic nitrogen and sulfur cycles • Rotifers and nematodes • Bacteria • Fungi • Protozoa • Algae • Three domains of living things • Woese • 16S or 18S RNA

4.1 Taxonomy of Microorganisms in Aquatic Environments

In this section, the classification of microorganisms will be discussed and emphasis will be laid on those microorganisms found in aquatic systems. The following are the organisms to be discussed:

Bacteria
Archeae
Eukarya
Fungi
   Algae
   Protozoa
Viruses

4.1.1 Nature of Modern Taxonomy

Modern taxonomy is the science of biological classification. It consists of three sections:

(a) *Classification*: The theory and process of arranging organisms into taxonomic groups or taxa (singular, taxon), on the basis of shared properties.

(b) *Nomenclature*: The assignment of names to taxonomic groups.

(c) *Identification*: The determination of the taxon to which a particular organism belongs, based on the properties of the organism.

N. Okafor, *Environmental Microbiology of Aquatic and Waste Systems*,
DOI 10.1007/978-94-007-1460-1_4, © Springer Science+Business Media B.V. 2011

Taxonomy is important because it:

1. Allows the orderly organization of huge amounts of information regarding organisms
2. Enables predictions about their properties and formation of hypothesis about them
3. Facilitates the accurate characterization and identification of "unknown" organisms
4. Places organisms in meaningful manageable groups and thus facilitates scientific communication

4.1.2 Evolution of the Classification of Living Things

Landmarks in the evolution and development of biological classification may be ascribed to the contributions of the following:

1. *Linnaeus (1707–1778)*

   The Swedish naturalist, Carolus Linnaeus, is credited with introducing the earliest organized classification of living things in his *Systema Naturae* or natural system. He divided living things into plants and animals. Based on morphology and motility, the distinction between the two groups of organisms was clear: plants were green and did not move; on the other hand, animals were not green, but moved.

2. *Ernst Haeckel (1834–1919)*

   Soon after the discovery of the microscope, previously invisible microscopic organisms were observed, some of which had properties common to both plants and animals. Some such as *Euglena* were green like, but they also moved about like animals. Because the clear-cut criteria which separated plants from animals were absent in these "new" organisms, the German biologist who was a contemporary of Charles Darwin, in 1866 coined the name Protista for a third kingdom, in addition to the Plant and Animal Kingdoms.

3. *Robert Harding Whittaker (1920–1980)*

   Whittaker was an American. Born in Wichita, Kansas, he worked in various places including the University of California, Irvine, and Cornell University. In 1968, he proposed the five-kingdom taxonomic classification of living things into the Animalia, Plantae, Fungi, Protista (Algae and Protozoa), and Monera (Bacteria). His categorization of living things was based on three criteria: cell-type (whether prokaryotic or eukaryotic); organizational level (unicellular or multi-cellular); nutritional type (autotrophy or heterotrophy).

4. *Carl R. Woese (1928–)*

   The current classification of living things is based on the work of Carl Robert Woese of the University of

Illinois. While earlier classifications were based mainly on morphological characteristics and cell-type, following our greater understanding of living things at the molecular level, Woese's classification is based on the sequence of the gene of the ribosomal RNA (rRNA) in the 16S of the small subunit of the prokaryotic ribosome, or the 18S of the small subunit of the eukaryotic ribosome (Petti et al. 2006).

The sequence of the rRNA in the 16S or 18S of the small subunit of the ribosome is used for the following reasons:

(a) The ribosome is an important organelle in all living things where it is used for a basic function for the support of life, namely, protein synthesis.

(b) The 16S (prokaryote) or 18 S (eukaryote) rRNA is an essential component of the ribosome.

(c) The function of 16S or 18 S rRNA is identical in all ribosomes.

(d) The sequences of the 16S or 18 S rRNA are ancient (or highly conserved) and change only slowly with evolutionary time.

(e) Organisms can generally inherit genes in two ways: From parent to offspring (vertical gene transfer), or by horizontal or lateral gene transfer, in which genes jump between unrelated organisms, a common phenomenon in prokaryotes. There is little or no lateral gene transfer in the sequences in the 16S or 18 S RNA of the ribosomal small units.

All the above properties make the sequence of the rRNA in the 16S or 18S of the small subunit of the ribosome useful as molecular chronometers for measuring evolutionary changes among organisms. Using this method, living things are now divided into three domains: Archae, Bacteria, and Eukarya. A diagrammatic representation of the three domains is given in Fig. 4.1, and their distinguishing properties are given in Table 4.1 (Woese 1987, 2000, 2002).

4.1.3 Determining Taxonomic Groups Within Domains

The smallest unit of biological classification is the *Species*. Species sharing similar properties are put in a *Genus*. Genera (plural of genus) sharing similar characteristics are put in a *Family*. Families with similar properties are arranged in an *Order*. Orders with similar properties are classified as into a *Class*. Classes which share similar properties are grouped into a *Phylum*. Phyla (plural of phylum) with similar properties are put in a *Kingdom* and similar kingdoms are in a *Domain*.

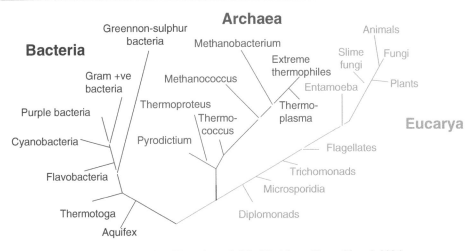

Fig. 4.1   Three domains of living things based on Woese's work (Modified from Ciccarelli et al. 2006)

Table 4.1   Summary of differences among the three domains of living things (Modified from Madigan and Martinko 2006)

S. No.	Characteristic	Bacteria	Archae	Eukarya
1.	Prokaryotic cell structure	+	+	−
2.	DNA present in closed circular form	+	+	−
3.	Histone proteins present[a]	−	+	+
4.	Nuclear membrane	−	−	+
5.	Muramic acid in cell wall	+	−	−
6.	Membrane lipids: Fatty acids or branched hydrocarbons	Fatty acids	Branched hydrocarbons	Fatty acids
	Unbranched fatty acid chains attached to glycerol by ester linkages	+	−	+
7.	Ribosome size	70 S	70 S	80 S
8.	Initiator t-RNA	Formyl-methionine	Methionine	Methionine
9.	Introns in most genes[b]	−	−	+
10.	Operons[c]	+	+	−
11.	Plasmids	+	+	Rare
12.	Ribosome sensitive to diphtheria toxin	−	+	+
13.	Sensitivity to streptomycin, chloramphenicol, and kanamycin	+	−	−
Physiological/special structures				
14.	Methanogenesis	+	+	−
15.	Denitrification	+	+	−
16.	Nitrogen fixation	+	+	−
17.	Chlorophyll-based photosynthesis	+	−	+ (Plants)
18.	Gas vesicles	+	+	−
19.	Chemolithotrophy	+	+	−
20.	Storage granules of poly-β-hydroxyalkanoates	+	+	−
21.	Growth above 80°C	+	+	−
22.	Growth above 100°C	−	+	−

[a]Histone proteins are present in eukaryotic chromosomes; histones and DNA give structure to chromosomes in eukaryotes; proteins in archeae chromosomes are different
[b]Noncoding sequences within genes
[c]Operons: Typically present in prokaryotes, these are clusters of genes controlled by a single operator
[d]TATA box (also called Hogness Box): an AT-rich region of the DNA with the sequence TATAT/AAT/A located before the initiation site
[e]Transcription factor is a protein that binds DNA at a specific promoter or enhancer region or site, where it regulates transcription

Several mnemonics exist to help with remembering the correct order of the listing of taxonomic groups. Two are given below:

Domain	Kingdom	Phylum	Class	Order	Family	Genus	Species
David	Kindly	Pay	Cash	Or	Furnish	Good	Security
Dignified	Kings	Play	Chess	On	Fine	Green	Silk

The names of the ascending taxonomic groups to which the fruit fly, humans, peas, and the bacterium, *E coli* belong are given in Table 4.2.

4.1.3.1 Definition of Species

In the Domain Eukarya, a species is defined as a group of organisms which can mate and produce fertile offspring. Even though goats are of different kinds, they will mate and produce fertile offspring; similarly with dogs. The horse and the donkey are of different species, because although they can mate and produce offspring, the offspring are not fertile. This definition is occasionally complicated by the lateral transfer of genes.

In Bacteria and Archae, the definition is a little different. A species in these Domains is a collection of strains that share many stable properties and differ significantly from other groups.

4.1.3.2 Nomenclature of Biological Objects

Biological objects, including microorganisms are named in the binomial system devised by Carolus Linnaeus. The genus name is written first and begins with an uppercase (capital) letter; the other half of the name is written in lowercase (small) letters and is the species name. The two are written in italics or underlined if written in long hand. When written formally, the name of the author who first described the organism is included and the year of the publication is given; the names are usually written in Latin or latinized. The name of a hypothetical Bacillus discovered in water by John Smith and published in 2007 could be *Bacillus aquanensis* Smith 2007. Usually only the genus and species names are given; the author and year of publication are omitted (van Regenmortel 1999).

4.1.3.3 Criteria and Methods for the Identification and Classification of Bacteria and Archae: Morphological, Physiological, Nucleic Acid, and Chemical Properties

Whereas members of the Domain Eukarya are classified largely on their morphological characteristics which are adequately diverse, morphological types are very limited in the Domains Bacteria and Archae. Therefore, while morphological properties are used, other characteristics are employed in addition to morphology. The properties used for classifying and identifying unknowns among organisms in the Domains Bacteria and Archae are given in Table 4.3. The principles of methods used are described briefly below.

Morphological and Physiological Methods

1. *Nutritional types of Bacteria*

   Living things are classified into major nutritional types on the basis of the following attributes:

   (a) Carbon source utilized

   A carbon skeleton is required for the compounds used for growth and development such as carbohydrates, amino acids, fats, etc. The organism is *autotrophic* if it manufactures its food and obtains its carbon through fixing CO_2 such as is the case with plants, algae, and some bacteria. When the organism cannot manufacture its own food from CO_2 but must utilize food already manufactured from CO_2, in the form of carbohydrates, proteins etc., it is *heterotrophic*. This is the case with animals and most bacteria.

Table 4.2 Taxonomic groups of some organisms (From Anonymous 2008a)

Rank	Human	Fruit fly	Pea	*E. coli*
Domain	Eukarya	Eukarya	Eukarya	Bacteria
Kingdom	Animalia	Animalia	Plantae	Bacteria
Phylum (animals) or division (plants)	Chordata	Arthropoda	Magnoliophyta	Proteobacteria
Class	Mammalia	Insecta	Magnoliopsida	Gammaproteobacteria
Order	Primates	Diptera	Fabales	Enterobacteriales
Family	Hominidae	Drosophilidae	Fabaceae	Enterobacteriaceae
Genus	*Homo*	*Drosophila*	*Pisum*	*Escherichia*
Species	*H. sapiens*	*D. melanogaster*	*P. sativum*	*E. coli*

Table 4.3 Some properties used for bacterial classification and identification

S. No.	Property
1.	Nutritional type
	(i) Autotrophy
	(ii) Heterotrophy
2.	Energy release
	(i) Lithotrophy
	(ii) Organotrophy
3.	Cell wall: Gram reaction
	(i) Gram negative
	(ii) Gram positive
4.	Cell morphology
	(i) Cell shapes
	(ii) Cell aggregation
	(iii) Flagellation – motility
	(iv) Spore formation and location
	(v) Special staining, e.g., Ziehl–Nielsen
5.	Physiological properties
	(i) Utilization of various sugars
	(ii) Utilization of various polysaccharides
	(iii) Utilization of various nitrogenous substrates
	(iv) Oxygen requirement
	(v) Temperature requirements
	(vi) pH requirement
	(vii) Production of special enzymes, e.g., catalase, coagulase, optochin, oxidase
6.	Antigenic properties
7.	Molecular (nucleic acid) methods
	(i) G + C composition
	(ii) DNA:DNA hybridization
	(iii) Ribotyping
	(iv) Fluorescent in-situ hybridization (FISH)
8.	Chemical analysis (Chemotaxonomy)
	(i) Lipid analysis
	(ii) Protein analysis

(b) Source of reducing equivalent

During the generation of energy in the cell, electrons are transferred from one compound to another. An organism is said to be *organotrophic* when it uses organic compounds as a source of electrons. When the source of electrons is inorganic, it is said to be *lithotrophic*.

(c) Source of energy

Some organisms derive energy for the generation of ATP used for the biosynthesis of new compounds and other cellular activities from sunlight; such organisms are *phototrophic*. When the generation of ATP occurs through energy obtained from chemical reactions, the organism is said to be *chemotrophic*.

The carbon source utilized, the source of reducing equivalent, and the source of energy determine the nutritional type of bacteria, and a wide variety of combinations of these three is possible. Table 4.4 gives a selection of the possible permutations.

2. *Cell wall: Gram reaction*

The Gram stain was devised by the German doctor, Christian Gram in 1884 and divides bacteria into two groups: Gram positive and Gram negative. On account of the greater thickness of peptidoglycan in the Gram positive wall (see Fig. 4.2), the iodine-crystal violet stain in the Gram stain is retained when decolorized with dilute acid, whereas it is removed in the Gram negative cell wall. The Gram stain also divides all bacteria into two groups regarding their susceptibility to the classical antibiotic penicillin: Gram-positive bacteria, being susceptible, while Gram negative bacteria are not (Fig. 4.3).

3. *Cell morphology*

(a) Individual cell shapes

Cell shapes in bacteria are limited and are spheres (coccus– cocci, plural), rods, spiral, or comma or vibrio (see Fig. 4.4).

Table 4.4 Nutritional types of living things

S/No	Nutritional type	Energy source	Carbon source	Reducing equivalent	Example
1	Photoautotrophs	Light	CO_2	Organotrophic	Plants, Cyanobacteria
2	Photoautotrophs	Light	CO_2	Lithotrophic $H_2S \rightarrow S$	Sulfur bacteria e.g., *Beggitoa* sp.
3	Chemoautotrophs	Chemotrophic	CO_2	Oxidation of sulfur	*Thiobacillus oxidans*
4	Photoheterotrophs	Light	Organic compounds	Organotrophic	Purple non-sulfur bacteria
5	Chemoheterotrophs	Chemotrophic	Organic compounds	Organotrophic	Animals, fungi, protozoa, most bacteria

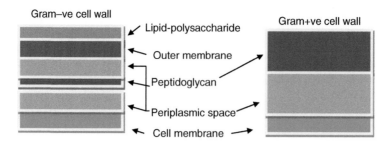

Fig. 4.2 Diagram illustrating the generalized structure of the bacterial cell wall (Note the comparative thicknesses of the peptidoglycan layers in the Gm+ve and Gm−ve walls)

Note that the peptidoglycan layer is very thick in Gram +ve walls, but very thin in Gram −ve bacterial cells. This thick peptidoglycan enables Gram +ve walls retain crystal violet, the primary stain in the Gram, when decolorized with dilute acid.

Crystal violet is not retained in Gram −ve bacteria because the peptidoglycan layer is thin. The Gram −ve wall would be colorless after decolorization with dilute acid. However in the Gram stain, after decolourization, the cells are counterstained with a red stain, safranin. On account of this the Gram negative cells appear red in the Gram stain, while Gram +ve cells are violet (see text and Fig. 4.3)

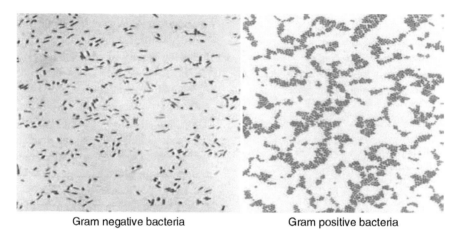

Gram negative bacteria   Gram positive bacteria

Fig. 4.3 Gram staining

(b) Cell aggregates

Cocci can occur in pairs (diplococci), in chains (streptococci), or in clumps (staphylococci). Rods may occur in short chains of two or three or in long chains or filaments.

(c) Flagellation

The flagella may be at one end (polar) and may occur singly or as a tuft. The flagella may occur all around the cell when it is peritrichous (see Fig. 4.5).

(d) Spores and location of spores

Spores are bodies resistant to heat and other adverse conditions which may be terminal or placed mid-way in the cell; in either position, it may be less than the diameter of the cell or may be wider. The terminal wider spore gives the shape of a drumstick, and is diagnostic of the

anaerobic rod-like spore-former, *Clostridium tetanii*, the causative agent of tetanus (see Fig. 4.6).

(e) Acid-fast (Ziel–Nieelsen) stain

If the bacterium is suspected to belong to *Mycobacterium* spp., or any of the other acid-fast bacteria it might be stained with hot basic fuchsin; acid fast bacteria retain the dye when decolorized.

4. *Utilization of various substrates*

Utilization of various sugars, carbohydrates, and nitrogenous sources

The ability of the organism to produce acid and/or gas from a medium containing a particular substrate is diagnostic of its ability to utilize it. The utilization of a wide range of sugars and other carbohydrates, and nitrogen sources including urea is tested by the presence of gas in the small (Durham) tube placed in the

Fig. 4.4 Bacterial cell shapes

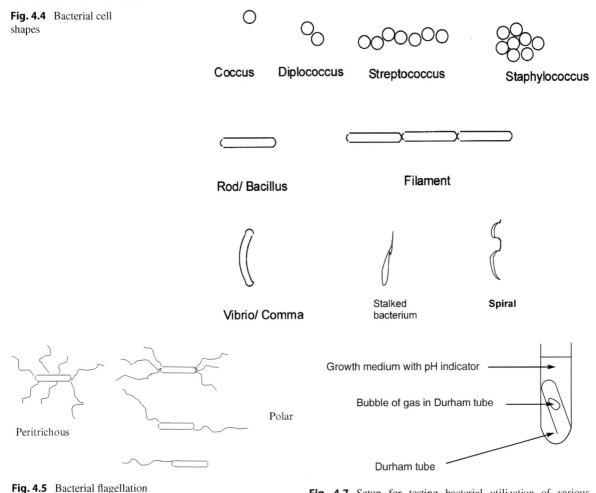

Coccus   Diplococcus   Streptococcus   Staphylococcus

Rod/ Bacillus   Filament

Vibrio/ Comma

Stalked bacterium   **Spiral**

Growth medium with pH indicator

Bubble of gas in Durham tube

Peritrichous

Polar

Durham tube

Fig. 4.5 Bacterial flagellation

Fig. 4.7 Setup for testing bacterial utilization of various substrates

Fig. 4.6 Spore locations in the bacterial cell

test tube containing the medium; a change in the color of indicator would indicate acid production by the organism (see Fig. 4.7).

5. *Determination of optimum growth conditions*

Optimum pH, temperature, and oxygen requirements are determined by growing the organism under different conditions of pH and temperature and finding the best condition. For oxygen requirement, the organism may be grown in an agar stab and sealed with sterile molten petroleum jelly to determine if it will grow under anaerobic conditions.

6. *Secretion of special enzymes*

The secretion of unique enzymes is diagnostic. Some of the following enzymes are diagnostic (see Fig. 4.7a).

Coagulase: Coagulase is an enzyme produced by *Staphylococcus aureus* that converts fibrinogen to fibrin. In the laboratory, it is used to distinguish between different types of *Staphylococcus* isolates. Coagulase negativity excludes *S. aureus*. The coagulase test is used to differentiate *Staphylococcus aureus* from the other species of *Staphylococcus*. The test uses rabbit plasma that has been inoculated with a staphylococcal colony. The tube is then incubated at 37°C for about 90 min. If positive (i.e., the suspect colony is *S. aureus*), the serum will coagulate, resulting in a clot. If negative (i.e., if the tested colony is *S. epidermidis*), the plasma remains liquid.

Catalase: Catalase is a common enzyme found in living organisms. Its functions include catalyzing the decomposition of hydrogen peroxide to water and oxygen.

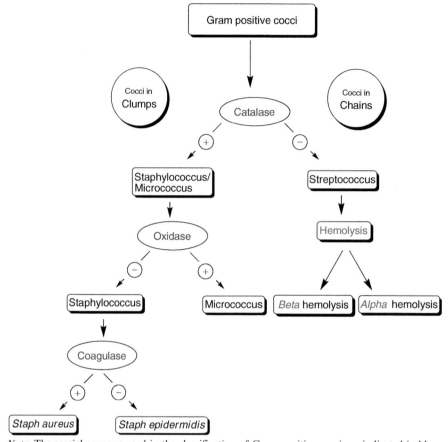

Note: The special enzymes used in the classification of Gram positive cocci are indicated in blue

Key: (−) = Negative for the property in question   (+) = Positive for the property in question

Fig. 4.7a The use of Special Enzymes secreted by some Gram positive cocci in their Classification

$$2H_2O_2 \rightarrow 2H_2O + O_2$$

In microbiology, the *catalase test* is used to differentiate between staphylococci and micrococci, which are catalase-positive, from streptococci and enterococci, which are catalase-negative

Optochin: Optochin (ethyl hydrocuprein hydrochloride) is used for the presumptive identification of *Streptococcus pneumoniae*, which is optochin sensitive, from *Streptococcus viridans* which is resistant. Bacteria that are optochin sensitive will not continue to grow (i.e., *Streptococcus pneumoniae* will die), while bacteria that are not optochin sensitive will be unaffected (i.e., *Streptococcus viridans* will survive).

Oxidase: An oxidase is any enzyme that catalyzes an oxidation/reduction reaction involving molecular oxygen (O_2) as the electron acceptor. In these reactions, oxygen is reduced to water (H_2O) or hydrogen peroxide (H_2O_2). The oxidases are a subclass of the oxidoreductases. In microbiology, the oxidase test is used as a phenotypic character for the identification of bacterial strains; it determines whether a given bacterium produces cytochrome oxidases (and therefore utilizes oxygen with an electron transfer chain) (see Fig. 4.7a).

7. *Serology*

Bacterial species and serotypes can be identified by specific antigen/antibody reactions. Antigens are substances that induce the production of antibodies in an animal body. Bacteria and bacterial components

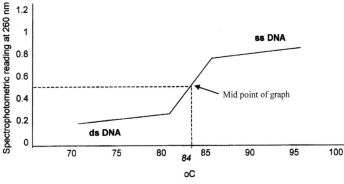

Fig. 4.8 Determination of temperature of melting (Tm) of DNA

ss DNA = Single stranded DNA
ds DNA= Double stranded DNA

serve as excellent antigens. The test includes production of antibodies in an animal host and testing of the antiserum by either the agglutination or precipitation test. In the agglutination test, a drop of the culture of a particular bacterium is mixed on a slide with the anti-serum of an individual infected by it and examined under a microscope. If clumping occurs, the test bacterium is considered to be the same or closely related to the bacterium used as the antigen.

Nucleic Acid Methods

The methods for the characterization and identification of bacteria which have been discussed so far are based on phenotypic properties, i.e., the outward manifestation of the innate (genetic) attributes of the organism. The properties to be discussed in this section are those of the nucleic acids of the organisms. They alone however do not define the organism and must be taken along with the phenotypic properties. They are very useful in refining the description of an organism. They are particularly useful in identifying strains within a species.

(a) *G + C ratio*

The G + C ratio is the percentage of guanine + cytosine in an organism's DNA. Several methods exist for determining this ratio. One method is to determine the Tm or temperature of melting of the DNA.

At room temperature, DNA is double stranded. However, as its temperature is raised gradually, the two strands separate and the rapidity of separation with increasing temperature depends on the amount of G and C in the organism's DNA. G and C are linked by triple bonds and are therefore less likely to separate than A and T bonds, which have double bonds. The higher the G + C content, the higher the temperature at which the DNA separates completely.

DNA begins to separate at 70–75°C and separates completely at about 90°C, when it is said to have melted. When cooled slowly, it begins to anneal (i.e., to reform itself into double strands). In annealing, the strands do not return to their previous "partners" but will anneal with any strand with complimentary bases no matter the source, including those coming from the same organism, other organisms, or even those synthesized in the laboratory. This phenomenon of annealing with complimentary strands from any source is important in other procedures such as in the identification of unknowns, the Polymerase Chain Reaction (PCR), etc.

When the Tm method is used to determine the G + C composition, the temperature of the double (ds) DNA is raised slowly and subjected to spectrophotometric reading at 260 nm. The graph of the spectrophotometric readings is plotted against the change in temperature (see Fig. 4.8). The Tm is the midpoint of the resulting graph. Two organisms with similar phenotypic properties and the same G + C ratio are likely to belong to the species.

(b) *DNA–DNA hybridization*

This technique measures the degree of genetic similarity between pools of DNA sequences. It is usually used to determine the genetic distance between two species.

It was seen above that when melted DNA is allowed to cool slowly, the single-stranded DNA will anneal with any single-stranded DNA no matter its source, as long the bases are complimentary. To determine how closely related an unknown organism is with a known one, DNA from the two

Table 4.5 Some signature sequences unique to the domains (Modified from Madigan and Martinko 2006)

S/No	Oligonucleotide sequence	Occurrence in (%)		
		Archaea	Bacteria	Eukarya
1	CACACCCG	100	0	0
2	CAACCYCR	0	>95	0
3	UCCCUG	>95	0	100
4	UACACACCG	0	>99	100

Y = Any pyridine
R = Any purine

organisms are mixed and heated slowly and allowed to anneal slowly. The unknown or the known is labeled with a fluorescent dye or with radioactive phosphorous and measured at 260 nm in the spectrophotometer. The extent of the taxonomic relatedness is reflected in the extent of the annealing. If the two organisms are of the same species, there will be complete annealment. Some authors have suggested that organisms of the same species will have 90% annealing, while those of the same genus will have about 75% annealing.

(c) *Ribotyping*

Ribotyping is an RNA-based molecular characterization of bacteria. In ribotyping, bacteria genomic DNAs are digested and separated by gel electrophoresis. Universal probes that target specific conserved domains of ribosomal RNA coding sequences are used to detect the band patterns. Ribosomal genes are known to be highly conserved in microbes, meaning that the genetic information coding for rRNA will vary much less within bacteria of the same strain than it will between bacterial strains. This characteristic allows for a greater ability to distinguish between different bacterial strains.

In ribotyping, restriction enzymes (i.e., enzymes which cut DNA at specific positions) are used to cut the genes coding for rRNA into pieces, and gel electrophoresis is used to separate the pieces by size. Genetic probes then visualize locations of different-size fragments of DNA in the gel, which appear as bands. The banding pattern of DNA fragments corresponding to the relevant rRNA is known as the ribotype.

A probe is a strand of nucleic acid which is synthesized in the laboratory and can be labeled with a dye or radioactively. Probes are used to hybridize to a complementary nucleic acid from a mixture. Probes can be general or specific. Thus, it is possible to design probes which will bind to sequences in the ribosomal RNA of all organisms irrespective of Domain. On the other hand, specific probes can be designed which will react only with nucleic acid of Bacteria, Archae, or Eukarya because of the unique sequences found in these groups. Even within species in the various domains, signature sequences exist which will enable the identification of the species using probes (see Table 4.5). Ribotyping is so specific that it has been nicknamed "molecular finger printing."

(d) *FISH – Fluorescent In Situ Hybridization*

This is a special type of ribotyping. In FISH, the whole organism is used without need to isolate the organism's DNA. The cells are treated with chemicals which make the cell walls and cell membranes permeable, thus permitting the entry of probes labeled with fluorescent dyes. After hybridization of the ribosomes with the dye, the entire organism fluoresces and can be seen under the light microscope. FISH is widely used in ecological and clinical studies. It can be used for the rapid identification of bacterial pathogens in clinical specimens; ordinary procedures take about 48 h, but FISH can be completed in a few hours.

Chemical Analysis of Microbial Components for Taxonomic Purposes (Chemotaxonomy)

(a) *Protein analyses*

Proteins are isolated from the whole bacterium, the cell membrane, or the ribosome. The proteins are run in a two-dimensional gel electrophoresis on polyacrylamide gel. The first run separates the proteins on the basis of their molecular weights and the second on the basis of their iso-electric points (Ochiai and Kawamoto 1995). The resulting protein pattern is diagnostic of a particular organism. If many samples are examined, the

Table 4.6 Principal gas chromatography (GC) fatty acid methyl ester (FAME) products of *B. pseudomallei* and *B. thailandensis* (From Inglis et al. 2003. With permission)

FAME peak	% of total FAME content for:	
	B. pseudomallei (n = 87)	*B. thailandensis* (n = 13)
18:1 w7c	32	32
16:0	23	25
17:0 cyclo	5.7	5.5
16:0 3OH	4.1	4.6
19:0 cyclo w8c	3.7	3.8
14:0	3.5	2.6
18:0	0.9	1.1
14:0 2OH[a]	0.58	Not detected

[a]http://www.pubmedcentral.nih.gov/articlerender.fcgi?artid=254375&rendertype=table&id=t1#t1fn3

patterns can be scanned with a computer. Patterns from unknown organisms are compared with patterns of known organisms to determine the relatedness of the known to the unknown.

(b) *Fatty acid analyses – fatty acid methyl ester (FAME)*

This method is widely used in clinical, food, and water microbiology for the identification of bacteria. Fatty acids from the cell membrane of bacteria as well from the outer membrane of Gram negative bacteria are extracted and converted to their methyl esters. The esters are then run in a gas chromatograph. The patterns of the gas chromatograms are diagnostic and can be used to identify unknowns. For example, *Burkholderia pseudomallei*, the cause of melioidosis, has been distinguished from the closely related but non-pathogenic *Burkholderia thailandensis* by gas chromatography (GC) analysis of fatty acid derivatives. A 2-hydroxymyristic acid derivative (14:0 2OH) was present in 95% of *B. pseudomallei* isolates but absent from all *B. thailandensis* isolates (see Table 4.6) (Inglis et al. 2003; Banowetz et al. 2006).

4.1.4 Bacteria

4.1.4.1 Taxonomic Groups Among Bacteria

Bacterial groups are described in two compendia, *Bergey's Manual of Determinative Bacteriology* and *Bergey's Manual of Systematic Bacteriology*. The first manual (on *Determinative Bacteriology*) is designed to facilitate the identification of a bacterium whose identity is unknown. It was first published in 1923 and the current edition, published in 1994, is the ninth. The companion volume (on *Systematic Bacteriology*) records the accepted published descriptions of bacteria, and classifies them into taxonomic groups. The first edition was produced in four volumes and published between 1984 and 1989. The bacterial classification in the latest (second) edition of *Bergey's Manual of Sytematic Bacteriology* is based on 16S RNA sequences, following the work of Carl Woese, and organizes the Domain Bacteria into 18 groups (or *phyla*; singular, *phylum*). It is to be published in five volumes: Volume 1 which deals with the *Archae* and the deeply branching and phototrophic bacteria was published in 2001; Volume 2 published in 2005 deals with the *Proteobacteria* and has three parts; Volume 3 (2009) and Volume 4 (2009) will deal with Firmicutes and The *Bacteroidetes*, *Planctomycetes*, *Chlamydiae*, etc. respectively; Volume 5 will be published in 2010 and deals with the *Actinobacteria* (Bergey's Manual Trust 2009; Garrity 2001–2006).

The manuals are named after Dr. D H Bergey who was the first Chairman of the Board set up by the then Society of American Bacteriologists (now American Society for Microbiology) to publish the books. The publication of *Bergey Manuals* is now managed by the Bergey's Manual Trust. Of the 18 phyla in the bacteria (see Figs. 4.9 and 4.10), the Aquiflex is evolutionarily the most primitive, while the most advanced is the Proteobacteria. In the following discussion, emphasis will be laid on the bacteria which are aquatic.

1. *Aquifex*

The two species generally classified in *Aquifex* are *A. pyrophilus* and *A. aeolicus*. Both are highly thermophilic, growing best in water temperature of 85–95°C. They are among the most thermophilic bacteria known. They can grow on hydrogen, oxygen, carbon dioxide, SO_2, S_2O_3, or NO_3 and mineral salts, functioning as a chemolithoautotroph (an organism which uses an inorganic carbon source for biosynthesis and an inorganic chemical energy source). As a hyperthermophilic bacterium, *Aquifex aeolicus* grows in extremely hot tempuratures such as near volcanoes or hot springs. They grow optimally at temperatures around 85°C but can grow at temperatures up to 95°C. It needs oxygen to carry on its metabolic machinery, but it can function in relatively low levels of oxygen. The genus *Aquifex* consists of Gram negative rods.

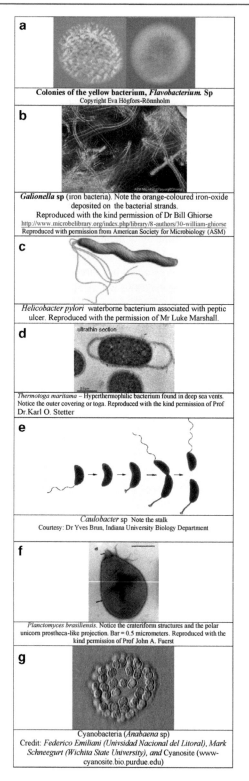

Colonies of the yellow bacterium, *Flavobacterium.* Sp
Copyright Eva Högfors-Rönnholm

Galionella sp (iron bacteria). Note the orange-coloured iron-oxide
deposited on the bacterial strands.
Reproduced with the kind permission of Dr Bill Ghiorse
http://www.microbelibrary.org/index.php/library/8-authors/30-william-ghiorse
Reproduced with permission from American Society for Microbiology (ASM)

Helicobacter pylori waterborne bacterium associated with peptic
ulcer. Reproduced with the permission of Mr Luke Marshall.

Thermotoga maritama – Hyperthermophilic bacterium found in deep sea vents.
Notice the outer covering or toga. Reproduced with the kind permission of Prof
Dr.Karl O. Stetter

Caulobacter sp Note the stalk
Courtesy: Dr Yves Brun, Indiana University Biology Department

Planctomyces brasiliensis. Notice the crateriform structures and the polar
unicorn prostheca-like projection. Bar = 0.5 micrometers. Reproduced with the
kind permission of Prof John A. Fuerst

Cyanobacteria (*Anabaena* sp)
Credit: *Federico Emiliani (Univsidad Nacional del Litoral), Mark
Schneegurt (Wichita State University), and* Cyanosite (www-
cyanosite.bio.purdue.edu)

Fig. 4.9 Illustrations of some bacteria (All items in table reproduced with permission)

2. *Thermodesulfobacterium*

Thermodesulfobacterium is a thermophilic sulfate reducer. Sulfate reducers include a wide range of morphological types, including rods, vibrios, ovals, spheres, and even tear-dropped or onion-shaped cells. Some are motile, others are not. Most sulfate-reducing bacteria are mesophilic, but a few are thermophiles, among which is the Gram negative and anaerobic *Thermodesulfobacterium*. The bacterium is non-spore-forming. It is an aquatic organism and has been isolated from volcanic hot springs, deep-sea hydrothermal sulfides, and other marine environments. In marine sediments and in aerobic wastewater treatment systems, sulfate reduction accounts for up to 50% of the mineralization of organic matter. Furthermore, sulfate reduction strongly stimulates microbially enhanced corrosion of metals. Sulfate Reducing Bacteria (SRB) are discussed in more detail later.

3. *Thermotoga*

Thermotoga is typically a rod-shaped cell enveloped in an outer cell membrane (the "toga" or jacket). *Thermotoga* enzymes are known for being active at high temperatures. Enzymes from *Thermotoga* spp. are extremely thermostable and therefore, useful for many industrial processes such as in the chemical and food industries. The organisms are thermophilic or hyperthermophilic, growing best around 80°C and in the neutral pH range. The salt tolerance of *Thermotoga* species varies greatly; while some display an extremely high salt tolerance, others are restricted to low-salinity habitats. This aerobic Gram-negative organism is typically non-spore-forming and metabolizes several carbohydrates, both simple and complex, including glucose, sucrose, starch, cellulose, and xylan. It can grow by anaerobic respiration using H_2 as electron donor and Fe^{3+} as electron acceptor. It is found in hot springs and in the hydrothermic vents of ocean floors. *Thermotoga maritime* has been widely studied.

4. *Green non-sulfur bacteria (Chloroflexi)*

The Green non-sulfur bacteria are now known as Chloroflexi are typically filamentous, and can move about by bacterial gliding. They are facultatively aerobic and have a different method of carbon fixation (photoheterotrophy) from other photosynthetic bacteria. Like green plants, they also carry out pho-

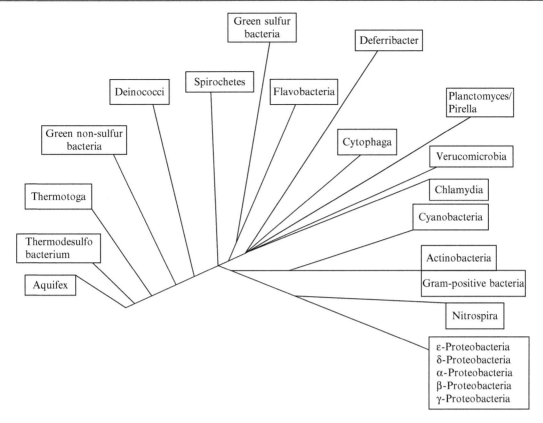

Fig. 4.10 Bacterial groups according to the 16S RNA classification

tosynthesis, but there are differences between the two; for instance, unlike plants, they do not produce oxygen during photosynthesis. The process of photosynthesis in the bacteria and in higher plants is discussed more fully below.

5. *Deinecoccus-Thermus*

The *Deinococcus-Thermus* are a small group of Gram negative bacteria comprised of cocci which are highly resistant to environmental hazards because they are able to quickly repair damage to their DNA. There are two main groups. The Deinococcales include a single genus, *Deinococcus*, with several species that are resistant to radiation; they have become famous for their ability to "eat" nuclear waste and other toxic materials, survive in the vacuum of atmosphere space, and survive extremes of heat and cold. *Thermus* spp. include several genera resistant to heat. *Deinococcus radiodurans* is an extremophilic bacterium, and is the most radioresistant organism known. It can survive heat, cold, dehydration,

vacuum, and acid, and because of its resistance to more than one extreme condition, *D. radiodurans* is known as a polyextremophile.

Thermus aquaticus is important in the development of the polymerase chain reaction (PCR) where repeated cycles of heating DNA to near boiling make it advantageous to use a thermostable DNA polymerase enzyme. These bacteria have thick cell walls that give them Gram-positive stains, but they include a second membrane and so are closer in structure to those of Gram-negative bacteria.

6. *Spirochetes*

Spirochetes are Gram-negative bacteria, which have long, helically coiled cells. Spirochetes are chemoheterotrophic in nature, with lengths between 5 and 250 µm and diameters around 0.1–0.6 µm. Spirochetes are distinguished from other bacterial phyla by the presence of flagella, sometimes called *axial filaments*, running lengthwise between the cell membrane and an outer membrane. These cause a twisting motion which allows the spirochete to move about. The spirochete shape may also be described as consisting of an axial filament around which the cell is wound giving spirochetes their characteristic corkscrew shapes. Most spirochetes are free-living and anaerobic, but they also include the following disease-causing members:

- *Leptospira* species, which causes leptospirosis (also known as Weil's disease)
- *Borrelia burgdorferi*, which causes Lyme disease
- *Borrelia recurrentis*, which causes relapsing fever
- *Treponema pallidum*, which causes syphilis

7. *Green sulfur bacteria*

Green Sulfur Bacteria are found in anaerobic environments such as muds, anaerobic, and sulfide-containing fresh or marine waters, and wetlands. These anoxygenic phototrophic bacteria live in environments where light and reduced sulfur compounds are present. They are found most often under the *Purple Sulfur* bacterial layer. Green sulfur bacteria are capable of using sulfide or elemental S as the electron donor. The elemental S arises from H_2S oxidation and is deposited extracellularly, before the oxidation of sulfate. There are four genera of green sulfur bacteria, *Chlorobium, Prosthecochloris* (with stalks or prostheca), *Pelodictyon* (with vacuoles), and *Clathrochloris* (motile).

The *Green Sulfur Bacteria* strains are green because of the presence of bacteriochlorophylls (bchls) "c" and "d" and small traces of bchl "a" located in chlorobium vesicles attached to the cytoplasmic membrane. Some are brown and they contain bacteriochlorophyll "e." These brown strains are found in the deeper layers of wetlands and water. Both of the two groups can be found also living in extreme conditions of salinity and high temperatures. The morphology of both color types is most often either straight or curved rods.

They are non-motile phototrophic short to long rods which utilize H_2S as electron donor oxidizing it to SO_2 and to SO_4^{2+}. The sulfur so produced lies outside the cells. Light energy absorbed by Bacteriochlorophylls c, d, or e is channeled to Bacteriochlorophyll a, which actually carries out photosynthetic energy conversion, and ATP synthesis takes place. A well-known member is *Chlorobium tepidum*.

In marine environments, they are found in the water column where hydrogen sulfide diffuses up from anaerobic sediments and where oxygen diffuses down from surface waters where oxygenic photosynthesis is taking place. In the Black Sea, the largest anoxic water body in the world, they are found at a depth of 100 m (Manske et al. 2008). They also live in special tissues in invertebrates such as *Riftia pachyptila* (vestimentiferan tube worms) and *Calyptogena magnifica* ("giant" white clams) that live around deep sea hydrothermal vents. There they provide energy, by oxidizing reduced sulfur compounds, and organic matter, by converting carbon dioxide to organic compounds, which the invertebrates use. They are sometimes abundant in coastal waters, and several members of the group have gas vacuoles in their cells to help them float.

8. *Flavobacteria*

Flavobacteria are Gram negative rods that are motile by gliding and found in aquatic environments, both freshwater and marine, and in the soil. Colonies are usually yellow to orange in color, hence their name. Flavobacteria are a group of commensal bacteria and opportunistic pathogens. *Flavobacterium psychrophilum* causes the septicemic diseases of rainbow trout fry syndrome and bacterial cold water disease. They decompose several polysaccharides including agar but not cellulose. The type species is *F. aquatile*.

9. *Defferibacter*

These are thermophilic, anaerobic, chemolithoautotrophic Gram negative straight to bent rods. They can use a wide range of electron acceptors including Fe^{3+} and Mn^{2+}. They are found in a wide range of aquatic environments including deep-sea hydrothermal vents. A well-known member is *Deferribacter desulfuricans*

10. *Cytophaga*

Cytophaga are unicellular, Gram-negative gliding bacteria. They are rod-shaped, but specific strains differ in diameter and length with some being pleomorphic (many shaped). The type species is *C. johnsonae*, which has a moderately long thin shape. Many strains are red, yellow, or orange because of unique pigments synthesized by the group. *Cytophaga* strains tend to be versatile in making these and one strain may synthesize 25 different structural varieties of pigment. The main habitats of *Cytophaga* are soils at or close to neutral pH, decaying plant material, and dung of animals. In freshwater environments, they are found on riverbanks and lake shores, in estuaries, bottom sediments, and algal mats. They are also common in sewage treatment plants, especially at the latter stages where only recalcitrant molecules remain. *Cytophaga* tend to degrade polymers such as cellulose and have been shown to be the major cellulose degraders in some lakes. A few species have been isolated from the oral cavity of humans where they appear to be part of the normal flora, but can occasionally cause septicemias. Some *Cytophaga* strains are pathogens of fish.

11. *Planctomyces/Pirella*

    Planctomyces, *Pirella*, *Gemmata,* and *Isosphaera* form a phylogenetically related group of microorganisms that have many unusual properties. They are the only bacteria, other than the confusing case of the *Chlamydia*, whose cells lack peptidoglycan. Cells of this group divide by budding. Some members of the group produce long appendages, called stalks, and new cells are motile, developing stalks as they mature. Some members of this group have structures resembling nuclear membranes (Bauld and Staley 1976): others have fimbrin.

    Cells of this group can be pigmented (light rose, bright red, or yellow to ochre) or non-pigmented. An example of the species is *Planctomyces limnophilus*, which is ovoid, has a diameter of 1.5 µm, and forms red pigmented colonies. It grows slowly at temperatures between 17°C and 39°C and takes at least a week to form colonies. Stalks of the organism are very thin and cannot be seen by light microscopy. These stalks appear to be made of thin fibers twisted into a bundle that emanates from one pole of the ovoid cell. Cells multiply by budding and new cells are motile and stalkless, but eventually grow stalks as part of a maturation process similar to that seen for *Caulobacter.*

    These microbes are common inhabitants of freshwater lakes, marine habitats, and salt ponds, but most have been difficult to isolate in pure culture. For example, three of the four species in *Planctomyces* have only been observed in lake water and never isolated.

12. *Verrucomicrobia*

    Verrucomicrobia, with the best example as *Verrucomicrobia spinosum*, has been isolated from freshwater, soil environments, and human feces. It produces cytoplasmic appendages called prostheca. Prostheca are like warts and the name of the group comes from the Greek word for warts. Both mother and daughter contain prostheca at the time of the cell division.

13. *Chlamydia*

    Chlamydia are obligate intracellular pathogens with poor metabolic capabilities. They cannot synthesize biomolecules such as amino acids which they obtain from their hosts. Many Chlamydiae coexist in an asymptomatic state within specific hosts, and it is widely believed that these hosts provide a natural reservoir for these species.

    Chlamydiae exist in two states: a metabolically inert *elementary body* (EB) and a metabolically active *reticulate body* (RB) found only inside cells. EB is similar to the virions of viruses and enters the body by phagocytosis. Once ingested and inside the cell, EB divides and becomes RB. After it has killed the cell, it becomes EB again and is ready to be transmitted. Chlamydiae are spread by aerosol or by contact and require no alternate vector.

    Diseases caused by Chlymidia include sexually transmitted infections (STIs) (*Chlamydia trachomatosis*), pneumonia (*Chlamydia pneumoniae*), and bird pneumonia (*Chlamydia psittaci*).

14. *Cyanobacteria*

    Cyanobacteria (Greek: κυανός (*kyanós*) = blue + bacterium) obtain their energy through photosynthesis. They are often referred to as blue-green algae, because they were once thought to be algae. They are a significant component of the marine nitrogen cycle system and an important primary producer in many areas of the ocean. Their ability to perform oxygenic (plant-like) photosynthesis is thought to have converted the reducing atmosphere of the early earth into an oxidizing one, which dramatically changed the life forms on Earth and provoked an explosion of biodiversity.

    Cyanobacteria are found in almost every conceivable habitat, from oceans to freshwater to bare rock to soil. Most are found in freshwater, while others are marine, occur in damp soil, or even temporarily moistened rocks in deserts. A few are endosymbionts in lichens, plants, various protists, or sponges and provide energy for the host. Some live in the fur of sloths, providing a form of camouflage.

    Cyanobacteria include unicellular and colonial species. Colonies may form filaments, sheets, or even hollow balls. Some filamentous colonies show the ability to differentiate into several different cell types: vegetative cells, the normal, photosynthetic cells that are formed under favorable growing conditions; akinetes, the climate-resistant spores that may form when environmental conditions become harsh; and thick-walled heterocysts, which contain the enzyme nitrogenase, vital for nitrogen fixation. Heterocysts may also form under the appropriate environmental conditions (anoxic) wherever nitrogen is necessary. Heterocyst-forming species are specialized for nitrogen fixation and are able to fix nitrogen gas, which cannot be used by plants, into ammonia (NH_3), nitrites (NO_2^-), or nitrates (NO_3^-), which can be absorbed by plants and converted to protein and nucleic acids. The rice paddies of Asia, which produce about 75% of the world's rice, do so because of the high populations of nitrogen-fixing cyanobacteria in the rice paddy fields.

Photosynthesis in cyanobacteria generally uses water as an electron donor and produces oxygen as a by-product, though some may also use hydrogen sulfide as is the case among other photosynthetic bacteria. Carbon dioxide is reduced to form carbohydrates via the Calvin cycle. In most forms, the photosynthetic machinery is embedded into folds of the cell membrane, similar to thylakoids found in the chloroplasts of higher plants.

The cyanobacteria are traditionally classified by morphology into five sections, I–V: Chroococcales, Pleurocapsales, Oscillatoriales, Nostocales, and Stigonematales. The latter two contain hetrocysts. The members of Chroococcales are unicellular and usually aggregated in colonies. In Pleurocapsales, the cells have the ability to form internal spores (baeocytes). In Oscillatorialles, the cells are singly arranged and do not form specialized cells, (akinets and heterocysts). In Nostocalles and Stigonematalles, the cells have the ability to develop heterocysts under certain conditions.

15. *Gram positive bacteria (including Mycoplasmas and Actinobacteria)*
Like the Proteobacteria, the Gram positive bacteria are very diverse; they contain many bacteria encountered in everyday life as agents of disease and inputs of production in industry or as important organisms in food microbiology. Some of them (the Mycoplasma) lack cell walls.

Gram-positive bacteria fall into two major phylogenetic divisions, "low-G + C" and "high-G + C.":
(a) Low G + C group: G + C below 50%;
(b) High G + C group: G + C higher than 50%
(a) Low G + C Group: G + C Below 50%
Non-sporulating Low G + C Group

Staphylococcus: The staphylococci have spherical cells often found in groups resembling clusters of grapes. Bacteria of this genus were originally grouped with other spherical microorganisms, especially of the genus *Micrococcus*, since these two genera often shared similar habitats. However, physiological studies and phylogenetic analysis have shown that these two genera are very different from one another. The differences between staphylococci and micrococci are discussed below.

Lactic Acid Bacteria: The lactic acid bacteria are Gram-positive rods and cocci that produce lactic acid as their primary end product. An important group characteristic is the absence of cytochromes, porphyrins and respiratory enzymes. They are therefore incapable of oxidative phosphorylation or any type of respiration and are totally dependent on fermentation. Lactic acid bacteria do, however, contain mechanisms to deal with the toxic byproducts of oxygen, which categorizes them as aerotolerant anaerobes. They include *Streptococcus, Leuconostoc, Pediococcus, Lactococcus, Enterococcus,* and *Lactobacillus.* Lactic acid bacteria are primarily differentiated based on the types of end products they form. Homofermentative lactic acid bacteria produce only lactic acid as an end product, while heterofermentative lactic acid bacteria produce lactate, ethanol, and CO_2 as well (Axelssson and Ahrne 2000; Narayanan et al. 2004).

Sporulating Low G + C Group
Bacillus: These are spore-forming aerobic rods. *Bacillus* is a genus of rod-shaped, beta-hemolytic Gram-positive bacteria. *Bacillus* species are catalase-positive obligate or facultative aerobes. Ubiquitous in nature, *Bacillus* includes both free-living in soil, water and air, as well as some pathogenic species. Under

stressful environmental conditions, the cells produce endospores resistant to heat, radiation, chemicals and other unfavorable conditions.

Clostridium: These are Gram-positive spore-forming obligately anaerobic rods. Individual cells are rod-shaped, and the name comes from the Greek for spindle. *Clostridium* includes common free-living bacteria as well as important pathogens, including *C. botulinum*, an organism producing a very potent toxin in food; *C. difficile*, which can overgrow other bacteria in the gut during antibiotic therapy; *C. tetani*, the causative organism of tetanus; *C. perfringens*, formerly called *C. welchii*, which causes a wide range of symptoms, from food poisoning to gas gangrene. Because *C. perfringens* produces much gas, it is also used as a replacement for yeasts in breadmaking. *C. sordellii* has been linked to the deaths of more than a dozen women. They are important in the anaerobic conditions of muds.

Heliobacteria: Heliobacteria are strictly anaerobic, spore-forming photoheterotrophic members of the Firmicutes. 16s rRNA studies put them among the *Firmicutes* (*Bacillus* and *Clostridium*) but they do not stain Gram-positively like the other members. They have no outer membrane and like certain other firmicutes (clostridia), they form heat resistant endospores. They are the only firmicutes known to conduct photosynthesis. Soluble periplasmic components appear absent in heliobacteria and photosynthesis takes place at the cell membrane, which does not form folds or compartments as it does in purple phototrophic bacteria. A particularity of heliobacterial photosynthesis is the occurrence of a unique Bacteriochlorophll (BChl) g. BChl *g* is chemically closer to Chl *a* than to BChl *a*. Correspondingly, heliobacteria appear to be more closely related to oxygenic photosynthesis than the green sulfur bacteria (based on 16S-rRNA phylogeny as well as on trees built from sequences of the photosynthetic reaction center). A small group, it is the only known phototrophic one among the Gram positives. Heliobacteria consist of three genera, *Heliobacterium* (3 spp.), *Heliobacillus* (1 sp.), and *Heliophilum* (1 sp.). They cannot tolerate sulfide, all known species can fix nitrogen. They are common in the waterlogged soils of paddy fields.

(b)   High G + C Group: G + C Above 50%
These include Actinomycetes, Mycobacteria, Micrococcus, and Corynebacterium:

Actinomycetes
Actinomycetes are filamentous and spore-forming (non heat resistant spores), found in soil. They are very important as antibiotic producers. Typical example is *Streptomyces* sp. They include some of the most common soil life, playing an important role in decomposition of organic materials, such as cellulose and chitin and thereby playing a vital part in organic matter turnover and carbon cycle. Actinomycetes of the family Actinoplanaceae, especially *Actinoplanes*, are readily isolated from the flowing waters of rivers and streams, where they are important in the decomposition of wood and other cellulolisic materials.

Mycobacterium
This is a slow-growing acid-fast strain (Ziel–Nielsen stain) implicated in diseases (*M. leprae*, leprosy; *M. tuberculosis*, tuberculosis). Many are however free-living and inhabit aquatic environments. These environmental or waterborne mycobacteria (WBM) inhabit a diverse range of natural environments and are a frequent cause of opportunistic infection in human beings and livestock. Several hospital and community outbreaks of mycobacterial infections, including infections as diverse as life-threatening pneumonia in patients with artificial ventilation, cystic fibrosis, and chronic granulomatous disease; outbreaks of skin infection following liposuction; furunculosis after domestic footbaths; mastitis after body piercing; and abscess formation in intravenous drug users.

Corynebacteria
Corynebacterium is a genus of Gram-positive, facultatively anaerobic, nonmotile, rod-shaped actinobacteria. Most do not cause disease, but are part of normal human skin flora. *Corynebacterium diphtheriae* is the cause of diphtheria

Table 4.7 Properties of Micrococci and Staphylococci

Species	Cells arrangement	G + C ratio	Oxygen requirement
Micrococcus	Clusters, tetrads (fours)	66–73	Strictly aerobic
Staphylococcus	Clusters, pairs	30–39	Microaerophilic

in humans. The genes encoding exotoxins that are the cause of diphtheria (caused by Corynebacterium) as well as cholera and some other bacterial diseases are mobile in aquatic and terrestrial environments and have been found in sediments and in river water using PCR.

Micrococcus

These are cocci in bunches and very similar to staphylococci. They are distinguished from each other according to the properties shown in Table 4.7.

16. *Nitrospira*

Nitrospira are nitrite-oxidizing bacteria that are important in marine habitats. In aquaria, for example, if the ammonia/nitrite/nitrate cycle is exhausted, the ecosystem suffers and fish can get sick or die. Therefore, nitrite-oxidizing bacteria as well as the other bacteria in this system are important for healthy marine ecosystems. In addition, *Nitrospira*-like bacteria are the main nitrite oxidizers in wastewater treatment plants.

17. *Proteobacteria (including purple bacteria)*

The *Proteobacteria* are a major group of bacteria. They include a wide variety of pathogens, such as *Escherichia*, *Salmonella*, *Vibrio*, and *Helicobacter*. Others are important agriculturally or industrally; still others are free-living, and include many of the bacteria responsible for nitrogen fixation. The group is defined primarily in terms of ribosomal RNA (rRNA) sequences, and is named after the Greek god Proteus, who could change his shape, because of the great diversity of forms found within the group. Proteus is also the name of a bacterial genus within the Proteobacteria.

All Proteobacteria are Gram-negative, with an outer membrane mainly composed of lipopolysaccharides. Many move with flagella, but some are nonmotile or move by bacterial gliding. The latter include the myxobacteria, a unique group of bacteria that can aggregate to form multicellular fruiting bodies.

There is also a wide variety in the types of metabolism. Most members are facultatively or obligately anaerobic and heterotrophic, but there are numerous exceptions. A variety of genera, which are not closely related to each other, convert energy from light through photosynthesis. These are called purple bacteria, referring to their mostly reddish pigmentation.

The Proteobacteria are divided into five sections, referred to by the Greek letters alpha through epsilon, again based on rRNA sequences.

Alpha (α) Proteobacteria

The Alphaproteobacteria comprise the most phototrophic genera, but also several genera metabolizing C1-compounds (compounds with a single carbon atom e.g., *Methylobacterium*, symbionts of plants (e.g., Rhizobia) and animals, and a group of intracellular pathogens, the Rickettsiaceae. Moreover, the precursors of the mitochondria of eukaryotic cells are thought to have originated in this bacterial group.

Beta (β) Proteobacteria

The Betaproteobacteria consist of several groups of aerobic or facultative bacteria which are often highly versatile in their degradation capacities, but also contain chemolithotrophic genera (e.g., the ammonia-oxidizing genus *Nitrosomonas*) and some phototrophs (genera *Rhodocyclus* and *Rubrivivax*). Beta Proteobacteria play an important role in nitrogen fixation in various types of plants, oxidizing ammonium to produce nitrite- an important chemical for plant function. Many of them are found in environmental samples, such as waste water or soil. Pathogenic species within this class are the *Neisseriaceae* (gonorrhea and meningoencephalitis) and species of the genus *Burkholderia*.

Gamma (γ) Proteobacteria

The Gammaproteobacteria comprise several medically and scientifically important groups of bacteria, such as the Enterobacteriaceae, Vibrionaceae, and Pseudomonadaceae. Many important pathogens belong to this class, e.g., *Salmonella* (enteritis and typhoid fever), *Yersinia* (plague), *Vibrio* (cholera), *Pseudomonas aeruginosa* (lung infections in hospitalized or cystic fibrosis patients), and *E coli*.

Fig. 4.11 Overview
of photosynthesis
(Modified from Moroney
and Ynalvez 2009)

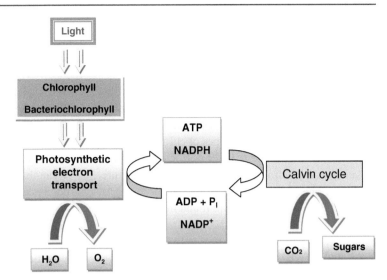

Delta (δ) Proteobacteria)
The Deltaproteobacteria comprise a group of pre-dominantly aerobic genera, the fruiting-body-forming myxobacteria, and a branch of strictly anaerobic genera, which contains most of the known sulfate-reducing bactria, (*Desulfovibrio, Desulfobacter, Desulfococcus, Desulfonema*, etc.) and sulfur-reducing bacteria (e.g., *Desulfuromonas*) alongside several other anaerobic bacteria with different physiology (e.g., ferric iron-reducing *Geobacter* and *Pelobacter* and *Syntrophus* species, which live symbiotically together).

Epsilon (ε) Proteobacteria
The Epsilonproteobacteria consist of only a few genera, mainly the curved to spiral-shaped *Wolinella, Helicobacter*, and *Campylobacter*. Most of the known species inhabit the digestive tract of animals and humans and serve as symbionts (*Wolinella* in cows) or pathogens (*Helicobacter* in the stomach and *Campylobacter* in the duodenum in humans). There have also been numerous environmental sequences of epsilons recovered from hydrothermal vent and cold seep habitats.

4.1.4.2 Aspects of the Physiology and Ecology of Microorganisms in the Aquatic Environment

This section will discuss the physiology of some of the activities of aquatic microorganisms which contribute to their ecology in bodies of water as well as to their economic importance. Items to be discussed

are photosynthesis, nitrogen economy, especially nitrogen fixation, sulfate reduction, and iron in the aquatic environment.

Photosynthesis

Photosynthesis is the conversion of CO_2 to carbohydrates using light energy. This process has been described as the most important biological reaction on earth, since it is the means by which the energy of the sun is harnessed by living things, through their consumption of the products of photosynthesis. Photosynthesis is carried out by plants, algae, and some bacteria, but not by Archae. It is an important factor affecting the ecology of microorganisms in aquatic environments (Achenbach et al. 2001). Photosynthesis is generally better known in plants than in bacteria; plant photosynthesis will therefore be discussed as a basis for understanding bacterial photosynthesis (see Fig. 4.11).

Photosynthesis is hinged on three items: (a) *Photosynthetic pigments*, (b) the *light* or light-dependent reactions of photosynthesis, and (c) the *dark* or light-independent reactions of photosynthesis.

The Pigments of Photosynthesis

A pigment is any substance that absorbs light. The color of the pigment comes from the wavelengths of light reflected by the pigment (in other words, those not absorbed). Chlorophyll, the green pigment common to all photosynthetic cells, absorbs all wavelengths of visible light except green, which it reflects, and thus is detected by human eyes as green. Black pigments absorb all of the wavelengths that strike them. White pigments/lighter colors reflect all or almost all of the light energy striking them. Pigments have their own characteristic absorption spectra, the absorption pattern of a given pigment.

Chlorophyll (chl) found in plants, algae, and cyanobacteria, is very similar to bacteriochlorophyll (bchl) found in bacteria, other than cyanobacteria (see Fig. 4.12). There are several types of chlorophylls and of bacteriochlorophylls, (named a, b, c, d, e, and g) differing from each in slight differences in structure. Bchls "a" and "b" are found in the purple bacteria; while bchls "c," "d," and "e" are found in Green sulfur bacteria; bchl "g" is found in *Heliobacteria*. In higher plants, photosynthesis takes place only in chl "a"; all other chlorophylls along with carotenoids are accessories and gather light which is channeled to chl "a." Similarly, in bacteria, bchl "a" is the site of photosynthesis; all the other bacteriochlorophylls are accessories and gather light which is channeled to bchl "a."

Accessory pigments include carotenoids found in higher plants and cyanobacteria and phycobilins found in the algae. Pigments have their own characteristic absorption spectra. Figure 4.13 shows the wavelength of various chlorophylls and accessory pigments.

The Light Reactions

In higher plants, the light dependent reactions, take place on membranous structures known as thylakoids found in chloroplasts in complex processes, that are not yet fully understood. The process is much simplified as described below.

In plants, light is absorbed by complexes formed between protein and chlorophyll molecules known as photosystems, Photosystem I (PSI) and Photosystem II (PSII). PSII absorbs light energy (photons) at a wavelength of 680 nm and is called P680 while PSI it absorbs photons at 700 nm and is called P700.

When a pigment absorbs light energy, one of three things will occur: Energy may be dissipated as heat; it may be re-emitted immediately as a longer wavelength, a phenomenon known as fluorescence; or the energy may trigger a chemical reaction, as in photosynthesis. In plant photosynthesis, the action begins at the PSII chlorophyll–protein complex which becomes excited and loses an electron; this electron is passed through a series of enzymes until it is transferred to water, causing it to lose electrons:

$$2H_2O \rightarrow 4H^+ + 4e^- + O_2$$

The electron released from the splitting of water is transferred to PSI, which can itself capture light energy; this energy is transferred by enzymes used to reduce $NADP^+$ to NADPH and ATP the other energy currency of cells, thus $ADP + Pi \Rightarrow ATP$.

While the photosynthetic process in cyanobacteria is similar to that of plants, green bacteria and purple sulfur bacteria have photosynthetic processes different from the process in plants.

Cyanobacteria do not have chloroplasts, but have structures on their cell membranes which are similar to thylakoids. They have photosystems similar to PS II and PS I found in the chloroplasts of higher plants. They can produce NADPH and ATP in the way as higher plants and they are the only bacteria which produce O_2 during photosynthesis. However, instead of carotenoids or chlorophyll "b" which act as accessory pigments in higher plants, they have phycobilins.

Purple Bacteria: Purple bacteria and green sulfur bacteria have only one type of photosystem. The single photosystem in purple bacteria is structurally related to PS II in cyanobacteria and plant chloroplasts; it, however, has a P870 molecule, i.e., it absorbs light at 870 nm and can make ATP in the transfer of electrons.

In order to make NADPH, purple bacteria use an external electron donor (hydrogen, hydrogen sulfide, sulfur, sulfite, or organic molecules such as succinate and lactate) to feed electrons into a reverse electron transport chain.

Green Sulfur Bacteria: These bacteria contain a photosystem that is analogous to PS I (P840) in chloroplasts. It makes ATP through the transfer of electrons. Electrons are removed from an excited chlorophyll molecule and used to reduce NAD^+ to NADH. The electrons removed from P840 must be replaced. This is accomplished by removing electrons from H_2S, which is oxidized to sulfur which appear as globules in the cells (hence the name "green sulfur bacteria").

$$H_2S \Rightarrow 2H + S^0.$$

The Dark or Light-Independent Reactions of Photosynthesis

The energy rich ATP and NADPH molecules formed in the light dependent phase of photosynthesis are used

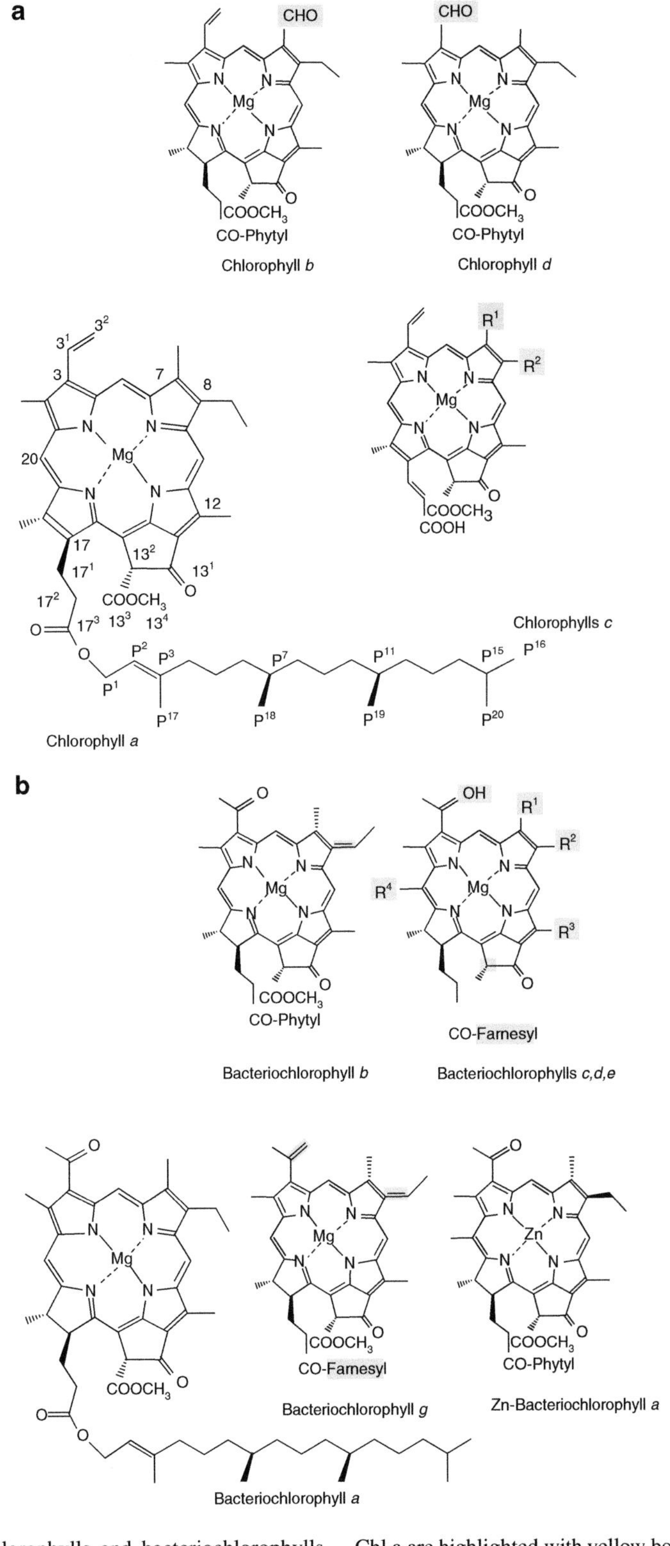

Fig. 4.12 Structure of chlorophylls and bacteriochlorophylls (From *Encyclopedia of Life Sciences*; Fujita 2005. With permission) Structure of various Chls (a) and BChls (b). (a) Chls a, b, c and d. Differences in side-chains of Chls b, c and d from Chl a are highlighted with yellow boxes. (b) BChls a, b, c, d, e, g and Zn-BChl a. Differences in sidechains, ring oxidation state, and the central metal ion of BChls b, c, d, e, g and Zn-BChl a are highlighted with yellow boxes.

Fig. 4.13 Wavelengths
of chorophylls and
photosythetic accessory
pigments (Modified from
Photosynthesis: Light
energy transduced to
chemical energy; http://
phototroph.blogspot.
com/2006/11/pigments-
and-absorption-spectra.
html)

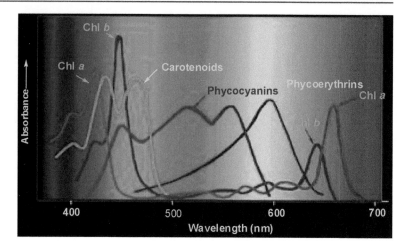

for the production of energy rich carbohydrates (sugars) in the Calvin cycle (see Fig. 4.11).

The fixation or reduction of carbon dioxide is a light-independent process in which carbon dioxide combines with a five-carbon sugar, ribulose 1,5-bis-phosphate (RuBP), to yield two molecules of a three-carbon compound, glycerate 3-phosphate (GP), also known as 3-phosphoglycerate (PGA). GP, in the presence of ATP and NADPH from the light-dependent stages, is reduced to glyceraldehyde 3-phosphate (G3P) and enters the citric acid cycle.

The processes of photosynthesis can be represented by the general formula:

$$CO_2 + 2H_2A \Rightarrow CH_2O + H_2O + 2A,$$

where H_2A is the source of the reducing power for the conversion of CO_2 to carbohydrates.

In higher plants, algae and cyanobacteria, where water is the source of the reducing power photosynthesis, can be represented thus:

$$CO_2 + 2H_2O \Rightarrow CH_2O + H_2O + O_2.$$

In bacteria, other than cyanobacteria, where water is not the source of reducing power and hence oxygen is not involved (anoxygenic), for example, the green sulfur bacteria, where hydrogen sulfide is utilized, the photosynthetic equation is given thus:

$$CO_2 + 2H_2S = CH_2O + H_2O + 2S.$$

Summary: Differences Between Photosynthesis in Plants and in the Bacteria

Like green plants, some bacteria are photosynthetic, using the energy of sunlight to reduce carbon dioxide to carbohydrate. There are a number of differences between the two groups which are summarized below:

1. *Chlorophyll and bacteriochlorophyll*

   Chlorophyll, the photosynthetic pigment in plants, is replaced in bacteria by bacteriochlorophyll (except in the Cyanobacteria). Both types of pigments are similar and differ only in some side chains (see Fig. 4.12).

2. *Sites for photosynthesis in green plants and bacteria*

   In higher plants, photosynthesis takes places in membraneous structures known as thylakoids which are located in organelles known as chloroplasts. In bacteria, the site for photosynthesis varies from one group of bacteria to the other. In the cyanobacteria, although chloroplasts are absent, photosynthesis occurs in thylakoid-like structures; in helicobacteria, it takes place on the cell membranes; in the purple bacteria, it takes place in invaginations of the cell membrane; in the green bacteria, it takes place on the cell membrane as well as in special membrane folding known as chlorosomes.

3. *Oxygenic and anoxygenic photosynthesis*

   In higher plants, algae and cyanobacteria, the light energy excites the molecules of chlorophyll leading to release of energy which splits the water molecule and to the release of oxygen as a by-product, and finally the provision of H for fixing the CO_2. In most bacteria (apart from cyanobacteria), oxygen is not released because water does not provide the H which converts the CO_2 to carbohydrates. Rather, light energy excites bacteriochlorophyll leading to energy which splits H from H_2S. In the dark, many photosynthetic bacteria can produce energy by the transfer of electron, or anaerobically.

Aspects of the Physiology of Photosynthetic Bacteria

The photosynthetic bacteria can be divided into two groups: The anaerobic photosynthetic groups and the aerobic photosynthetic bacteria.

1. *The anaerobic photosynthetic bacteria (AnPB)*

   The bacterial order *Rhodospirillales* contains three photosynthetic families:

   (a) *Rhodospirillaceae*: Purple non-sulfur bacteria, e.g., *Rhodospirillum*. These cells contain bacteriochlorophyll "a" or "b" located on specialized membranes continuous with the cytoplasmic membrane. They are not able to use elemental sulfur as electron donor and typically use an organic electron donor, such as succinate or malate, but can also use hydrogen gas.

   (b) *Chromatiaceae*: These include purple sulfur bacteria, e.g., *Chromatium*. They are able to use sulfur and sulfide as the sole photosynthetic electron donor and sulfur can be oxidized to sulfate. They can use inorganic sulfur compounds, such as hydrogen sulfide as an electron donor. Purple sulfur bacteria must fix CO_2 to survive, whereas non-sulfur purple bacteria can grow aerobically in the dark by respiration on an organic carbon source. They store elemental sulfur inside their cells, and these appear globules within their cells, hence their name, purple *sulfur* bacteria.

   (c) *Chlorobiaceae*: These are green sulfur bacteria; their cells contain bacteriochlorophyll "c" or "d" located in chlorobium vesicles attached to the cytoplasmic membrane.

   (d) *Heliobacteria*: The heliobacteria are anaerobic and phototrophic, converting light energy into chemical energy by photosynthesis using a PSI type reaction center (RC) (P798). The primary pigment involved is bacteriochlorophyll *g*, which is unique to the group and has a unique absorption spectrum. On account of this, the heliobacteria occupy their own special environmental niche. Phototrophy takes place on the cell membrane, which does not form folds or compartments as it does in purple phototrophic bacteria.

Using 16 S RNA analysis, they are placed among the Firmicutes, Gram positive bacteria; although, they do not stain Gram positive, but they form heat resistant endospores. Heliobacteria are the only firmicutes known to conduct photosynthesis. They are *photoheterotrophic*, i.e., they require organic carbon sources. They do not fix carbon dioxide, they lack rubisco, and do not have Calvin cycle.

They are found in soils, especially water logged soils such as in paddy fields. They are also strong nitrogen fixers.

2. *The aerobic photosynthetic bacteria (APB)*

   The cyanobacteria are photosynthetic and aerobic, but recently another photosynthetic aerobic group was discovered. It was previously generally believed that anoxygenic photosynthesis was an anaerobic growth mode of either obligately anaerobic, or facultatively anaerobic bacteria capable of switching between respiration under aerobic conditions and phototrophy under anaerobic conditions. Recently (1979), the first reported member of the aerobic phototrophic bacteria, *Erythrobacter longus*, discovered in the Bay of Japan, changed our previous knowledge of the phototrophic bacteria. APBs have since been found in a wide variety of both marine and freshwater habitats, including acid mine drainage sites, soils, saline lakes, and soda lakes. (Rathgeber et al. 2004). Other genera of APBs found in freshwater and marine environments include the following: *Erythrobacter,, Roseobacter,, Porphyrobacter,, Acidiphilium Erythromonas, Erythromicrobium, Roseococcus,* and *Sandaracinobacter.*

   The distinguishing features of APBs are:

   (a) They produce their photosynthetic apparatus only in the presence of oxygen and the absence of light.

   (b) The presence of bacteriochlorophyll a (BChl a) incorporated into light harvesting (LH) and reaction center (RC) complexes capable of transforming light into electrochemical energy under aerobic conditions.

   (c) A relatively low amount of photosynthetic units per cell.

   (d) Extreme inhibition of BChl synthesis by light.

   (e) An abundance of carotenoid pigments.

   (f) Apparent lack of intracytoplasmic photosynthetic membranes.

   (g) The inability to grow phototrophically under anaerobic conditions.

The APBs produce a photosynthetic apparatus similar to that of purple phototrophic bacteria. However, this apparatus, in contrast to that of the anaerobic photosynthetic bacteria is produced only under aerobic conditions. In facultatively anaerobic organisms, the photosynthetic apparatus is synthesized under conditions of oxygen shortage and absence of light.

They are a phylogenetically diverse group interspersed predominantly throughout the α-Proteobacteria, closely related to anoxygenic phototrophic purple non-sulfur bacteria as well as chemotrophic species. Recently, however, more and more APBs have been placed in the β-Proteobacteria.

Nitrogen Economy in Aquatic Systems

Nitrogen Fixation

Nitrogen is important in microorganisms for the manufacture of proteins and nucleic acids, both of which are essential for the continued existence of all living things. Although the element is abundant in the atmosphere, constituting about 80%, the ability of making atmospheric nitrogen available to living things is present only in a few organisms. Biological nitrogen fixation can be represented by the following equation, in which 2 moles of ammonia are produced from 1 mole of nitrogen gas, at the expense of 16 moles of ATP and a supply of electrons and protons (hydrogen ions):

microorganisms including aerobic and anaerobic ones. (Naqvi 2006). Among aerobes, nitrogen fixers include all members of *Azotobacter* and *Beijerinckia,*, some *Klebsiella* and some cyanobacteria. Under anaerobic conditions, such as, occur in sediments or in the deeper regions of water columns, the following organisms fix nitrogen: Some *Clostridium* spp., *Desulfovibrio*, purple sulfur bacteria, purple non-sulfur bacteria, and green sulfur bacteria.

The nitrogenase enzyme complex is highly sensitive to oxygen and it is inactivated when exposed to oxygen, because oxygen reacts with the iron component of the proteins. Aerobic organisms including cyanobacteria, which produce oxygen during photosynthesis, combat the problem of nitrogenase inactivation by different methods. Cyanobacteria for example have special cells, heterocysts, where nitrogen fixation occurs and in which nitrogenase is protected because they contain only photosystem I whereas the other cells have both photosystem I and photosystem II (which generates oxygen when light energy is used to split water to supply H_2 in photosynthesis.). For the same reason, also, *Azotobacter* and *Rhizobium* produce large amounts of extracellular polysaccharide, which helps limit the diffusion of oxygen to the cells. Furthermore, *Rhizobium* root nodules contain oxygen-scavenging molecules such as leghemoglobin, which regulate the supply of oxygen to the nodule tissues in

$$N_2 + 8H + + 8e^- + 16\ ATP = 2NH_3 + H_2 + 16ADP + 16\ Pi.$$

This reaction is performed exclusively by prokaryotes using a nitrogenase enzyme complex. This enzyme consists of two proteins – an iron protein and a molybdenum-iron protein – as shown Fig. 4.14.

The reactions occur while N_2 is bound to the nitrogenase enzyme complex. The Fe protein is first reduced by electrons donated by ferredoxin. Then the reduced Fe protein binds ATP and reduces the molybdenum–iron protein, which donates electrons to N_2, producing HN=NH. In two further cycles of this process (each requiring electrons donated by ferredoxin), HN=NH is reduced to $H_2N–NH_2$, and this in turn is reduced to $2NH_3$. Depending on the type of microorganism, the reduced ferredoxin, which supplies electrons for this process, is generated by photosynthesis, respiration, or fermentation (Anonymous 2010e).

Nitrogen fixation may be done by bacteria living symbiotically with higher plants such as *Rhizobium* spp. and legumes or by free-living organisms. In the aquatic environment, the nitrogen fixers are free-living

the same way as hemoglobin regulates the supply of oxygen to mammalian tissues.

Other microbial activities which participate in regulating the nitrogen economy of aquatic systems are *nitrification* and *denitrification*.

Nitrification

Nitrification is the conversion of ammonium to nitrate by the nitrifying bacteria. These bacteria are chemo-autotrophs which obtain energy by oxidizing ammonium, while using CO_2 as their source of carbon to synthesize organic compounds. The nitrifying bacteria are found in most soils and waters of moderate pH, but are not active in highly acidic soils. They almost always are found as mixed-species communities or *consortia*. Some of them – e.g., *Nitrosomonas* convert ammonium to nitrite (NO_2^-) while others – e.g., *Nitrobacter* convert nitrite to nitrate (NO_3^-). The nitrifying bacteria are so numerous in waters rich in ammonium such as sewage

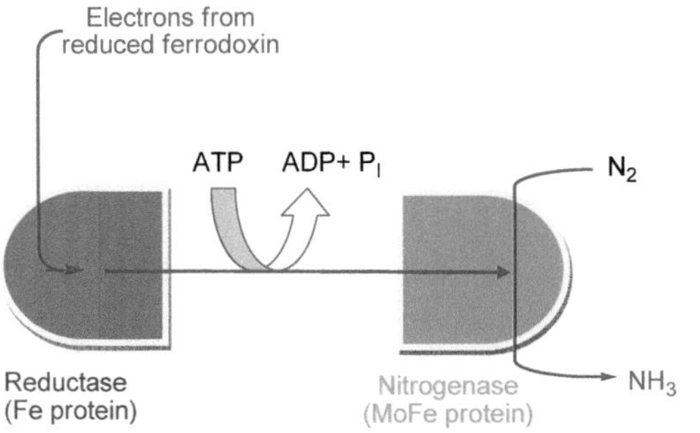

Fig. 4.14 The enzymes of nitrogen fixation (Reproduced from Berg et al. 2002. With permission)

Note: Ferrodoxins are a group of red-brown proteins containing iron and sulfur, which act as electron carriers during photosynthesis, nitrogen fixation, or oxidation-reduction reactions in green plants, algae, and anaerobic bacteria.

Nitrogenase is a two-protein complex. One component, **nitrogenase reductase** is an iron-containing protein that accepts electrons from ferredoxin, a strong reductant, and then delivers them to the other component, **nitrogenase**, which contains **Iron (Fe) and molybdenum (Mo)**.

The overall reaction in nitrogen fixation via nitrogenase is:
$$8H^+ + N_2 + 8e^- + 16ATP + 16H_2O \rightarrow$$
$$2NH_3 + H_2 + 16ADP + 16P_i + 16H_+$$

effluents that they readily convert the ammonium compounds therein into nitrates. The nitrates can accumulate in groundwater, and may ultimately enter drinking water. Regulations in many countries control the amount of nitrate in drinking water, because in the anaerobic conditions of the animal alimentary canal, nitrates can be reduced to highly reactive nitrites by microorganisms. Nitrites are absorbed from the gut and bind to hemoglobin, reducing its oxygen-carrying capacity. In young babies, this can lead to a respiratory illness known as *blue baby syndrome*. Nitrites can also react with amino compounds, forming nitrosamines which are highly carcinogenic.

Denitrification

Denitrification is the conversion of nitrate to gaseous compounds (nitric oxide, nitrous oxide, and N_2) by microorganisms. Denitrification goes through some combination of the following intermediate forms:

$$NO_3^- \rightarrow NO_2^- \rightarrow NO \rightarrow N_2O \rightarrow N_2 \uparrow$$
nitrate   nitrite   nitric oxide   nitrous oxide   gas

The denitrification process can be expressed in terms of electron transfer thus:

$$2NO_3^- + 10e^- + 12H^+ \rightarrow N_2 + 6H_2O.$$

Denitrification is brought by a large number of different bacteria which are mainly heterotrophic. They complete the nitrogen cycle by returning N_2 to the atmosphere. Denitrification occurs under special conditions in both soil and aquatic conditions, including marine environments. Denitrification occurs when oxygen supply is low such as in ground water, wetlands in seafloors, and other poorly aerated parts of aquatic systems. The conditions which encourage denitrification are those in which there is a supply of oxidizable organic matter, and absence of oxygen and the availability of reducible nitrogen sources. Under such conditions, the terminal electron acceptor for the denitrifying bacteria is not oxygen but the nitrogen compounds given in the formula above. The organisms prefer nitrates and the other compounds in the equation, in the order they occur in the equation above and ending with nitrous oxide. When the terminal electron acceptor is an inorganic compound such as those in formula above, the condition is also termed respiration as is also the case with oxygen.

A mixture of gaseous nitrogen products is often produced because of the stepwise use of nitrate, nitrite, nitric oxide, and nitrous oxide as electron acceptors in anaerobic respiration. The commonest denitrifying bacteria include several species of *Pseudomonas*, *Alkaligenes*, *Bacillus*,, and *Paracoccus denitrificans*. Autotrophic denitrifiers (e.g., *Thiobacillus denitrificans*) have also been identified. In general, however, several species of bacteria are involved in the complete reduction of nitrate to molecular nitrogen, and more than one enzymatic pathway have also been identified.

Anammox

In some organisms, direct reduction from nitrate to ammonium (also known as dissimilatory nitrate reduction to ammonium or DNRA) may also occur; although, this is less common than denitrification. Anammox, an abbreviation for ANaerobic AMMonium OXidation, is a globally important microbial process of the nitrogen cycle. It takes place in many natural environments.

The bacteria mediating this process were identified only 20 years ago. They belong to the bacterial phylum *Planctomycetes*, of which *Planctomyces* and *Pirellula* are the best known genera. Four genera of anammox bacteria have been identified: *Brocadia*, *Kuenenia*, *Anammoxoglobus*, *Jettenia* (all freshwater species), and *Scalindua* (marine species).

Fig. 4.15 Nitrogen cycle
in the marine environment
(From Codispoti et al. 2001.
With permission)
"X" and "Y" are intracel-
lular intermediates that do
not accumulate in water
column

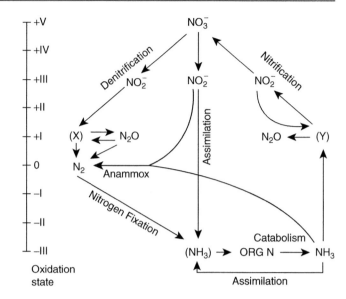

The anammox bacteria are characterized by several
striking properties:

(a) They all possess one anammoxosome, a membrane
    bound compartment inside the cytoplasm which is
    the site of anammox catabolism.
(b) Further, the membranes of these bacteria mainly
    consist of ladderane lipids which are rare in living
    organisms.
(c) Hydrazine (normally used as a high-energy rocket
    fuel, and poisonous to most living organisms) is a
    by-product of these organisms.
(d) Finally, the organisms grow very slowly, the gen-
    eration or doubling time being nearly 2 weeks.

The anammox process was originally found to
occur only from 20°C to 43°C, but more recently, anammox has been observed at temperatures from 36°C to
52°C in hot springs and 60°C to 85°C at hydrothermal
vents located in the ocean floor.

Reduction under anaerobic conditions can also occur
through anaerobic ammonia oxidation (Anammox) thus:

$$NH_4^+ + NO_2^- \rightarrow N_2 + 2H_2O.$$

Because denitrifying bacteria are principally heterotrophic, in some wastewater treatment plants, small
amounts of methanol are added to the wastewater to
provide a carbon source for the bacteria.

Nitrogen fixation, nitrification, and denitrification
are interlinked in the nitrogen cycle. The nitrogen
cycle in the marine environment is given in Fig. 4.15.

The Sulfur Cycle in the Aquatic System and Bacteria

In the environment, through changes brought about
mostly by bacteria, sulfur changes from one form to
the other: From hydrogen sulfide (H_2S) to sulfate via
elemental sulfur (S^0) and sulfate is changed again to
hydrogen sulfide. Hydrogen sulfide is also evolved
from hot springs and volcanoes, and occurs when dead
animals, the excreta of animals, and dead plants are
decomposed by bacteria. The compound is oxidized to
sulfuric acid by the sulfur-oxidizing bacteria and photosynthetic sulfur bacteria via elemental sulfur. The
change of hydrogen sulfide to elemental sulfur occurs
also abiotically in the presence of molecular oxygen.

Dimethyl sulfide ($(CH_3)_2S$ is produced by marine
algae and marine cyanobacteria and contributes to the
typical smell of thee sea. Dimethyl sulfide is degraded
by bacteria such as *Thiobacillus* and *Hyphomicrobium*,
leading to the formation of acid. The various transformations are summarized in Fig. 4.16. A major group of
bacteria important in the global economy of sulfur,
especially in aquatic environments are the sulfate reducing bacteria. They will be discussed briefly below.

The sulfate reducing bacteria (SRB) are ubiquitous
anaerobes found in diverse environments. They include
several groups of bacteria that use sulfate as an oxidizing agent, reducing it to sulfide (Fig 4.17) (Luptakova
2007). They can also utilize other sulfur compounds,
including sulfite, thiosulfate, and elemental sulfur in
a type of metabolism known as dissimilatory, because

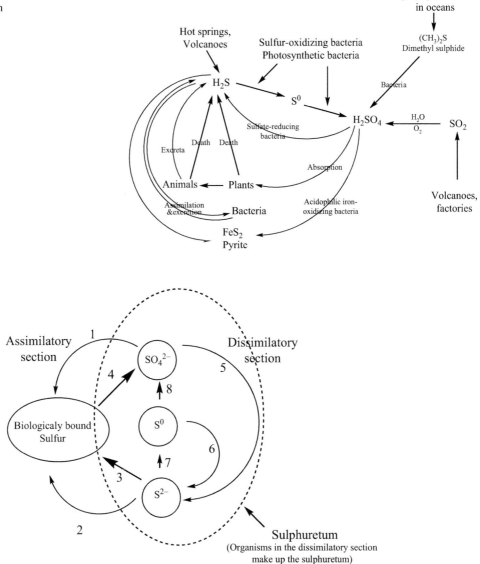

Fig. 4.16 General circulation of sulfur on Earth (Modified from Yamanaka 2008)

Fig. 4.17 The biological sulfur cycle (After Luptakova 2007). Key: 1 – Assimilatory sulfate reduction by plants, fungi and bacteria; 2 – Death and decomposition by fungi and bacteria; 3 – Sulfide assimilation by bacteria and some plants; 4 – Excretion of sulfate by animals; 5 – Dissimilatory sulfate-reducing bacteria; 6 – Dissimilatory sulfur-reducing bacteria; 7 – Phototrophic and chemotrophic sulfide-oxidizing bacteria; 8 – Phototrophic and chemotrophic sulfur-oxidizing bacteria

sulfur is not converted into organic compounds. The rotten egg odor of hydrogen sulfide in the environment usually indicates the presence of sulfate-reducing bacteria in nature. Sulfate-reducing bacteria are responsible for the rotten egg odors of salt marshes, mud flats, and intestinal gas. They slowly degrade materials that are rich in cellulose in anaerobic environments. Apart from soil, sulfate reducing bacteria are found in various habitats such as seas and oceans, mud and sediments of freshwaters (rivers, lakes), waters rich in decaying organic material, thermal or nonthermal sulfur springs, mining waters from sulfide deposits, waters from deposits of mineral oil and natural gas, industrial waste waters from metallurgical industry, as well as in the gastrointestinal tract of man and animals (Barton and Hamilton 2007).

Based on the energy source of sulfate reducing bacteria, there are two types of anaerobic respiration of sulfates: autotrophic and heterotrophic.

1. *Autotrophic reduction of sulfates*: In this case, the energy source is gaseous hydrogen; the reaction proceeds in several stages and the whole process can be expressed by:

$$4H_2 + SO_4^{2-} \xrightarrow{\text{SRB}} S^{2-} + 4H_2O \qquad (4.1)$$

2. *Heterotrophic reduction of sulfates*: The energy sources in heterotrophic reduction are simple organic substances (lactate, fumarate, pyruvate, some alcohols, etc.). The organic substrate may be incompletely or completely oxidized as shown in the two reactions given below:

(a) Incomplete heterotrophic oxidation of the organic substrate (acetate in this example):

$$2\,CH_3CHOHCOO^- + SO_4^{2-} \xrightarrow{\text{SRB}} 2\,CH_3COO^- + 2\,HCO_3^- + H_2S \qquad (4.2)$$

(b) Complete heterotrophic oxidation of organic substrate in which the final products are CO_2 and H_2O (Eq. 4.3):

$$4\,CH_3COOONa + 5\,MgSO_4 \xrightarrow{\text{SRB}} 5\,MgCO_3 + 2\,Na_2CO_3 + 5\,H_2S + 5\,CO_2 + H_2O \qquad (4.3)$$

During anaerobic respiration of sulfates, SRBs produce large amounts of gaseous hydrogen sulfide (H_2S) which react easily in the water medium with heavy metal cations forming fairly insoluble metallic sulfides (Eq. 4.4):

$$Me^{2+} + H_2S \rightarrow MeS + 2H^+ \,(Me^{2+}\text{-metal cation}). \quad (4.4)$$

SRBs are of great economic importance especially in the oil industry. They are ubiquitous in oil-bearing shale and strata and therefore play an important economic role in many aspects of oil technology. They are:

1. Responsible for extensive corrosion of drilling and pumping machinery and storage tanks
2. Contaminate resulting crude oil and thereby increase undesirably the sulfur content of the oil through the H_2S which they release into it
3. Important in secondary oil recovery processes, where bacterial growth in injection waters can plug machinery used in these processes
4. Speculated to play a role in biogenesis of oil hydrocarbons

For all of these reasons, SRB are of vital importance in petroleum producing and processing industries.

Apart from the above, SRB are responsible for the corrosion of buried tanks and tanks made of iron; in some industries, such as the paper industry, they cause undesirable blackening of paper due to iron sulfides in the processing water.

In nature, sulfur circulates permanently because it is continuously oxidized or reduced by chemical or biological processes. In such a biogeochemical sulfur cycle (Fig. 4.16), the biological transformations may have either assimilatory or dissimilatory metabolic functions. SRB play an important in this cycle. Figure 4.16 shows the global sulfur cycle, including biological and nonbiological activities. The biological component of sulfur transformation is given in Fig. 4.17. Most plants, fungi, and bacteria are capable of performing an assimilatory reduction of sulfate to sulfide which is necessary for the biosynthesis of sulfur containing cell compounds. On the other hand, the energy producing dissimilatory sulfur metabolism is restricted to a few groups of bacteria. The bacteria which participate in the dissimilatory section of the biological sulfur cycle are collectively known as the sulfuretum.

These groups include:

(a) Anaerobic dissimilatory sulfate reducers (*Desulfovibrio, Desulfotomaculum, Desulfomonas*)
(b) Anaerobic dissimilatory sulfur reducers (*Desulfuromonas, Beggiatoa*)
(c) Anaerobic phototrophic sulfur oxidisers (some cyanobacteria and most anoxygenic phototrophic bacteria)
(d) Anaerobic chemotrophic sulfur oxidisers (*Thiobacillus denitrificans, Thiomicrospira denitrificans*)

Iron Bacteria

Iron bacteria are chemoautotrophs which derive energy by oxidizing dissolved ferrous iron, and sometimes manganese and aluminum. The resulting ferric oxide is insoluble, and appears as brown gelatinous slime that will stain plumbing fixtures, and clothing or utensils washed with the water carrying the oxide.

$$Fe^{2+} + 2H^+ + 0.5O_2 \Rightarrow Fe^{3+} + 0.5H_2O.$$

Iron bacteria grow in waters containing as low as 0.1 mg/l of iron. They produce the brownish scale that forms inside the tanks of flush toilets. They complete the oxidation of partially oxidized iron compounds and are able to couple the energy produced to the synthesis of carbohydrate.

Many different bacteria can be involved in producing oxidized iron seen as "rusty" sediments in water. The true iron bacteria are those whose metabolism has been described above. The genera involved are *Leptothrix,, Clonothrix, and Gallionella* and *Sphaerotilus*. They are usually stalked, filamentous, and difficult or impossible to cultivate. They are sheathed and the outer portion of the sheaths is covered with slime in which oxides of iron are deposited giving them the colors ranging from red to brown. This sheath makes them somewhat resistant to disinfectants.

Typical symptoms of iron bacterial growths in water supplies are:

(a) Discoloration of the waters (yellow to rust-red or brown)
(b) Reduction in flow rates through the system caused by coatings of iron bacteria inside the pipes
(c) Development of thick red or brown coatings on the sides of reservoirs, tanks, and cisterns; sometimes, sloughing off to form either fluffy specks in the water or gelatinous clumps of red to brown filamentous growths
(d) Rapid Clogging of Filter screens
(e) Heavy surface and sedimented growths of a red or brown color sometimes iridescent (ochre) in water

Iron bacteria do not cause disease and their nuisance value is mainly esthetic. They cause economic loss due to stained porcelain fixtures, fouled laundry, etc.

Iron bacteria are not active at temperatures of about 5°C or lower and they require water with iron content of at least 0.2 mg/l. They thrive in situations where there is good aeration, some source of nutrition, and some heat such as provided by water pumps, and a regular supply of water with dissolved iron. They are susceptible to ultraviolet of the sun and hence are found deep in the ground or hidden in pipes.

Heavy growths of iron bacteria form a substrate for other bacteria which may then degrade these materials anaerobically to form acidic products and hydrogen sulfide. The growth of iron bacteria can controlled through the use of chlorine.

It should be pointed out that passing that "rust" is not always solely due to bacterial activity but could be due to physicochemical reactions, especially where the geological formations contain iron oxides in the form of different iron minerals: Siderite (iron carbonate), pyrite or greigite (iron sulfide) and hematite (iron oxide or hydroxide). Ground water is low in oxygen and has pH near neutrality. The dissolved iron oxides can rise to as high as 5 mg/l under these conditions. When the water is pumped from underground, it is exposed to air and the dissolved oxides are quickly oxidized and sediment as fine rusty colored powder. Oxidizing agents such as chlorine and potassium permanganate accelerate the oxidation of the oxides and deposition of rust.

During water purification, the aeration of the raw also hastens the deposition of the oxides. Manganese oxides are frequently common in waters with iron oxides. They form black deposit when oxidized.

4.1.5 Archae

4.1.5.1 General Properties of Archaea

Like the Domain Bacteria, the Domain Archaea consist of single-celled organisms lacking nuclear membranes, and are therefore prokaryotes. A *single* organism from this domain is called an *archaeon*, just as a single member in the Domain Bacteria is a bacterium. As seen in Table 4.1 the properties of Archaea make them closer, evolutionarily, to Eukaryotes than they are to Bacteria. Thus their genetic transcription and translation do not show many typical bacterial features, and are in many aspects similar to those of eukaryotes. Many archaeal tRNA and rRNA genes harbor unique archaeal introns which are neither like eukaryotic introns, nor like bacterial introns. Several other characteristics also set the Archaea apart.

1. With the exception of one group of methanogens, Archaea lack a peptidoglycan wall. Even in this case, the peptidoglycan is very different from the type found in bacteria.
2. Archaeans also have flagella that are notably different in composition and development from the superficially similar flagella of bacteria. Flagella from both domains consist of filaments extending outside of the cell, and rotate to propel the cell. Recent studies show that there are many detailed differences between the archaeal and bacterial flagella:
   (a) Bacterial flagella are motorized by a flow of H^+ ions, wheras archaeal flagella move by the action of ATP.
   (b) Bacterial cells often have many flagellar filaments, each of which rotates independently; the archaeal

flagellum is composed of a bundle of many filaments that rotate as a single assembly.

(c) Bacterial flagella grow by the addition of flagel-
lin (the protein in the flagella) subunits at the
tip; archaeal flagella grow by the addition of
subunits to the base.

(d) Bacterial flagella are thicker than archaeal
flagella, and have a hole through which flagel-
lin flows to be added at the tip, whereas archeal
flagella are too thin for such a hole.

Many Archaea are inhabitants of aquatic environments, both marine and freshwater.

4.1.5.2 Taxonomic Groups Among Archeae

Archaea are divided into two main groups based on rRNA
trees, the. Two other groups have recently been tentatively
added: *Koracheota* and *Nanoarheota*. The discussion
will be on the first two, and better known, groups.

Euryarchaeota

Members of this group can be arranged as follows:

1. *Extremely halohilic Archaea*: Members of this group
   survive in hypersaline environments, high levels of
   salt, such as are found in Great Salt Lake in Utah, and
   the Dead Sea. All are known as extremely Halophilic
   Archaea stain Gram negative. There are ten genera
   and 20 species of extreme halophiles, five of these
   genera contain only one species each: *Halobacterium*;
   Halobaculum;   *Natrosobacterium*;   Natrialba;
   Natrosomonas. The other genera are: *Natrarococcus*
   (two species); *Haloarcula* (two species); *Halococcus*
   (two species); *Haloferax* (four species); *Halorubrum*
   (five species). The key genera in this group are
   Halobacterium,, Haloferax, and *Natronobacterium.*

2. *Methane producing Archaea*: Nearly half of the
   known species of Archaea are unique in being capable of producing methane energy from selected low
   molecular weight carbon compounds and hydrogen
   as part of their normal biochemical pathways.
   Methanogens are anaerobic and are the most common and widely dispersed of the Archaea being
   found in anoxic sediments and swamps, lakes,
   marshes, paddy fields, landfills, hydrothermal vents,
   and sewage works as well as in the rumen of cattle,
   sheep, and camels, the cecae of horses and rabbits,
   the large intestine of dogs and humans, and in the
   hindgut of insects such as termites and cockroaches.

   In their natural habitats, methanogens depend on
   substrate supply from associated anaerobic microbial

communities or geological sources, and depending
on the substrates they utilize, three types of methanogenic pathways are recognized (see Fig. 4.18):

(a) *Hydrogenotrophic methanogens* which grow
with hydrogen (H_2) as the electron donor and
carbon dioxide (CO_2) as the electron acceptor.
Some hydrogenotrophs also use formate, which
is the source of both CO_2 and H_2.

(b) *Acetoclastic methanogens* which cleave acetate
into a methyl and a carbonyl group. Oxidation of
the carbonyl group into CO_2 provides potential
for reduction of the methyl group into CH_4.

(c) *Methylotrophic methanogens* grow on methy-
lated compounds such as methanol, methy-
lamines, and methyl sulfides, which act as
both electron donor and acceptor or are
reduced with H_2.

The important genera among methane producing
Archaea are *Methanobacterium*, *Methanosarcina*
and *Methanocaldococcus*. Methanogens utilize a
wide variety of substrates for producing methane.
These include CO_2, alcohols, methyl substrates,
methanol (CH_3OH), methylamine (CH_3NH^{3+}), and
trimethylamine (($CH_3)_3NH^+$) and acetic compounds
such as acetate (CH_3COO^-) and pyruvate.

3. *Thermophilic and Extremely Acidophilic (Thermo-
plasmatales)*: This is a small group of extreme acidophilic organisms. They containing four species in
two genera, they are unusual in their ability to tolerate acid conditions. The two *Picrophilus* species are
the most acidophilic organisms known. They have
an optimal pH requirement of 0.7, can still grow at
a pH of -0.06 and die at pH values of less than 4.0.
Both *Picrophilus* species were found in acid solfatoras in Japan. Solfatoras are craters, often near volcanoes, spewing out steam, and gases such as CO_2,
SO_2, and HCl. When sulfurous gases are spewed
out from such craters they are solfatoras (from the
Italian for sulfur). The two species of *Thermoplasma*
grow optimally at pH 2.0. *Thermoplasma* spp. are
also very unusual in that they do not have a cell
wall. *T. volcanium* has been isolated from a number
of solfatoras around the world. The cell membrane
of *Thermoplasma* is composed of a lipopolysaccha-
ride-like compound consisting of lipid with man-
nose and glucose units and called a lipoglycan.
Examples of this group are Thermoplasma and
Ferroplasma. These Archae lack cellwalls and in
this regard are like Mycoplasmas. They not only

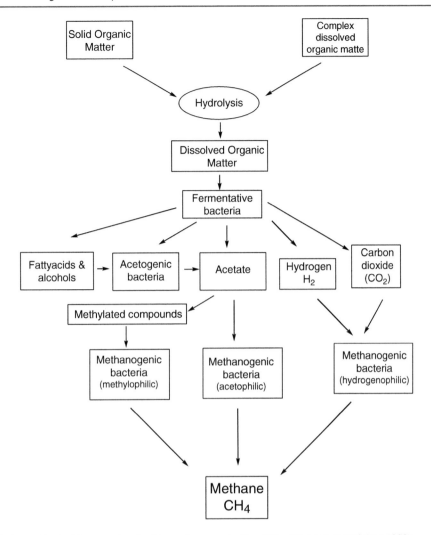

Fig. 4.18 Substrates and bacterial groups involved in methane production (After Christen and Kjelsen 1989)

survive without cellwalls but they also survive high temperatures and low acid conditions. For these conditions, these organisms have special polysaccharide structures in their cell membranes, a lipopolyssacharide.

4. *Hyperthermophilic Archae*: Well-known members of this group are *Thermococcus*, *Pyrococcus*, and *Methanopyrus*. Members of this group have optimal temperatures of 80°C and many grow at temperatures higher than that of boiling water. Thus *Thermococcus* and *Pyrococcus* (cocci with a tuft of flagella on one side) grow at between 70°c and 106°C with an optimum at 100°C (*Pyrococcus*). Proteins, starch or maltose are oxidized as electron donors and S^0 is the terminal acceptor and is reduced to H_2S.

Crenarchaeota

Crenarchaeota has the distinction of including microbial species with the highest known growth temperatures of any organisms. As a rule, they grow best between 80°C and 100°C and several species will not grow below 80°C. Several species also prefer to live under very acidic conditions in dilute solutions of hot sulfuric acid. Approximately 15 genera are known, and most of the hyperthermophilic species have been isolated from marine or terrestrial volcanic environments, such as hot springs and shallow or deep-sea hydrothermal vents. Recent analyses of genetic sequences obtained directly from environmental samples, however, indicate the existence of low temperature Crenarchaeota, which have not yet been cultivated. The most spectacular feature of the Crenarchaeota, however, is their tolerance

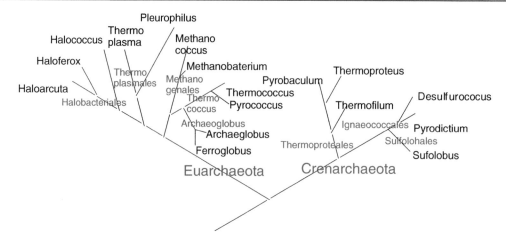

Fig. 4.19 Phylogenetic tree of the Archae (Modified from Ciccarelli et al. 2006)

   Note: The genera of Archeae are given above. They are placed in the two phyla of Archeae which have been cultivated, Euarchaeota and Crenarchaeota. A third phylum, the

Korarcheota are only known from their DNA sequences and have not yet been cultivated. The notations in *red* are orders in which the genera are grouped. Thus the halophilic archeae are grouped among the Halobacteriales the methane producers are in Methagenales

of, and even preference for, extremes of acidity and temperature. While many prefer neutral to slightly acidic pH ranges, members of the Crenarchaeal order Sulfolobales flourish at pH 1–2 and die above pH 7. Optimum growth temperatures range from 75°C to 105°C and the maximum temperature of growth can be as high as 113°C (*Pyrobolus*). Most species are unable to grow below 70°C, although they can survive for long periods at lower temperatures. Crenarchaeota contains representations of organisms which live in a wide variety of environments including terrestrial environments (hot springs, geothermal power plants) or in marine (submarine hot vents, deep oil wells, marine smokers up to 400°C). Some exist in environments of over 100°C, while others live at ice cold conditions. For substrates, they utilize a wide range of gases: CO_2, CO, CH_4, S_2O_3, N_2, NH_4 (Fig. 4.19).

(a) *Hyperthermophiles in underwater volcanic environments*

   Temperatures as high as 100°C occur around terrestrial volcanoes and *Sulfolobus* and *Thermoproteus*, both hyperthermophiles, have been isolated from such environments.

(b) *Hyperthermophiles in land volcanic environments*

   Archae with the highest optimum temperature of growth known occur in underwater vents and near underwater volcanoes. *Pyrodictium* sp. and *Pyrolobus* sp. have optimum temperatures of growth of 100°C and 106°C respectively and are

found in such environments. *Desulforococcus* (90°C, optimum) and *Staphylothermus* (95°C, optimum) are also found in that environment.

4.1.6 Microbial Taxonomic Groups Among Eucharia

The Domain Eukarya includes plants, animals, algae, fungi, and protozoa. The last three are regarded as microorganisms, although some of them are quite large. In Fig. 4.20, green algae, brown algae, red algae, and diatoms are Algae. Diplomonads, Trichomonads, Ciliates, Flagellates and Slime molds are Protozoa.

4.1.6.1 Protozoa

Protozoa are thought to be the evolutionary ancestors of all multi-cellular organisms, including plants, fungi, and animals. The basis for this assumption is that Protozoa contain members which like plants are phototrophic, as well as members which like animals and fungi are heterotrophic. Furthermore Protozoa contain species with intermediate (mixotrophic) trophic capabilities, i.e., they have the capability for both auto- and heterotrophic existence. Thus, many dinoflagellates are auxotrophic because they take up vitamins produced by other organisms; on the other hand, other Protozoa such as some euglenoids alter from one form of trophic existence system to another.

Fig. 4.20 Phylogenetic tree among the Eukarya (Redrawn from Ciccarelli et al. 2006)

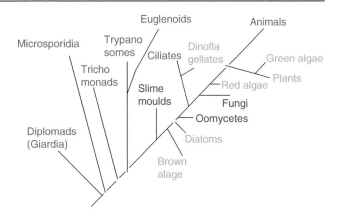

Protozoa are classified in many ways, primarily on morphological characteristics, and the one adopted here groups them into five: Mastigophora, Sarcodina, Ciliata, Sporozoa, and Suctoria.

1. *Flagellata (Mastigophora)*

   These possess flagella and are subdivided into "phytoflagellata" and "zooflagellata," depending on whether they are plant-like (with chlorophyll) or animal-like (without chlorophyll). They usually multiply by longitudinal binary fission. Many flagellates are able to feed autotrophically as well as heterotrophically, and are important primary producers in lakes and oceans; yet, they can also feed like animals, ingesting or absorbing food synthesized by other organisms.

   Many are free-living, but some are parasitic. Examples of parasitic Mastigophora are *Trypanosoma gambiense* and *T. rhodesiense* which cause African sleeping sickness and is transmitted by tsetse flies. *T. cruzi* is the cause of Chagas' disease, prevalent in South and Central America, which affects the nervous system and heart; it is transmitted by the bite of assassin bugs. *Giardiasis* is caused by the mastigophoran *Giardia lamblia.*

2. *Rhizopoda (Sarcodina)*

   These Protozoa use pseudopodia (false feet) for locomotion and for catching preys. Members of the group Sarcodina move by pseudopodia; although, flagella may be present in the reproductive stages. Cytoplasmic streaming assists movement. Asexual reproduction occurs by fission of the cell.

   Sarcodina includes two marine groups known as foraminiferans and radiolarians. Both groups were present on earth when the oil fields were in formative stages, and marine geologists use them as potential markers for oil fields. Some amoebae live in shells

from which the pseudopodia are extruded. Some members of the group such as *Entamoeba histolytica,* are pathogenic, causing amoebic dysentery in humans. This organism can cause painful lesions of the intestine and is contracted in polluted water (Fig. 4.21).

3. *Ciliata*

   Ciliates possess cilia (short and highly coordinated flagellae), a somatic (macro) nucleus, and genetic (micro) nucleus, and a contractile vacuole is usually present. They move by means of cilia. Conjugation may be used for sexual reproduction and binary fission also occurs. The distinctive rows of cilia vibrate in synchrony and propel the organism in one direction. One of the best known members of the group is *Paramecium*; another of the free-living members of this group, is *Tetrahymena.*

   Ciliates form an extremely large group are distinguished by the possession of cilia, two different types of nuclei and transverse fission of the organism when it divides, unlike flagellates and sarcodina which divide longitudinally.

4. *Sporozoa*

   Members of the group *Sporozoa* form spores at one stage in their life cycle. Sporozoa are endoparasites which have spores. Most of them spend at least part of their life-cycle inside a host cell. Reproduction is a complex phenomenon in this group. Members of the group display no means of locomotion in the adult form. Their motile stages move by bending, creeping, and gliding and usually have an apical complex at their anterior end which help them penetrate their hosts. The group includes *Plasmodium,* the agent of malaria, *Toxoplasma*, the agent of toxoplasmosis, and *Pneumocystis carinii,* the cause of a serious pneumonia in AIDS patients.

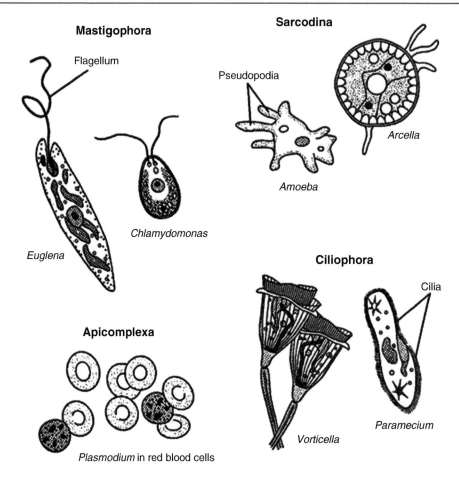

Fig. 4.21 Protozoan groups. Illustrations of some Protozoa (From http://www.cliffsnotes.com/study_guide/topicArticleId-8524,articleId-8461.html; Anonymous 2010b. With permission)

5. *Suctoria*

The juvenile forms are ciliated and motile, while the adult forms are sessile and capture food by tentacles. They feed by extracellular digestion and lack cilia in the adult phase. The adult have structures called haptocysts at the tip which attach to the prey. The prey's cytoplasm is then sucked directly into a food vacuole inside the cell, where its contents are digested and absorbed. Most suctoria are around 15–30 µm in size, with a non-contractile stalk and often a shell. Suctoria reproduce primarily by budding, producing swarmers which lack both tentacles and stalks but have cilia. Once the swarmers (motile young) have found a place to attach themselves, they quickly develop stalks and tentacles and lose their cilia. Because of the presence of cilia in the young of suctoria, some authors group the suctoria among ciliates.

Suctoria are found in both freshwater and marine environments, and some which live on the surface of aquatic animals, and typically feed on ciliates. Some marine species form symbiotic relationships with crustaceans and even some fish. One species, *Ephelota gemmipara* lives on the external parasite of salmon, *Lepeophtheirus salmonis* (salmon louse).

4.1.6.2 Fungi

Fungi are eukaryotic microorganisms which

(a) Are non-photosynthetic and hence do not contain chlorophyll

(b) Contain chitin and/or cellulose in their cell walls

(c) Are usually filamentous (called molds), but they may be unicellular (called yeasts)

(d) Reproduce asexually with spores

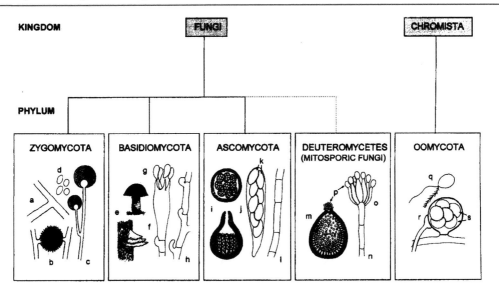

Fig. 4.22 Diagnostic features of fungi (From Guarro et al. 1999)

Zygomycota (Phycomycetes): (a) coenocytic hypha; (b) zygospore; (c) sporangiophore; (d) sporangiospores.

Basidiomycota (Basidiomycetes): (e) basidiomate; (f) basidium; (g) basidiospore; (h) hypha with clamp connections.

Ascomycota (Ascomycetes): (i) ascommata; (j) ascus; (k) ascospores; (l) septate hyphae.

Deuteromycota (Deuteromycetes): (m) pycnidium; (n) conidiophore; (o) conidiogenous cells; (p) conidia.

Oomycota (regarded as aquatic Phycomycetes) : (q) zoospore; (r) gametangia; (s) oospores

Taxonomy of Fungi

The classification of the fungi is based mainly on morphology of the hyphae, the structures housing the sexual structures, or the structure to which the sexual spores are attached (Samson and Pitt 1989; Guarro et al. 1999). The principal diagnostic characteristics are shown in Fig. 4.22 and are as follows:

(a) *Septation of the hyphae*

The septation, or lack thereof, of the hyphae is important in classifying fungi. Non-septate or coenocytic hyphae are found in Phycomycetes (Zygomycota). All other fungal groups have septate hyphae. Examples of Phycomycetes are *Mucor* spp. and *Rhizopus* spp., the bread mold.

(b) *The nature of the asexual spores of aquatic Phycomycetes*

Aquatic fungi are found among Phycomycetes. The asexual cells of many aquatic Phycomycetes are motile and are flagellated and help in the identification of the organisms (see Figs. 4.22 and 4.23). Many aquatic Phycomycetes, classified as Oomycetes, by some authors are pathogens of plants and fish. *Phytphthora infestans* which caused the famous potato blight and subsequent famine in Ireland which led to massive Irish immigration to the US belong to this group. Others are *Plasmopara viticola* (the cause of downy mildew of grapes), *Plasmopara halstedii* (sunflower downy mildew), and *Saprolegniales* spp., or water molds, which cause diseases of fish and other aquatic vertebrates. Some authors argue that the Oomycetes are so different from other fungi (so-called true fungi) that they should not be classified with them. The majority of authors classify them with the Phycomycetes, their peculiarities notwithstanding (see Table 4.8).

(c) *The presence of asci*

An ascus or sac (*plural*, asci) which contains ascosposcores (sexual spores typically eight in number housed in an ascus) is diagnostic of *Ascomycetes* (Ascomycota). The trivial name of this group of fungi is sac fungi.

(d) *The presence of Basidiomycetes*

The presence of basidiospores, typically four in number attached to a basidium, (a club-like structure) identifies Basidiomycetes. Some of the best known *Basidiomycetes* (Basidiomycota) are mushrooms. A microscopic examination of the "gills" on the underside of the mushrooms reveals the basidia carrying the basidiospores.

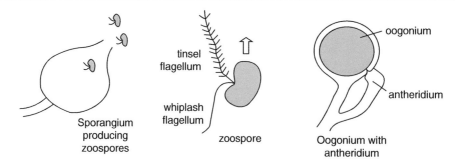

Fig. 4.23 Reproductive structures in aquatic fungi (From Rossman and Palm 2006. With permission)

Table 4.8 Properties of oomycetes and other ("true") fungi (Modified from Rossman and Palm 2006. With permission)

Character	Oomycota	True Fungi
Sexual reproduction	Heterogametangia. Fertilization of oospheres by nuclei from antheridia forming oospores	Oospores not produced; sexual reproduction results in zygospores, ascospores, or basidiospores
Nuclear state of vegetative mycelium	Diploid	Haploid or dikaryotic
Cell wall composition	Beta glucans, cellulose	Chitin. Cellulose rarely present
Type of flagella on zoospores, if produced	Heterokont, of two types, one whiplash, directed posteriorly, the other fibrous, ciliated, directed anteriorly	If flagellum produced, usually of only one type: posterior, whiplash
Mitochondria	With tubular cristae	With flattened cristae

(e) *Deutromycetes (Deuteromycota) or fungi imperfecti*

These are fungi whose perfect or sexual stages have not been discovered. Some of the best examples are *Penicillin* spp. (with broom-like structures) and *Aspergillus* spp. (with club-like structures).

4.1.6.3 Algae

Algae are photosynthetic eukaryotic organisms which lack the structures of vascular plants. Many authors classify them as microorganisms, but they are highly variable in size and range from microscopic sizes to the brown algae which could be up to 70 m long (Trainor 1978; Sze 1986) (Fig. 4.24).

Taxonomy of Algae

The classification of the algae is based on the following:

(a) *Pigmentation*

The various kinds of pigments in the algae, as well as the overall color of the alga, are used in classifying the organisms. All of the groups contain the following pigments which are soluble in organic solvents: Chlorophylls and several carotenoids, which include carotenes and xanthophylls. Chlorophylls "a" and "b," alpha or beta carotene, and some xanthophylls are common. The water soluble phycobiliproteins (phycobilins) are found in blue-green algae, red algae, and a small group of flagellates.

(b) *The reserve compound*

Reserve food material is usually stored within the cell and frequently within the plastid in which photosynthesis occurred. Starch, starchlike compounds, fats, or oils are the most common forms.

(c) *The nature of the zoospores*

Some organisms are motile during much of their lives, whereas other genera lack motility, or any motile reproductive stages. Adult algae are usually nonmotile; often, however, some reproductive stages (zoospores) are motile. The overall shape of the zoospores, the shape, number, and the insertion position of the flagella, and the presence or absence of hairs on the flagella are diagnostic (Fig. 4.24; Table 4.9).

(d) *Wall composition*

The cell wall may be a simple outer covering around the protoplast or an elaborately ornamented structure. The materials found in algal walls are cellulose, xylans, mannans, sulfated polysaccharides, alginic acid, protein, silicon dioxide, and calcium carbonate

(e) *Gross morphology of the alga*

The overall shape of the alga is diagnostic.

The various groups of algae are given in the Table 4.9.

Fig. 4.24   Illustrations of some algae (All items in the table reproduced with permission) a) Red algae are red because of the presence of the pigment phycoerythrin; this pigment reflects red light and absorbs blue light. Because blue light penetrates water to a greater depth than light of longer wavelengths, these pigments allow red algae to photosynthesize and live at somewhat greater depths than most other algae. The picture of the red alga, Dichotomaria marginata, shown here was taken and kindly supplied by Keoki Stender, University of Hawaii. b) Sargassum fluitans. Sargassum seaweed, Gulf Weed (brown alga). This is a major component of the algae in the Sargasso Sea in the pelagic Atlantic. It has long, serrated fronds with a distinctive mid-rib, and smooth berry-like spherical gas-filled bladders, pneumatocysts, which assist the floatation of the alga. The photo of Sargassum above was kindly supplied by the South Carolia Department of Natural Resource, courtesy of H. Scott Meister. c) Synura spp, a member of the Chrysophyceae (golden or golden-brown algae on account of their content of fucoxanthin which in the presence of chlorophyll makes them look brown or golden brown), forms swimming colonies from a variable number of cells joined together at their posterior ends in a spherical or elongated cluster. Synura is important because it gives drinking water a bitter taste and a "fishy" cod liver oil type of odour. Synura is freshwater; some marine members such as Olisthodiscus luteus, produce neurotoxins which may kill aquatic fauna and may affect humans through eating shell fish raw. Credit: Dr Graham Matthews, graham@gpmatthews.nildram.co.uk (http://www.gpmatthews.nildram.co.uk/microscopes/pondlife_plants01.html). Dr Graham Mathews is also Hon Secretary of the Secretary of the Quekett Microscopical Club) d) Bacillaropyceae (Diatoms) are one of the largest and ecologically most significant groups of organisms on Earth. Diatoms are microscopic algae which are easily recognizable because of their unique cell structure, silicified cell wall and life cycle. Diatoms are found anywhere there is water and light: in oceans, lakes and rivers; marshes, fens and bogs; damp moss and rock faces. They are an important part of the food chain in aquatic environments, especially in nutrient-rich areas of the world's oceans, where they occur in abundance. Photograph of diatoms kindly supplied by Dr David Carling) e) Dinoflagellates are minute marine unicellular algae with diverse morphology, the largest, Noctiluca, being as large as 2 mm in diameter!. Many are photosynthetic, while some are parasites of fish. In temperate climates they form blooms in summer months which may be golden or red. Their blooms produce neurotoxins marine animals eating and humans who consume them raw (such as shellfish)

Prorocentrum mexicanum   Protoperidinium crassipe

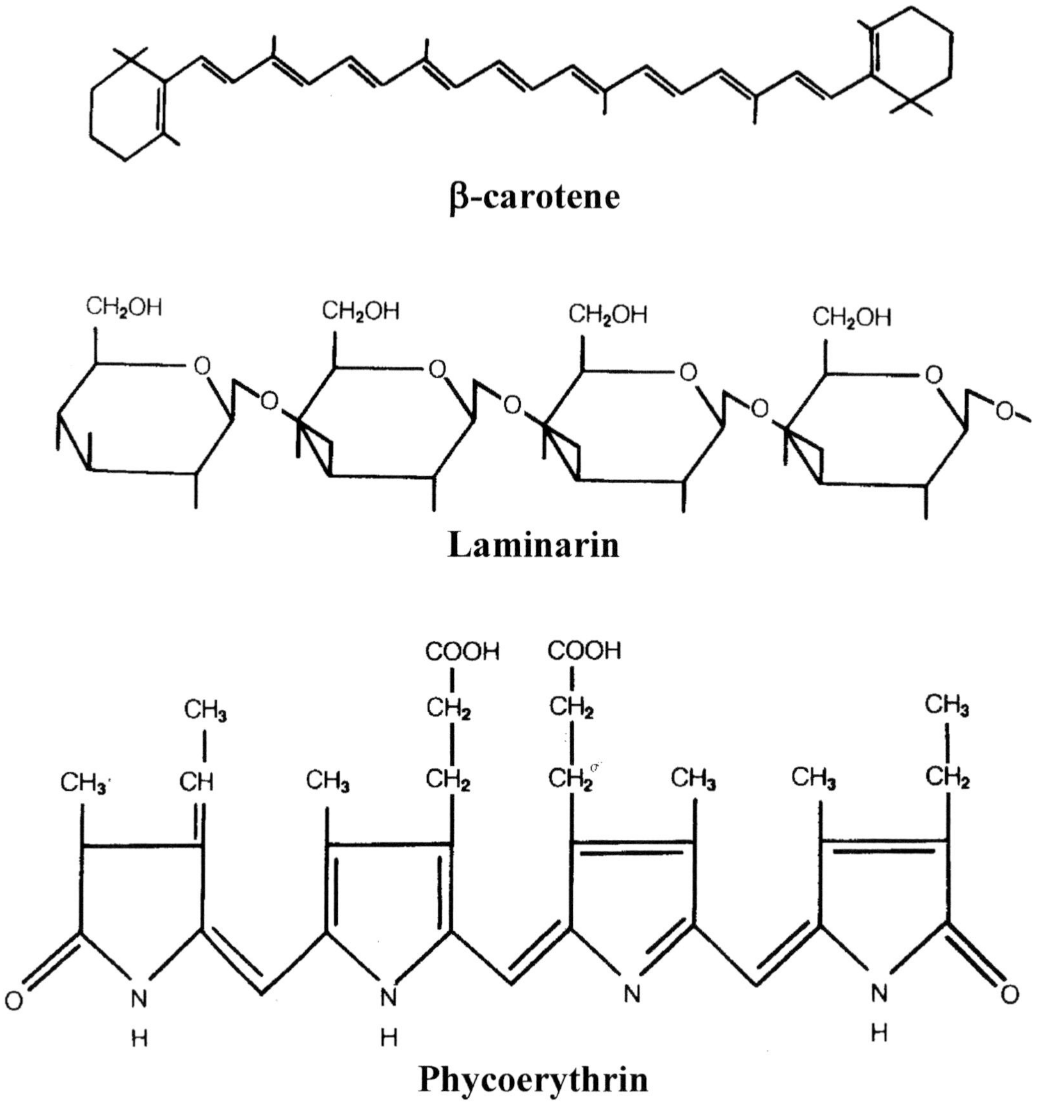

β-carotene

Laminarin

Phycoerythrin

Fig. 4.25 Some pigments and storage material of algae

4.1.7 Viruses

Until recently, viruses were not thought to be abundant or important in the aquatic environment. We now know that they are not only abundant, but that they profoundly influence the ecology and food status of the aquatic environment including seas and oceans.

Viruses are lifeless crystals of nucleic acid which are able to grow and reproduce only in living cells. They differ from cells in the following ways, and also have the following properties:

1. Whereas cells contain both DNA and RNA, viruses contain either DNA or RNA, never both.

2. Viruses have a nucleic acid inner core (the genome) and outer protein cover, the capsid (see Fig. 4.26).

3. Viruses enter only susceptible cells: thus there are viruses which will attack only plants while some will attack only animals. Even among plants and animals, some viruses will attack some members and not the others.

4. All living things are attacked by viruses, including the microorganisms: Bacteria, fungi, algae, and protozoa. Viruses attacking bacteria and fungi are bacteriophages and mycophages, respectively.

Viruses used to be classified on their diseases they cause and their sizes and shapes, but these criteria have

Table 4.9 The various algal groups (Compiled from Sze 1986 and Trainor 1978)

S. No.	Group	Pigments	Storage	Zoospores	Walls,	Morphology	Example	Habitat
1.	Chlorophyceae (Green algae)	Chlorophyll a,b, Xanthophyll	Starch	Variable, when present	Cellulose, chitin, calcium carbonate	Unicellular to multicellular	Spirogyra, Chlamydomonas	Freshwater, oceans, soils
2.	Rhodophyceae (Red algae)	Chlorophyll a,d Carotene a, b Phycobilins (masks other pigments)	Floridean starch (like glycogen)	None	Cellulose	Multicellular: large, up to 5 m long	Gelidium (agar, microbial cultivation) and Chondrus (carageenan – food thickening)	Few freshwater, mostly marine:warmer waters, especially along coastlines
3.	Phaemyceae (Brown algae)	Chlorophyll a,c Carotene a, b Xanthophylls	Laminarin, mannitol, fat	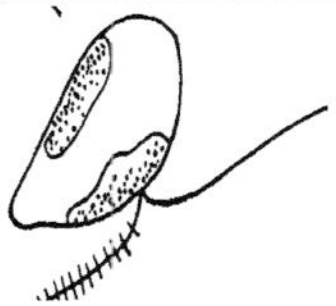	Alginic acid, cellulose	Multicellular: large, up to 5 m long	Ectocarpus	Mostly marine in north temperate regions
4.	Bacillariophyceae (Diatoms)	Chlorophyll a,c Xanthophylls	Glucans, Oil	No moving member at any stage	Celluose, pectin in above structure	Unicels, colonial or filametous	Tabelaria spp.(freshwater)	Marine and freshwater
5.	Xanthophyceae (Yellow green algae)	Chlorophyll a,c Carotene a, b Xanthophylls	Chrysolaminarin, Oil		Celluose, pectin in above structure	Unicells, colonies, branched and unbranched filaments	Vaucheria sp.	Mostly temperate freshwater
6.	Chrysophyceae (Golden algae)	Chlorophyll c Carotene a, b Xanthophylls	Chrysolaminarin	Adults nonmotile or motile with a flagellum	Naked or cellulose	Typically Unicellular; few colonies, branched and unbranched filaments	Synura sp. (imparts fish odor to water)	Cold freshwater in northern hemisphere
7.	Dinophyceae (Dinoflagellates; dark-brown algae)	Chlorophyll c Carotene b Xanthophylls	Starch or oils		Often only cell membrane; sometimes a pellicle Girdle in which one flagellum lodged; other flagellum trails	Motile adult flagellates, colonies or filamentous		Freshwater and marine

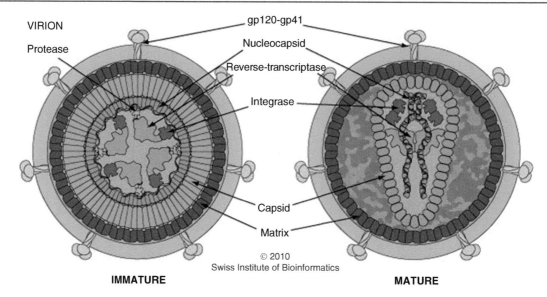

Fig. 4.26 Structure of HIV 1 virus (Reproduced with permission from the Swiss Institute of Bioinformatics [SIB]; Anonymous 2010c)

Note that the HIV 1 virus has an envelope (matrix in the diagram above). Not all viruses have envelopes; those which do not are said to be *naked* (see text). The gp structures are glycoprotein.

(Retroviral) integrase is an enzyme produced by a retrovirus that enables its genetic material to be integrated into the DNA of the infected cell. Note also the reverse transcriptase (produced by retroviruses such as HIV), a DNA polymerase enzyme that transcribes single-stranded RNA into double-stranded DNA

been abandoned owing to the small sizes of viruses and because the disease symptoms of different viruses were sometimes similar (Anonymous 2005; Sander 2007). The current classification of viruses is credited to David Baltimore, who won the Nobel Prize for his discovery of retroviviruses and reverse transcriptase. According to the Baltimore classification, viruses are grouped into seven based on their nucleic acid (DNA or RNA), strandedness (single-stranded or double stranded), and method of replication. The groups are numbered with Roman numerals thus:

Group I: double-stranded DNA viruses

Group II: single-stranded DNA viruses

Group III: double-stranded RNA viruses

Group IV: positive-sense single-stranded RNA viruses

Group V: negative-sense single-stranded RNA viruses

Group VI: reverse transcribing diploid single-stranded RNA viruses

Group VII: reverse transcribing circular double-stranded DNA viruses

Nomenclature of viruses: The nomenclature of viruses is based on a set of rules set up by the International Committee on the Taxonomy of Viruses (ICTV), which since the 1960s has been arranging viruses in these seven groups into taxonomic hierarchies.

The criteria for the taxonomic arrangements are:

1. *Morphology*
   (Helical, e.g., bacteriophage M13; icosohedral/polyhedral/cubic, e.g., poliovirus, enveloped – may have poyhedral (e.g., herpes simplex) or helical (e.g., influenza virus) capsids, complex, e.g., poxviruses)
2. *Nucleic acid type*,
3. Whether the virus is *naked or enveloped*
4. *Mode of replication*
5. *Host organisms*
6. The *type of disease* they cause

4.1.7.1 Viral Taxonomy and Nomenclature

Viral taxonomic nomenclature is modeled after that of cellular organisms. However viruses suffer from the absence of fossil record which will enable more phylogenetic relationships among the various groups. Consequently, the highest level in the viral taxonomic hierarchy is the order, thus:

Order (*-virales*)

Family (*-viridae*)

Subfamily (*-virinae*)

Genus (*-virus*)

Species (*-virus*)

Regarding nomenclature, the rules set up by the ICTV are as follows (Van Regenmortel 1999):

- The names of virus orders, families, subfamilies, genera, and species should be written in italics with the first letter capitalized.
- Other words are not capitalized unless they are proper nouns, e.g., *Tobacco mosaic virus*, *Poliovirus*, *Murray River encephalitis virus*.
- This format should only be used when official taxonomic entities are referred to - it is not possible to centrifuge the species, for example, *Poliovirus,* but it is possible to centrifuge poliovirus.
- Italics and capitalization are not used for vernacular forms (e.g., rhinoviruses, c.f. the genus *Rhinovirus*), for acronyms (e.g., HIV-1), nor for adjectival usage (e.g., poliovirus polymerase).

4.1.7.2 The Viral Groups
DNA Viruses (Groups I and II)
Group I: These are double-stranded DNA viruses and include such virus families as Herpesviridae (examples like HSV1 [oral herpes], HSV2 [genital herpes], VZV [chickenpox], EBV [Epstein–Barr virus], CMV [Cytomegalovirus]), Poxviridae (smallpox), and many tailed bacteriophages. The mimivirus is also placed into this group (see Table 4.10).

Group II: These viruses possess single-stranded DNA and include such virus families as Parvoviridae and the important bacteriophage M13 (see Table 4.11).

RNA Viruses (Groups III, IV and V)
Group III: These viruses possess double-stranded RNA genomes, e.g., rotavirus. These genomes are always segmented. Segmented virus genomes are those which are divided into two or more physically separate molecules of nucleic acid, all of which are then packaged into a single virus particle (see Table 4.12).

Group IV: These viruses possess positive-sense single-stranded RNA genomes. Many well known viruses are found in this group, including the picornaviruses (which is a family of viruses that includes well-known viruses like Hepatitis A virus, enteroviruses, rhinoviruses, poliovirus, and foot-and-mouth virus), SARS virus, hepatitis C virus, yellow fever virus, and rubella virus. Positive-sense viral RNA is identical to viral mRNA and thus can be immediately translated by the host cell (see Table 4.13).

Group V: These viruses possess negative-sense single-stranded RNA genomes. The deadly Ebola and Marburg viruses are well known members of this group, along with influenza virus, measles, mumps, and rabies. Negative-sense viral RNA is complementary to mRNA and thus must be converted to positive-sense RNA by an RNA polymerase before translation (see Table 4.14).

Reverse Transcribing Viruses (Groups VI and VII)
A *reverse transcribing virus* is any virus which replicates using reverse transcription, the formation of DNA from an RNA template. Both Group VI and Group VII viruses fall into this category. Group VI contains single-stranded RNA viruses which use a DNA intermediate to replicate, whereas Group VII contains double-stranded DNA viruses which use an RNA intermediate during genome replication. Thus a reverse transcriptase, also known as RNA-dependent DNA polymerase, is a DNA polymerase enzyme that transcribes single-stranded RNA into single-stranded DNA. Normal transcription involves the synthesis of RNA from DNA; hence, reverse transcription is the *reverse* of this.

Group VI: *RNA Reverse Transcribing Viruses* possess single-stranded RNA genomes and replicate using reverse transcriptase. The retroviruses are included in this group, of which HIV is a member. Members of Group VI use virally encoded reverse transcriptase, a RNA-dependent DNA polymerase, to produce DNA from the initial virion RNA genome. This DNA is often integrated into the host genome, as in the case of retroviruses and pseudoviruses, where it is replicated and transcribed by the host. Group VI includes the following in Table 4.15.

Group VII: *DNA Reverse Transcribing Viruses* possess double-stranded DNA genomes and replicate using reverse transcriptase. The hepatitis B virus can be found in this group. Group VII have DNA genomes contained within the invading virus particles. The DNA genome is transcribed into both mRNA, for use as a transcript in protein synthesis, and pre-genomic RNA, for use as the template during genome replication. Virally encoded reverse transcriptase uses the pre-genomic RNA as a template for the creation of genomic DNA. They are shown in Table 4.16.

The Structure of the DNA and RNA Viruses
The structures of the various viruses whether they are DNA or RNA are highly variable. The structure of the DNA viruses are shown in Table 4.17 and those of the RNA viruses are shown in Table 4.18.

Table 4.10 Group I: dsDNA Viruses (Reproduced from Van Regenmortel et al. 2005; http://www.microbiologybytes.com/virology/VirusGroups.html#VI; Anonymous 2005. With permission)

Order *Caudovirales* – tailed bacteriophages

Family				
Subfamily		Genus	Type species	Hosts
Myoviridae		T4-like viruses	*Enterobacteria phage T4*	Bacteria
		P1-like viruses	*Enterobacteria phage P1*	Bacteria
		P2-like viruses	*Enterobacteria phage P2*	Bacteria
		Mu-like viruses	*Enterobacteria phage Mu*	Bacteria
		SP01-like viruses	*Bacillus phage SP01*	Bacteria
		φH-like viruses	*Halobacterium virus φH*	Bacteria
Podoviridae		T7-like viruses	*Enterobacteria phage T7*	Bacteria
		P22-like viruses	*Enterobacteria phage P22*	Bacteria
		φ29-like viruses	*Bacillus phage φ29*	Bacteria
		N4-like viruses	*Enterobacteria phage N4*	Bacteria
Siphoviridae		λ-like viruses	*Enterobacteria phage λ*	Bacteria
		T1-like viruses	*Enterobacteria phage T1*	Bacteria
		T5-like viruses	*Enterobacteria phage T5*	Bacteria
		L5-like viruses	*Mycobacterium phage L5*	Bacteria
		c2-like viruses	*Lactococcus phage c2*	Bacteria
		ψM1-like viruses	*Methanobacterium virus ψM1*	Bacteria
		φC31-like viruses	*Streptomyces phage φC31*	Bacteria
		N15-like viruses	*Enterobacteria phage N15*	Bacteria
Ascoviridae		*Ascovirus*	*Spodoptera frugiperda ascovirus*	Invertebrates
Adenoviridae		*Atadenovirus*	*Ovine adenovirus D*	Vertebrates
		Aviadenovirus	*Fowl adenovirus A*	Vertebrates
		Mastadenovirus	*Human adenovirus C*	Vertebrates
		Siadenovirus	*Frog adenovirus*	Vertebrates
Asfarviridae		*Asfivirus*	*African swine fever virus*	Vertebrates
Baculoviridae		*Nucleopolyhedrovirus*	*Autographa californica nucleopolyhedrovirus*	Invertebrates
		Granulovirus	*Cydia pomonella granulovirus*	Invertebrates
Corticoviridae		*Corticovirus*	*Alteromonas phage PM2*	Bacteria
Fuselloviridae		*Fusellovirus*	*Sulfolobus virus SSV1*	Archaea
Guttaviridae		*Guttavirus*	*Sulfolobus virus SNDV*	Archaea
Herpesviridae:		*Ictalurivirus*	*Ictalurid herpesvirus 1*	Vertebrates
	Alphaherpesvirinae	*Mardivirus*	*Gallid herpesvirus 2*	Vertebrates
		Simplexvirus	*Human herpesvirus 1*	Vertebrates
		Varicellovirus	*Human herpesvirus 3*	Vertebrates
		Iltovirus	*Gallid herpesvirus 1*	Vertebrates
	Betaherpesvirinae	*Cytomegalovirus*	*Human herpesvirus 5*	Vertebrates
		Muromegalovirus	*Murine herpesvirus 1*	Vertebrates
		Roseolovirus	*Human herpesvirus 6*	Vertebrates
	Gammaherpesvirinae	*Lymphocryptovirus*	*Human herpesvirus 4*	Vertebrates
		Rhadinovirus	*Simian herpesvirus 2*	Vertebrates
Iridoviridae		*Iridovirus*	*Invertebrate iridescent virus 6*	Invertebrates
		Chloriridovirus	*Invertebrate iridescent virus 3*	Invertebrates
		Ranavirus	*Frog virus 3*	Vertebrates
		Lymphocystivirus	*Lymphocystis disease virus 1*	Vertebrates
		Megalocytivirus	*Infectious spleen and kidney necrosis virus*	Vertebrates

(continued)

Table 4.10 (continued)

Order *Caudovirales* – tailed bacteriophages

Family Subfamily	Genus	Type species	Hosts
Lipothrixviridae	*Alphalipothrixvirus*	*Thermoproteus virus 1*	Archaea
	Betalipothrixvirus	*Sulfolobus mislandicus filamentous virus*	Archaea
	Gammalipothrixvirus	*Acidianus filamentous virus 1*	Archaea
Nimaviridae	*Whispovirus*	*White spot syndrome virus 1*	Invertebrates
Mimivirus		*Acanthamoeba polyphaga mimivirus*	Protozoa, Vertebrates
Polyomaviridae	*Polyomavirus*	*Simian virus 40*	Vertebrates
Papillomaviridae	*Alphapapillomavirus*	*Human papillomavirus 32*	Vertebrates
	Betapapillomavirus	*Human papillomavirus 5*	Vertebrates
	Gammapapillomavirus	*Human papillomavirus 4*	Vertebrates
	Deltapapillomavirus	*European elk papillomavirus*	Vertebrates
	Epsilonpapillomavirus	*Bovin papillomavirus 5*	Vertebrates
	Zetapapillomavirus	*Equine papillomavirus 1*	Vertebrates
	Etapapillomavirus	*Fringilla coelebs papillomavirus*	Vertebrates
	Thetapapillomavirus	*Psittacus erithacus timneh papillomavirus*	Vertebrates
	Iotapapillomavirus	*Mastomys natalensis papillomavirus*	Vertebrates
	Kappapapillomavirus	*Cottontail rabbit papillomavirus*	Vertebrates
	Lambdapapillomavirus	*Canine oral papillomavirus*	Vertebrates
	Mupapillomavirus	*Human papillomavirus 1*	Vertebrates
	Nupapillomavirus	*Human papillomavirus 41*	Vertebrates
	Xipapillomavirus	*Bovine papillomavirus 3*	Vertebrates
	Omikronpapillomavirus	*Phocoena spinipinnis papillomavirus*	Vertebrates
	Pipapillomavirus	*Hamster oral papillomavirus*	Vertebrates
Phycodnaviridae	*Chlorovirus*	*Paramecium bursaria Chlorella virus 1*	Algae
	Prasinovirus	*Micromonas pusilla virus SP1*	Algae
	Prymnesiovirus	*Chryosochromomulina brevifilium virus PW1*	Algae
	Phaeovirus	*Extocarpus siliculosus virus 1*	Algae
	Coccolithovirus	*Emiliania huxleyi virus 86*	Algae
	Raphidovirus	*Heterosigms akashiwo virus 01*	Algae
Plasmaviridae	*Plasmavirus*	*Acholeplasma phage L2*	Mycoplasma
Polydnaviridae	*Ichnovirus*	*Campoletis sonorensis ichnovirus*	Invertebrates
	Bracovirus	*Cotesia melanoscela brachovirus*	Invertebrates
Poxviridae: *Chordopoxvirinae*	*Orthopoxvirus*	*Vaccinia virus*	Vertebrates
	Parapoxvirus	*Orf virus*	Vertebrates
	Avipoxvirus	*Fowlpox virus*	Vertebrates
	Capripoxvirus	*Sheeppox virus*	Vertebrates
	Leporipoxvirus	*Myxoma virus*	Vertebrates
	Suipoxvirus	*Swinepox virus*	Vertebrates
	Molluscipoxvirus	*Molluscum contagiosum virus*	Vertebrates
	Yatapoxvirus	*Yaba monkey tumor virus*	Vertebrates

(continued)

Table 4.10 (continued)

Order *Caudovirales* – tailed bacteriophages

Family Subfamily		Genus	Type species	Hosts
	Entomopoxvirinae	*Entomopoxvirus A*	*Melolontha melolontha entomopoxvirus*	Invertebrates
		Entomopoxvirus B	*Amsacta moorei entomopoxvirus*	Invertebrates
		Entomopoxvirus C	*Chironomus luridus entomopoxvirus*	Invertebrates
Rhizidovirus			*Rhizidomyces virus*	Fungi
Rudiviridae		*Rudivirus*	*Sulfolobus virus SIRV1*	Archaea
Tectiviridae		*Tectivirus*	*Enterobacteria phage PRD1*	Bacteria

Table 4.11 Group II: The ssDNA viruses (Reproduced from Van Regenmortel et al. 2005; http://www.microbiologybytes.com/virology/VirusGroups.html#VI; Anonymous 2005. With permission)

Group II: ssDNA viruses

Family Subfamily		Genus	Type species	Hosts
Anellovirus			*Torque teno virus*	Vertebrates
Circoviridae		*Circovirus*	*Porcine circovirus*	Vertebrates
		Gyrovirus	*Chicken anemia virus*	Vertebrates
Geminiviridae		*Mastrevirus*	*Maize streak virus*	Plants
		Curtovirus	*Beet curly top virus*	Plants
		Topocuvirus	*Tomato pseudo-curly top virus*	Plants
		Begomovirus	*Bean golden mosaic virus*	Plants
Inoviridae		*Inovirus*	*Enterobacteria phage M13*	Bacteria
		Plectrovirus	*Acholeplasma phage MV-L51*	Bacteria
Microviridae		*Microvirus*	*Enterobacteria ØX174*	Bacteria
		Spiromicrovirus	*Spiroplasma phage 4*	Spiroplasma
		Bdellomicrovirus	*Bdellovibrio phage MAC1*	Bacteria
		Chlamydiamicrovirus	*Chlamydia phage 1*	Bacteria
Nanoviridae		*Nanovirus*	*Subterranean clover stunt virus*	Plants
		Babuvirus	*Banana bunchy top virus*	Plants
Parvoviridae:	*Parvovirinae*	*Parvovirus*	*Mice minute virus*	Vertebrates
		Erythrovirus	*B19 virus*	Vertebrates
		Dependovirus	*Adeno-associated virus 2*	Vertebrates
		Amdovirus	*Aleutian mink disease virus*	Vertebrates
		Bocavirus	*Bovine parvovirus*	Vertebrates
	Densovirinae	*Densovirus*	*Junonia coenia densovirus*	Invertebrates
		Iteravirus	*Bombyx mori densovirus*	Invertebrates
		Brevidensovirus	*Aedes aegypti densovirus*	Invertebrates
		Pefudensovirus	*Periplanta fuliginosa densovirus*	Invertebrates
		Circovirus	*Porcine circovirus*	Vertebrates

4.1.7.3 Bacteriophages in the Aquatic Environment

Until recently, it was thought that aquatic environments, marine and freshwater, were devoid of viruses. New techniques now show them to be abundant in the aquatic environment, where they contribute to nutrient cycle by lysing microorganisms. All microorganisms, bacteria (bacteriophages), fungi (mycophages), algae (phycophages), and protozoa are attacked by viruses (phages, phago = eat, Greek) which attack (eat) them.

Table 4.12 Group III: dsRNA viruses (Reproduced from Van Regenmortel et al. 2005; http://www.microbiologybytes.com/virology/VirusGroups.html#VI; Anonymous 2005. With permission)

Family / Subfamily	Genus	Type species	Hosts
Birnaviridae	*Aquabirnavirus*	*Infectious pancreatic necrosis virus*	Vertebrates
	Avibirnavirus	*Infectious bursal disease virus*	Vertebrates
	Entomobirnavirus	*Drosophila X virus*	Invertebrates
Chrysoviridae	*Chrysovirus*	*Penicillium chrysogenum virus*	Fungi
Cystoviridae	*Cystovirus*	*Pseudomonas phage Ø6*	Bacteria
Endornavirus		*Vicia faba endornavirus*	Plants
Hypoviridae	*Hypovirus*	*Cryphonectria hypovirus 1-EP713*	Fungi
Partitiviridae	*Partitivirus*	*Atkinsonella hypoxylon virus*	Fungi
	Alphacryptovirus	*White clover cryptic virus 1*	Plants
	Betacryptovirus	*White clover cryptic virus 2*	Plants
Reoviridae	*Orthoreovirus*	*Mammalian orthoreovirus*	Vertebrates
	Orbivirus	*Bluetongue virus*	Vertebrates
	Rotavirus	*Rotavirus A*	Vertebrates
	Coltivirus	*Colorado tick fever virus*	Vertebrates
	Aquareovirus	*Golden shiner virus*	Vertebrates
	Seadornavirus	*Banna virus*	Vertebrates
	Cypovirus	*Cypovirus 1*	Invertebrates
	Idnoreovirus	*Idnoreovirus 1*	Invertebrates
	Fijivirus	*Fiji disease virus*	Plants
	Phytoreovirus	*Wound tumor virus*	Plants
	Oryzavirus	*Rice ragged stunt virus*	Plants
	Mycoreovirus	*Mycoreovirus 1*	Fungi
Totiviridae	*Totivirus*	*Saccharomyces cerevisiae virus L-A*	Fungi
	Giardiavirus	*Giardia lamblia virus*	Protozoa
	Leishmaniavirus	*Leishmania RNA virus 1-1*	Protozoa

Table 4.13 Group IV: (+)sense RNA viruses (Reproduced from Van Regenmortel et al. 2005; http://www.microbiologybytes.com/virology/VirusGroups.html#VI; Anonymous 2005. With permission)

Order *Nidovirales* – "Nested" viruses

Family / Subfamily	Genus	Type species	Hosts
Arteriviridae	*Arterivirus*	*Equine arteritis virus*	Vertebrates
Coronaviridae	*Coronavirus*	*Infectious bronchitis virus*	Vertebrates
	Torovirus	*Equine torovirus*	Vertebrates
Roniviridae	*Okavirus*	*Gill-associated virus*	Vertebrates
Astroviridae	*Avastrovirus*	*Turkey astrovirus*	Vertebrates
	Mamastrovirus	*Human astrovirus*	Vertebrates
Barnaviridae	*Barnavirus*	*Mushroom bacilliform virus*	Fungi
Benyvirus		*Beet necrotic yellow vein virus*	Plants
Bromoviridae	*Alfamovirus*	*Alfalfa mosaic virus*	Plants
	Bromovirus	*Brome mosaic virus*	Plants
	Cucumovirus	*Cucumber mosaic virus*	Plants
	Ilarvirus	*Tobacco streak virus*	Plants
	Oleavirus	*Olive latent virus 2*	Plants

(continued)

Table 4.13 (continued)

Order *Nidovirales* – "Nested" viruses			
Family			
Subfamily	**Genus**	**Type species**	**Hosts**
Caliciviridae	*Lagovirus*	*Rabbit haemorrhagic disease virus*	Vertebrates
	Norovirus	*Norwalk virus*	Vertebrates
	Sapovirus	*Sapporo virus*	Vertebrates
	Vesivirus	*Swine vesicular exanthema virus*	Vertebrates
Cheravirus		*Cherry rasp leaf virus*	Plants
Closteroviridae	*Ampelovirus*	*Grapevine leafroll-associated virus 3*	Plants
	Closterovirus	*Beet yellows virus*	Plants
Comoviridae	*Comovirus*	*Cowpea mosaic virus*	Plants
	Fabavirus	*Broad bean wilt virus 1*	Plants
	Nepovirus	*Tobacco ringspot virus*	Plants
Dicistroviridae	*Cripavirus*	*Cricket paralysis virus*	Invertebrates
Flaviviridae	*Flavivirus*	*Yellow fever virus*	Vertebrates
	Pestivirus	*Bovine diarrhea virus 1*	Vertebrates
	Hepacivirus	*Hepatitis C virus*	Vertebrates
Flexiviridae	*Potexvirus*	*Potato virus X*	Plants
	Mandarivirus	*Indian citrus ringspot virus*	Plants
	Allexivirus	*Shallot virus X*	Plants
	Carlavirus	*Carnation latent virus*	Plants
	Foveavirus	*Apple stem pitting virus*	Plants
	Capillovirus	*Apple stem grooving virus*	Plants
	Vitivirus	*Grapevine virus A*	Plants
	Trichovirus	*Apple chlorotic leaf spot virus*	Plants
Furovirus		*Soil-borne wheat mosaic virus*	Plants
Hepevirus		*Hepatitis E virus*	Vertebrates
Hordeivirus		*Barley stripe mosaic virus*	Plants
Idaeovirus		*Rasberry bushy dwarf virus*	Plants
Iflavirus		*Infectious flacherie virus*	Invertebrates
Leviviridae	*Levivirus*	*Enterobacteria phage MS2*	Bacteria
	Allolevivirus	*Enterobacteria phage Qβ*	Bacteria
Luteoviridae	*Luteovirus*	*Cereal yellow dwarf virus-PAV*	Plants
	Polerovirus	*Potato leafroll virus*	Plants
	Enamovirus	*Pea enation mosaic virus-1*	Plants
Machlomovirus		*Maize chlorotic mottle virus*	Plants
Marnaviridae	*Marnavirus*	*Heterosigma akashiwo RNA virus*	Fungi
Narnaviridae	*Narnavirus*	*Saccharomyces cerevisiae narnavirus 20S*	Fungi
	Mitovirus	*Cryphonectria parasitica mitovirus 1-NB631*	Fungi
Nodaviridae	*Alphanodoavirus*	*Nodamura virus*	Invertebrates
	Betanodovirus	*Striped jack nervous necrosis virus*	Vertebrates
Pecluvirus		*Peanut clump virus*	Plants
Ourmiavirus		*Ourmia melon virus*	Plants
Picornaviridae	*Enterovirus*	*Poliovirus*	Vertebrates
	Rhinovirus	*Human rhinovirus A*	Vertebrates
	Hepatovirus	*Hepatitis A virus*	Vertebrates
	Cardiovirus	*Encephalomyocarditis virus*	Vertebrates
	Aphthovirus	*Foot-and-mouth disease virus O*	Vertebrates
	Parechovirus	*Human parechovirus*	Vertebrates
	Erbovirus	*Equine rhinitis B virus*	Vertebrates
	Kobuvirus	*Aichi virus*	Vertebrates
	Teschovirus	*Porcine teschovirus*	Vertebrates

(continued)

Table 4.13 (continued)

Order *Nidovirales* – "Nested" viruses			
Family			
Subfamily	Genus	Type species	Hosts
Pomovirus		*Potato mop-top virus*	Plants
Potyviridae	*Potyvirus*	*Potato virus Y*	Plants
	Ipomovirus	*Sweet potato mild mottle virus*	Plants
	Macluravirus	*Maclura mosaic virus*	Plants
	Rymovirus	*Ryegrass mosaic virus*	Plants
	Tritimovirus	*Wheat streak mosaic virus*	Plants
	Bymovirus	*Barley yellow mosaic virus*	Plants
Sadwavirus		*Satsuma dwarf virus*	Plants
Sequiviridae	*Sequivirus*	*Parsnip yellow fleck virus*	Plants
	Waikavirus	*Rice tungro spherical virus*	Plants
Sobemovirus		*Southern bean mosaic virus*	Plants
Tetraviridae	*Betatetravirus*	*Nudaurelia capensis β virus*	Invertebrates
	Omegatetravirus	*Nudaurelia capensis ω virus*	Invertebrates
Tobamovirus		*Tobacco mosaic virus*	Plants
Tobravirus		*Tobacco rattle virus*	Plants
Tombusviridae	*Tombusvirus*	*Tomato bushy stunt virus*	Plants
	Avenavirus	*Oat chlorotic stunt virus*	Plants
	Aureusvirus	*Pothos latent virus*	Plants
	Carmovirus	*Carnation mottle virus*	Plants
	Dainthovirus	*Carnation ringspot virus*	Plants
	Machlomovirus	*Maize chlorotic mottle virus*	Plants
	Necrovirus	*Tobacco necrosis virus*	Plants
	Panicovirus	*Panicum mosaic virus*	Plants
Togaviridae	*Alphavirus*	*Sindbis virus*	Vertebrates
	Rubivirus	*Rubella virus*	Vertebrates
Tymoviridae	*Maculavirus*	*Grapevine fleck virus*	Plants
	Marafivirus	*Maize rayado fino virus*	Plants
	Tymovirus	*Turnip yellow mosaic virus*	Plants
Umbravirus		*Carrot mottle virus*	Plants

Table 4.14 Group V: (−) sense RNA viruses (Reproduced from Van Regenmortel et al. 2005; http://www.microbiologybytes.com/virology/VirusGroups.html#VI; Anonymous 2005. With permission)

Order Mononegavirales				
Family				
Subfamily		Genus	Type species	Hosts
Bornaviridae		*Bornavirus*	*Borna disease virus*	Vertebrates
Filoviridae		*Marburgvirus*	*Lake Victoria marburgvirus*	Vertebrates
		Ebolavirus	*Zaire ebolavirus*	Vertebrates
Paramyxoviridae	*Paramyxovirinae*	*Avulavirus*	*Newcastle disease virus*	Vertebrates
		Henipavirus	*Hendra virus*	Vertebrates
		Morbillivirus	*Measles virus*	Vertebrates
		Respirovirus	*Sendai virus*	Vertebrates
		Rubulavirus	*Mumps virus*	Vertebrates
	Pneumovirinae	*Pneumovirus*	*Human respiratory syncytial virus*	Vertebrates
		Metapneumovirus	*Avian pneumovirus*	Vertebrates

(continued)

Table 4.14   (continued)

Order Mononegavirales			
Family			
Subfamily	Genus	Type species	Hosts
Rhabdoviridae	Vesiculovirus	Vesicular stomatitis Indiana virus	Vertebrates, invertebrates
	Lyssavirus	Rabies virus	Vertebrates
	Ephemerovirus	Bovine ephemeral fever virus	Vertebrates, invertebrates
	Novirhabdovirus	Infectious hematopoetic necrosis virus	Vertebrates
	Cytorhabdovirus	Lettuce necrotic yellows virus	Plants, invertebrates
	Nucleorhabdovirus	Potato yellow dwarf virus	Plants, invertebrates
Arenaviridae	Arenavirus	Lymphocytic choriomeningitis virus	Vertebrates
Bunyaviridae	Orthobunyavirus	Bunyamwera virus	Vertebrates
	Hantavirus	Hantaan virus	Vertebrates
	Nairovirus	Nairobi sheep disease virus	Vertebrates
	Phlebovirus	Sandfly fever Sicilian virus	Vertebrates
	Tospovirus	Tomato spotted wilt virus	Plants
Deltavirus		Hepatitis delta virus	Vertebrates
Ophiovirus		Citrus psorosis virus	Plants
Orthomyxoviridae	Influenza A virus	Influenza A virus	Vertebrates
	Influenza B virus	Influenza B virus	Vertebrates
	Influenza C virus	Influenza C virus	Vertebrates
	Isavirus	Infectious salmon anemia virus	Vertebrates
	Thogotovirus	Thogoto virus	Vertebrates
Tenuivirus		Rice stripe virus	Plants
Varicosaivirus		Lettuce big-vein associated virus	Plants

Note: Families are in uppercase and genera in bold starting with capital letters. The Order is represented in this case and is in larger display than the family

RNA viruses are also designated according to the sense or polarity of their RNA into negative-sense and positive-sense, or **ambisense.** Positive-sense viral RNA is similar to mRNA and thus can be immediately translated by the host cell. Negative-sense viral RNA is complementary to mRNA and thus must be converted to positive-sense RNA by an RNA polymerase before translation. Ambisense RNA viruses resemble negative-sense RNA viruses, except they also translate genes from the positive strand. They differ from those of other negative-sense RNA viruses in that some proteins are coded in viral-complementary RNA sequences and others are coded in the viral RNA sequence

Bacteriophages are the most abundant among the phages and they have been more widely studied. Bacteriophages were first formally described by the French Canadian Felix d' Herelle in 1915, but the initial observations were made during 1896, followed by observations made by the British bacteriologist Frederick Twort in 1913 (see Fig. 4.26). On account of their importance in aquatic systems, the life cycle of bacterial viruses, the methods of isolating and enumerating them from water, their grouping, and their host range will be discussed below.

Life History of Bacteriophages

When bacteriophages enter susceptible bacteria, they take over the genetic apparatus of their hosts and force the hosts to produce more viruses of their type. When the virions (virus particles) mature, they produce enzymes which lyse the host cell wall releasing the virus particle to start life afresh. When they lyse the host they are in the lytic phase.

Sometimes they enter into a phase, the lysogenic phase, in which the phages remain in the cell and replicate with it. This phase is the lysogenic phase (see Fig. 4.27).

Table 4.15 Group VI: reverse transcribing Diploid single-stranded RNA viruses (Reproduced from Van Regenmortel et al. 2005; http://www.microbiologybytes.com/virology/VirusGroups.html#VI; Anonymous 2005. With permission)

Family Subfamily	Genus	Type species	Hosts
Retroviridae	*Alpharetrovirus*	*Avian leukosis virus*	Vertebrates
	Betaretrovirus	*Mouse mammary tumor virus*	Vertebrates
	Gammaretrovirus	*Murine leukemia virus*	Vertebrates
	Deltaretrovirus	*Bovine leukemia virus*	Vertebrates
	Epsilonretrovirus	*Walley dermal sarcoma virus*	Vertebrates
	Lentivirus	*Human immunodeficiency virus 1*	Vertebrates
	Spumavirus	*Human spumavirus*	Vertebrates
Metaviridae	*Metavirus*	*Saccharomyces cerevisiae Ty3 virus*	Fungi
	Errantivirus	*Drosophila melanogaster gypsy virus*	Invertebrates
Pseudoviridae	*Pseudovirus*	*Saccharomyces cerevisiae Ty1 virus*	Invertebrates
	Hemivirus	*Drosophila melanogaster copia virus*	Invertebrates

Table 4.16 Group VII: DNA reverse transcribing viruses (Reproduced from Van Regenmortel et al. 2005; http://www.microbiologybytes.com/virology/VirusGroups.html#VI; Anonymous 2005. With permission)

Family Subfamily	Genus	Type species	Hosts
Hepadnaviridae	*Orthohepadnavirus*	*Hepatitis B virus*	Vertebrates
	Avihepadnavirus	*Duck hepatitis B virus*	Vertebrates
Caulimoviridae	*Caulimovirus*	*Cauliflower mosaic virus*	Plants
	Badnavirus	*Commelina yellow mottle virus*	Plants
	Cavemovirus	*Cassava vein mosaic virus*	Plants
	Petuvirus	*Petunia vein clearing virus*	Plants
	Soymovirus	*Soybean chlorotic mottle virus*	Plants
	Tungrovirus	*Rice tungro bacilliform virus*	Plants

Table 4.17 The structures of DNAviruses (From http://en.wikipedia.org/wiki/Virus_classification; Anonymous 2010a)

S. No.	Virus family	Virus genus	Virion – naked/enveloped	Capsid symmetry	Type of nucleic acid
1.	Adenoviridae	Adenovirus	Naked	Icosahedral	ds
2.	Papovaviridae	Papillomavirus	Naked	Icosahedral	ds circular
3.	Parvoviridae	B 19 virus	Naked	Icosahedral	ss
4.	Herpesviridae	Herpes Simplex Virus, Varicella zoster virus, Cytomegalovirus, Epstein Barr virus	Enveloped	Icosahedral	ds
5.	Poxviridae	Small pox virus, Vaccinia virus	Complex coats	Complex	ds
6.	Hepadnaviridae	Hepatitis B virus	Enveloped	Icosahedral	ds circular
7.	Polyomaviridae	Polyoma virus (progressive multifocal leucoencephalopathy)	?	?	ds

Table 4.18 The structure of RNA viruses (From http://en.wikipedia.org/wiki/Virus_classification; Anonymous 2010a)

S. No.	Virus family	Virus genus	Virion – naked/enveloped	Capsid symmetry	Type of nucleic acid
1.	Reoviridae	Reovirus, Rotavirus	Naked	Icosahedral	ds
2.	Picornaviridae	Poliovirus, Rhinovirus, Hepatitis A virus	Naked	Icosahedral	ss
3.	Caliciviridae	Norwalk virus, Hepatitis E virus	Naked	Icosahedral	ss
4.	Togaviridae	Rubella virus	Enveloped	Icosahedral	ss
5.	Arenaviridae	Lymphocytic choriomeningitis virus	Enveloped	Complex	ss
6.	Retroviridae	HIV-1, HIV-2, HTLV-I	Enveloped	Complex	ss
7.	Flaviviridae	Dengue virus, hepatitis C virus, yellow fever virus	Enveloped	Complex	ss
8.	Orthomyxoviridae	Influenza virus	Enveloped	Helical	ss
9.	Paramyxoviridae	Measles virus, mumps virus, respiratory syncytial virus	Enveloped	Helical	ss
10.	Bunyaviridae	California encephalitis virus, Hantavirus	Enveloped	Helical	ss
11.	Rhabdoviridae	Rabies virus	Enveloped	Helical	ss
12.	Filoviridae	Ebola virus, Marburg virus	Enveloped	Helical	ss
13.	Coronaviridae	Corona virus	Enveloped	Complex	ss
14.	Astroviridae	Astro virus	Naked	Icosahedral	ss
15.	Bornaviridae	Borna disease virus	Enveloped	Helical	ss

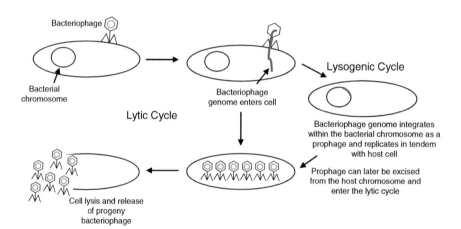

Fig. 4.27 (The lytic and lysogenic cycles of bacteriophage replication (from Zourob & Ripp, 2010, with kind permission from Springer Science+Business Media: Zourob, M. & Ripp, S. (2010). Bacteriophage-Based Biosensors. In M. Zourob (Ed.) Recognition Receptors in Biosensors (Fig. 11.3, p. 419). New York: Springer)

Methods for the Study of Bacteriophages

Because of their small size, bacteriophages are not usually studied directly as is the case with microorganisms. An electron microscope is necessary to study viruses, but not only are electron microscopes expensive, but they require skilled operators to handle them. Therefore, viruses are studied indirectly through their effects. The indirect methods used especially for animal and plant viruses, include

the changes (known as cytopathic effects, CPE) they bring about in the cells in the cell culture in which they are grown: lysis, altered shape, detachment from substrate, membrane fusion, altered membrane permeability.

Other methods for studying viruses, particularly of animals and plants, are serological methods based on the interaction between virus and antibody produced specifically against it, detection of viral nucleic acid, including the use of polymerase chain reaction (PCR) for the detection of DNA or RNA. These methods are used mostly for studying animal and plant viruses. For bacteriophages, the chief method of detection is the cytopathic effect (CPE).

Isolation and Enumeration of Bacteriophages

Bacteriophages are sometimes very abundant in water and because of the specificity of the bacteria they attack, it has been suggested that they can be used as indicators of fecal pollution of water.

Bacteriophages may be isolated and/or enumerated from water in the following ways (McLaughlin et al. 2006):

(a) *Enrichment*

This method is used if the aim is to isolate phages attacking a particular bacterium from the water sample. A pure culture of the bacterium whose phages are to be isolated is introduced into sample of the water to be assessed for phage load, say about 1 ml of a log phase culture of the bacterium is added to about 10–15 ml of the water and incubated under conditions which will encourage the growth of the bacterium, including shaking if necessary. At the end of 18–24 h growth, a quantity of chloroform is added to kill the bacteria, and the broth is filtered in a 0.45 μm filter to remove the debris. The filtrate is then serially diluted and about 0.5–1 ml of the dilutions mixed with molten agar at 45°C and poured in to plates and incubated. It is assumed that each zone of clearing (plaque) on the bacterial lawn indicates one bacteriophage, or more correctly, one plaque forming unit (PFU), since as the case with bacteria, a clearing could be formed by a clump of bacteriophages. To ensure the avoidance of clumps, the counting or selection should done with as high a dilution as possible; the filtrate can also be shaken to breakup clumps before introduction into the agar.

Some workers prefer to use the soft agar method. In this method, a small volume of a dilution of phage suspension and a small quantity of host cells grown to high cell density, sufficient to give 10^7–10^8 CFU/ml (colony forming units/ml), are mixed in about 2.5 ml of molten, "soft" agar at 46°C. The resulting suspension is then poured on to an appropriate basal agar medium. This poured mixture cools and forms a thin "top layer" which hardens and immobilizes the bacteria.

(b) *Direct plating out*

If the aim is to assess the diversity of phages present in the water body, several dilutions of the water are made. At each dilution, pour plates are made as described above, each plate with one of the variety of bacteria whose phages are being sought in the water.

(c) *Direct counting in a flow cytometer*

In this method, the water may be centrifuged to concentrate the virus. The phages are stained with highly fluorescent nucleic acid specific dyes such as SYBR Green I, SYBR Green II, OliGreen, or PicoGreen. Flow cytometry allows extremely rapid enumeration of single cells, primarily by optical means. Cells scatter light when passing through the laser beam and emit fluorescent light when excited by the laser. Flow cytometry has become an invaluable tool for both qualitative and quantitative analyses owing to its rapidity. It has been used as a rapid method for detecting viruses from different families. Flow cytometry appears faster and more accurate than any other method currently used for the direct detection and quantification of virus particles (Brussaard et al. 2000).

Although epifluorescence microscopy is commonly used for the enumeration of bacteria and other microorganisms in natural water samples including viruses, because of its simplicity and ready availability of the microscopes, distinguishing viruses based on differences in fluorescence intensity is difficult with the epifluorscence microscope; small bacteria may for example be counted for large viruses.

Bacteriophages and Their Bacterial Hosts

Figure 4.28 gives the names of the bacteriophages, the description of their virions and the hosts which they attack. It will be seen that in many cases, the same bacteriophages may sometimes attack numerous hosts, while in some cases, a bacteriophage is restricted to one host.

4.1.8 Small Multicellular Macroorganisms in Aquatic Systems

Two groups of small multicellular macroorganisms occur in water, namely, crustaceans (including rotifers) and nematodes.

1 CORTICOVIRIDAE

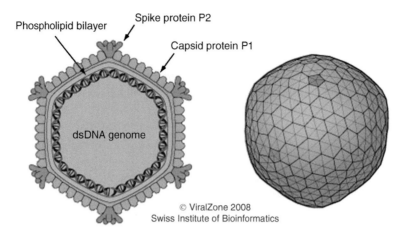

© ViralZone 2008
Swiss Institute of Bioinformatics

$T = 21$

Structure

Phages consist of a capsid and an internal lipid membrane. Virus capsid is **not enveloped**. Internal lipid membrane located between outer and inner protein shell. Capsid/nucleocapsid is round and exhibits icosahedral symmetry (T=12), or 13). The capsid is isometric and has a diameter of 60 nm (or more). The capsid shells of virions are composed of three layers. Capsids appear hexagonal in outline. The capsid surface structure reveals a regular pattern with distinctive features. The capsomer arrangement is clearly visible.

Surface projections are distinct, brush-like spikes protruding from the 12 vertices. Capsids all have the same appearance.

The genome is not segmented, constitutes 13% of the virus's weight and contains a single molecule of circular, supercoiled, double-stranded DNA of 9500-12000 nucleotides in length. The genome has a g + c content of 43%.

Host

Their hosts are members of the Phylum *Proteobacteria*.

2 CYSTOVIRIDAE

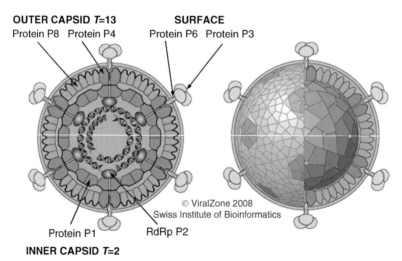

© ViralZone 2008
Swiss Institute of Bioinformatics

Structure

Enveloped, spherical virion of 85 nm in diameter. The virion has a double capsid structure: Outer capsid a has an icosahedral T=13 symmetry, inner capsid an icosahedral symmetry T=2.

All cystoviruses are distinguished by their three strands (analogous to chromosomes) of double stranded (ds) RNA, totalling ~14 kb in length and their protein and lipid outer layer. No other bacteriophage have any lipid in their outer coat, though the *Tectiviridae* and the *Corticoviridae* have lipids within their capsids.

Members of the *Cystoviridae* appear to be most closely related to the *Reoviridae* but also share homology with the

Totiviridae. Cystoviruses are the only bacteriophage that are more closely related to viruses of eukaryotes than to other phage.

Host

Most identified cystoviruses infect *Pseudomonas* species, but this is likely biased due to the method of screening and enrichment. The type species is *Pseudomonas phage Φ6*, but there are many other members of this family. Φ7, Φ8, Φ9, Φ10, Φ11, Φ12 and Φ13 have been identified and named, but other cystoviruses have also been isolated.

3   FUSSELLOVIRIDAE

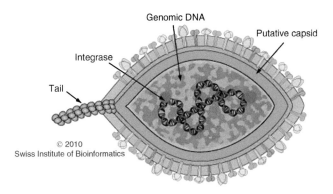

© 2010
Swiss Institute of Bioinformatics

Structure

Enveloped, lemon-shaped, with short tail fibers attached to one pole. The virion is about 60 nm in diameter and 100 nm in length.

The genome of a fuselloviridae is non-segmented and contains a single molecule of circular, double-stranded DNA. The DNA is positively supercoiled. The complete genome is 15500 nucleotides in length; encodes for 31 to 37 genes.

Fuselloviridae virions consist of an envelope and a nucleocapsid. The capsid is enveloped. Virions are spindle-shaped, flexible, and have protrusions that extend through the envelope. One pole has short tail-like fibers attached to it. The virions are 100 nm in length and 60 nm in diameter.

Host

Fuselloviridae infect the Archae *Sulfolobus* which inhabits high-temperature (>70°C), acidic (pH of <4.0) environments. Members of this family have been found in acidic hotsprings in Japan and Iceland. The Fuselloviridae family currently consists of only one virus, *Sulfolobus* spindle-shaped virus 1 (SSV1), and three tentative members (SSV2, SSV3, and the staaelite virus pSSVx, which stands for plasmid SSV x). SSV1, the type virus for the family, was the first high-temperature virus to be characterized.

4   INOVIRIDAE

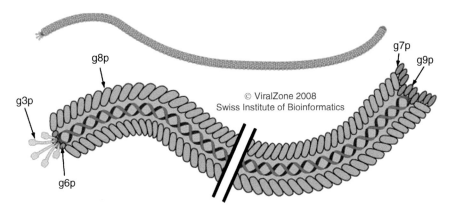

© ViralZone 2008
Swiss Institute of Bioinformatics

Structure

Non-enveloped, Rod of filaments ut 7nm in diameter and from 700 to 2000nm long.

Circular, single-stranded DNAof 4.5 to 8kb encoding for 4 to 10 proteins. Replication occurs via dsDNA intermediate and rolling circle.

Host

Members of the family infect their natural hosts without causing lysis, and the infected cells continue to divide and produce virus indefinitely. The hosts are plant and animal pathogens. In several systems the phage enter into lysogenic phases.

Inouirus hosts are all gram-negative bacteria (i.e., *Escherichia coli, Salmonella, Pseudomonas, Vibrio, Xanthomonas,* etc.).

Host ranges are determined primarily by host cell receptors, which are usually conjugative pili. Some pili are encoded chromosomally and some are encoded on plasmids of different incompatibility groups. Transmission of the plasmids to new bacterial species usually transfers phage sensitivity. Additional host range determinants include restriction-modification systems, host periplasmic proteins involved in viral ssDNA translocation into the cytoplasm, and host protein(s) involved in membrane assembly. Transfections of non-natural hosts with naked ssDNA or dsDNA are sometimes possible. When *Vibrio cholera* phage lysogens colonize the human intestine, states of elevated cholera toxin expression and release, and of progeny filamentous choleraphage extrusion, are induced. Thus *Inaoirus* lysogeny is a critical virulence factor in cholera pathogenesis.

5 LEVIVIRIDAE

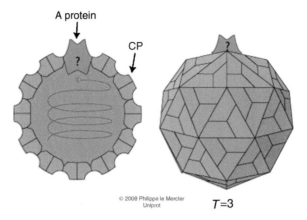

© 2008 Philippe le Mercier
Uniprot

T=3

Structure

Non-enveloped, spherical virion about 26nm in diameter with icosahedral symmetry (T=3) composed of 180 CP proteins and a single A protein. Linear, ssRNA(+) genome about 4 kb in size. The 5' end is capped. Encodes for 4 proteins.

Host

It attacks Enterobacteriacea, Acinetobacter, Caulobacter and Pseudomonas.

6 LIPOTHRIXVIRIDAE

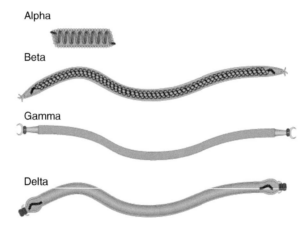

Alpha

Beta

Gamma

Delta

Structure

Enveloped, rod-shaped. The capsid is about 24-38 nm in diameter and 410-1950 nm in length.

Linear dsDNA genome of 15.9 to 56 kb. Extremities of the DNA are modified in an unknown manner.

The Lipothrixviridae family consists of a family of viruses that infect archaea. They share characteristics from the Rudiviridae family and both have are filamentous viruses with linear dsDNA genomes that infect thermophilic archaea in the kingdom Crenarchaeota. Lipothrixviridae are enveloped.

Host

Lipothrixviridae is a family crenarchaeal viruses. It is by far the most diverse family of crenarchaeal viruses, with six isolates divided into three genera: *Alphalipothrixvirus*, *Betalipothrixvirus*, and *Gammalipothrixvirus*. *Alphalipothrixvirus* contains TTV1, TTV2, and TTV3, isolated from acidic hot springs Iceland. *Betalipothrixvirus* contains SIFV, also isolated in Iceland. Finally, *Gammalipothrixvirus* is represented by AFV1, isolated from Yellowstone National Park.

Also Acidianus, Sulfolobus, Thermoproteus.

7   MICROVIRIDAE

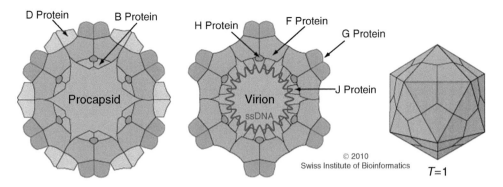

Structure

Non-enveloped, round, icosahedral symmetry (T=1), about 30 nm in diameter. The capsid consists of 12 pentagonal trumpet-shaped pentomers. The virion is composed of 60 copies each of the F, G, and J proteins, and 12 copies of the H protein. There are 12 spikes which are each composed of 5 G and one H proteins.

Host

Attacks Bdellovibrio, Chlamydia, Enterobacteria, Spiroplasma, Enterobacteria Clamydiamicrovirus and Bdellomicrovirus: intracellular parasitic **bacteria**. Spiromicrovirus: Spiroplasma.

8   Myoviridae

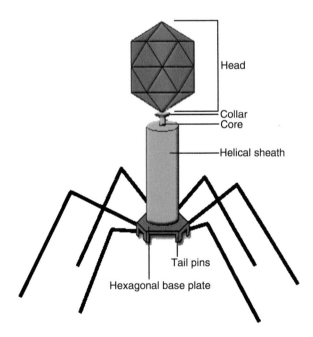

Structure

Myoviruses are not enveloped and consist of a head and a tail separated by a neck. The head has icosahedral symmetry, while the tail is tubular and has helical symmetry. The capsid that constitutes the head is made up of 152 capsomers. The head has a diameter of 50-110nm; the tail is 16-20nm in diameter. The tail consists of a central tube, a contractile sheath, a collar, a base plate, six tail pins and six long fibers. Tail structure is similar to tectiviridae, but differs in the fact that a myovirus' tail is permanent. Contractions of the tail require ATP. When the sheath is contracted, it measures 10-15 nm in lengthicosahedral capsid, circular ssDNA.

Host

Myoviruses, being bacteriophages, infect bacteria. The most commonly infected bacteria is *Escherichia coli*. Myoviruses are virulent phages, meaning they do not integrate their genetic material with their host cell's, and they usually kill their host cell. Others are Bdellovibrio, Chlamydia, Enterobacteria, Spiroplasma.

9   PLASMAVIRIDAE

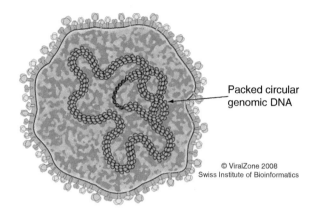

Packed circular genomic DNA

© ViralZone 2008
Swiss Institute of Bioinformatics

Structure

Pleomorphic, envelope, lipids, no capsid, circular supercoiled dsDNA. Enveloped, spherical to pleomorphic, no head-tail structure. The capsid is about 80 nm in diameter.

The Plasmaviridae is a family of bacteriophages, viruses that infects bacteria. Virions have an envelope, a nucleoprotein complex, and a capsid. They are 50-125 nm in diameter with a baggy or loose membrane.

The genome is condensed, non segmented and consists of a single molecule of circular, supercoiled double-stranded DNA,

12000 base pairs in length. The genome has a rather high G-C content of around 32 percent.

A productive infectious cycle begins before a lysogenic cycle establishes the virus in the infected bacteria. After initial infection of the viral genome the virus may become latent within the host. Lysogeny involves integration into the host chromosome.

Host

A well-known host is *Acholeplasma*.

10   PODOVIRIDAE

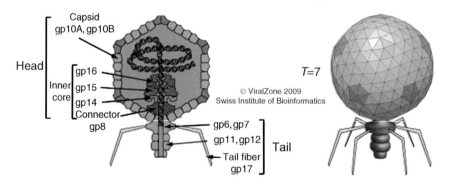

© ViralZone 2009
Swiss Institute of Bioinformatics

Structure

Nonenveloped, head-tail structure. Head is about 60 nm in diameter. The tail is non-contractile, has 6 short subterminal fibers. The capsid is icosahedral with a T=7 symmetry.

It has a linear, dsDNA genome of about 40-42 kb encoding for 55 genes.

Host

Wide range of bacterial hosts including Gram positive and Gram negative.

11 RUDIVIRIDAE

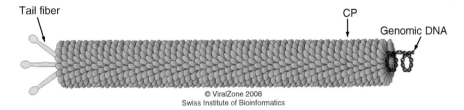

© ViralZone 2008
Swiss Institute of Bioinformatics

Structure

Non-enveloped, rod-shaped, rigid, with three tail fibers at either end. Linear dsDNA genome of 32-35 kb. At both ends, there are inverted terminal repeats as well as seven direct repeats. The two strands of the linear genomes are covalently linked.

Host

Wide range of bacteria, Gram positive and Gram negative.

12 SIPHOVIRIDAE

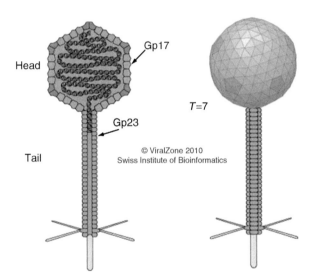

© ViralZone 2010
Swiss Institute of Bioinformatics

Siphoviridae are a family of double-stranded DNA viruses infecting only bacteria that are characterized by a long non-contractile tail and an isometric capsid (morphotype B1) or a prolate capsid (morphotype B2).

The *Siphoviridae* viruses have a capsid with a diameter of about 55-60 nm and a long tail that can reach up to 570 nm. Their double-stranded DNA is linear.

Nonenveloped, head-tail structure. The head is about 60 nm in diameter. The tail is non-contractile, has fibers, and is filamentous. The capsid is icosahedral with a T=7 symmetry. Linear, dsDNA genome of about 50 kb, containing about 70 genes.

Host

Wide range of bacteria, Gram positive and Gram negative.

13 TECTIVIRIDAE

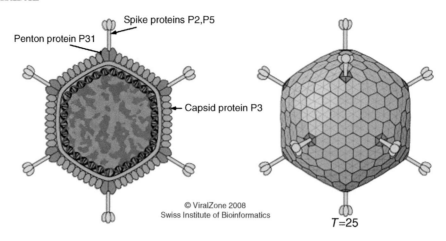

© ViralZone 2008
Swiss Institute of Bioinformatics

T=25

Structure

Non enveloped, icosahedral virion with a pseudo T=25 symmetry. Virion size is about 66 nm with apical spikes of 20 nm. The capsid encloses an inner membrane vesicle within which the genomic DNAis coiled.

Linear, dsDNA genome of about 15 kb flanked by inverted repeats. Encodes for 30 ORFs. Replication is protein-primed.

Host

A wide range of bacteria including Alicyclobacillus, Bacillus, Enterobacteria, Pseudomonas, Thermus.

Fig. 4.28 Bacteriophages and their hosts (All items in the table reproduced with permission; Anonymous 2010d)

4.1.8.1 Crustaceans (Including Rotifers)

Microscopic crustaceans like tiny lobsters are found in sewage works where they feed on bacteria and algae. Some of the species encountered are *Cyclops* spp., *Paracyclops* spp., and *Oaphia* spp.

Rotifers, small microscopic animals of the class *Rotifera*, are found in water in great abundance. They are usually less than 1 mm, most usually in the range of 500 µm in length. Three orders are known: *Seisonidae* is marine, while the other two, *Bdelloidea* and *Monongononta*, are freshwater and found in reservoirs, streams, and sewage treatment plants.

Rotifers may be sessile or planktonic, although they can swim with their cilia; their usual locomotive method is by crawling. They are common in structures with large exposed surfaces such as in trickling filters sewage treatment plants. They are found in oligotrpic waters, i.e., waters low in organic matter, for example in sewage effluents and reservoirs after protozoa have died off, and thus are indicators of post-eutrophication waters. Indeed, they have been used as indicators of water quality. Various species are identified with different levels of water quality. *Brachionus angularis*, *Trichocerca cylindrica*, *Polyurthra euryptera*,

Pompholyx sulcata, Rotaria rotatoria, Filinia longiseta have been designated as indicators of heavy pollution (eutrophic) while *Ascomorpha ovalis, Asplanchna herricki, Synchaeta grandis, Ploesoma hudsoni, Anuraeopsis fissa, Monostyla bulla,* and *M. hamata* are indicators of fresh and clean waters (oligotrophic). A variety of rotifers including *Brachionus, Keratella* spec, are inhabitants of moderately clean (mesotrophic) waters (Saksena 2006).

Rotifers have been also been used to detect the oocysts of *Cryptosporidium*, in water samples. The fluorescent in-situ hybridization (FISH) technique (see Sect. 4.1.3.3.2d) applied to rotifers has enabled the detection of biological contamination of surface water through an assessment of the dispersive stages of the parasite (see Table 4.19).

Other crustaceans found in water are *Daphnia, Cyclops, Synchaeta.*

4.1.8.2 Nematodes

Nematodes are invertebrate roundworms that inhabit marine, freshwater, and terrestrial environments. They comprise the phylum Nematoda (or Nemata) which includes parasites of plants and of animals,

Table 4.19 Rotifers common in aquatic environments (Modified from Saksena 2006)

S. No.	Taxonomic group	Common aquatic habitat	Food/feeding type	Examples
1.	Seisonidea	Marine epizoic	Sessile parasite	*Seison*
2.	Bdelloidiae	Freshwater planktonic, freshwater sessile, freshwater creeping	Suspension feeders	*Philodina*
3.	Monogonata	Freshwater planktonic	Suspension feeders	*Brachionus*
				Keratella
				Polyarthra
				Floscularia
			Raptorial predator	*Asplanchna*

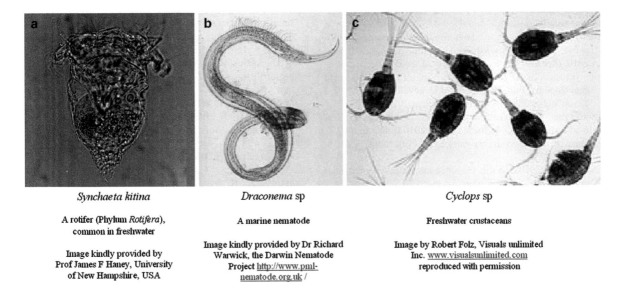

Synchaeta kitina

A rotifer (Phylum *Rotifera*),
common in freshwater

Image kindly provided by
Prof James F Haney, University
of New Hampshire, USA

Draconema sp

A marine nematode

Image kindly provided by Dr Richard
Warwick, the Darwin Nematode
Project http://www.pml-
nematode.org.uk /

Cyclops sp

Freshwater crustaceans

Image by Robert Folz, Visuals unlimited
Inc. www.visualsunlimited.com
reproduced with permission

Fig. 4.29 Marine and freshwater crustaceans and nematode (All items in the table reproduced with permission)

including humans, as well as species that feed on bacteria, fungi, algae, and on other nematodes. Four out of every five multicellular animals on the planet are nematodes. The majority of nematodes are microscopic, averaging less than a millimeter in length, but some of the animal parasites are quite large and readily visible to the naked eye. Nematodes that feed on other organisms are important participants in the cycling of minerals and nutrients in the ecosystem that is fundamental to other biological activity. Some of these nematodes may have major roles in decomposition, including biodegradation of toxic compounds. In fact, the incidence of certain nematode species is sometimes used as an indicator of environmental quality (Fig. 4.29).

Nematodes are, by nature, aquatic organisms. It is estimated that about 50% of nematode species inhabit marine environments, although many of these have yet to be described and characterized. The remainder of the species inhabit soil and freshwater. In the soil, their aquatic requirements are satisfied by inhabiting the water films around soil particles. Parasitic nematodes are biologically active when bathed in moisture films supplied by water in the tissues or body fluids of the host.

Nematodes in freshwater aquatic systems also serve as a nutrient source for invertebrates, small vertebrates, and fungi. The source of food for these nematodes is primarily bacteria, but algae and fungi are also consumed.

The marine environment provides habitat for an enormous diversity of nematodes, from surface, littoral, and estuarine zones to the ocean depths. One interesting group of deep sea nematodes are the *Rhaptothyridae*, which have no mouth and a very reduced alimentary tract. The digestive tract is filled with symbiotic chemoaototrophic bacteria. A similar relationship exists in the mouthless genus *Astomonema*.

Nematodes are widely distributed in aquatic and soil habitats and are particularly common in waters rich in organic matter such as sewage. They feed on bacteria and algae etc. Some nematodes encountered are *Rhabditis* spp., *Pelodra* spp., and *Diplogaster* spp.

References

Achenbach, L. A., Carey, J., & Madigan, M. T. (2001). Photosynthetic and phylogenetic primers for detection of anoxygenic phototrophs in natural environments. *Applied and Environmental Microbiology, 67,* 2922–2926.

Anonymous. (2005). Virus taxonomy. http://www.microbiologybytes.com/virology/VirusGroups.html. Accessed 10 Sept 2010. (See also Van Regenmortel et al. 2005).

Anonymous. (2008a). Taxon (2008, August 21) New world encyclopedia. http://www.newworldencyclopedia.org/entry/Taxon?oldid=788257>. Retrieved 6 Sept 2010.

Anonymous. (2008b). Lytic cycle. (2008, December 22) New world encyclopedia. Retrieved 6 Sept 2010 from http://www.newworldencyclopedia.org/entry/Lytic_cycle?oldid=886635. Accessed 25.

Anonymous. (2010a). Virus classification. http://en.wikipedia.org/wiki/Virus_classification. Accessed 20 Sept 2010.

Anonymous. (2010b). Illustrations of some Protozoa. http://www.cliffsnotes.com/study_guide/topicArticleId-8524,articleId-8461.html. Accessed 8 Aug 2010.

Anonymous. (2010c). Human immunodeficiency virus 1. Viralzone. http://www.expasy.org/viralzone/all_by_species/7.html. Accessed 2 Oct 2010.

Anonymous. (2010d). Corticoviridae. Viralzone. http://www.expasy.org/viralzone/all_by_species/14.html. Accessed 2 Oct 2010.

Anonymous. (2010e). Bacterial nitrogenase. http://www.cliffsnotes.com/study_guide/Bacterial-Nitrogenase.topicArticleId-24594,articleId-24526.html. Accessed 24 Dec 2010.

Axelssson, L., & Ahrne, S. (2000). Lactic acid bacteria. In F. G. Priest & M. Goodfellow (Eds.), *Applied microbial systematics* (pp. 367–388). The Netherlands: Dordrecht.

Banowetz, G. M., Whittaker, G. W., Dierksen, K. P., Azevedo, M. D., Kennedy, A. C., Griffith, S. M., & Steiner, J. J. (2006). Fatty acid methyl ester analysis to identify sources of soil in surface water. *Journal of Environmental Quality, 35,* 133–140.

Barton, L. L., & Hamilton, W. A. (2007). *Sulphate-reducing bacteria.* Cambridge: Cambridge University Press.

Bauld, J., & Staley, J. T. (1976). *Planctomyces mavis sp. nov.:* a marine isolate of the *Planctomyces-Blastocaulis* group of budding bacteria. *Journal of General Microbiology, 97,* 45–55.

Berg, J. M., Tymoczko, J. L., & Stryer, L. (2002). *Biochemistry* (5th ed.). New York: W. H. Freeman.

Bergey's manual trust (2009). http://www.bergeys.org/pubinfo.html. Accessed 2 Aug 2009.

Brussaard, C. P. D., Marie, D., & Bratbak, G. (2000). Flow cytometric detection of viruses. *Journal of Virological Methods, 85,* 175–182.

Christensen, T. H., & Kjeldsen, P. (1989). Basic biochemical processes in landfills. Chapter 2.1 in *Sanitary Landfilling: Process, Technology and Environmental Impact,* Christensen, T. H., Cossu, R., & Stegmann, R. (Eds.). London: Academic Press. p 29.

Ciccarelli, F. D., Doerks, T., von Mering, C., Creevey, C. J., Snel, B., & Bork, P. (2006). Toward automatic reconstruction of a highly resolved tree of life. *Science, 311,* 1283–1288.

Codispoti, L. A., Brandes, J. A., Christensen, J. P., Devol, A. H., Naqvi, S. W. A., Paerl, H. W., & Yoshinari, T. (2001). The oceanic fixed nitrogen and nitrous oxide budgets: moving targets as we enter the Anthropocene? *Scientia Marina, 65,* 85–105.

Fujita, Y. (2005). *Chlorophylls* (Encyclopedia of life sciences). New York: Wiley.

Garrity, G. M. (2001–2006). *Bergey's manual of systematic bacteriology* (2nd ed.). New York: Springer.

Guarro, J., Gené, J., & Stchigel, A. M. (1999). Developments in fungal taxonomy. *Clinical Microbiology Reviews, 12,* 454–500.

Inglis, T. J. J., Aravena-Roman, M., Ching, S., Croft, K., Wuthiekanun, V., & Mee, B. J. (2003). Cellular fatty acid profile distinguishes *Burkholderia pseudomallei* from avirulent *Burkholderia thailandensis. Journal of Clinical Microbiology, 41,* 4812–4814.

Luptakova, A. (2007). Importance of sulphate-reducing bacteria in environment. *Nova Biotechnologica, 7,* 17–22.

Madigan, M. T., & Martinko, J. M. (2006). *Brock biology of microorganisms* (11th ed.). Upper Saddle River: Pearson Prentice Hall.

Manske, A. K., Henbge, U., Glaeser, J., & Overman, J. (2008). Subfossil 16S r RNA gene sequences of green sulfur bacteria in the Black Sea and their implications for past photic zone anoxia. *Applied and Environmental Microbiology, 74,* 624–632.

McLaughlin, M. R., Balaa, M. F., Sims, J., & King, R. (2006). Isolation of Salmonella bacteriophages from swine effluent lagoons. *Journal of Environmental Quality, 35,* 522–528.

Moroney, J. V., & Ynalvez, R. A. (2009). *Algal photosynthesis* (Encyclopedia of Life Sciences). Chichester: Wiley.

Naqvi, W. (2006). Marine nitrogen cycle. In C. J. Cleveland (Ed.), *Encyclopedia of Earth.* Washington, DC: Environmental Information Coalition, National Council for Science and the Environment.

Narayanan, N., Pradip, K., Roychoudhury, P.K., Srivastava, A. (2004). L (+) lactic acid fermentation and its product polymcrization. *Electronic Journal of Biotechnology, 7*(3). http://www.ejbiotechnology.info/content/vol2/issue3/full/3/index.html. ISSN 0717-3458. Accessed 15 Aug 2004.

Ochiai, K., & Kawamoto, I. (1995). Two-dimensional gel electrophoresis of ribosomal proteins as a novel approach to bacterial taxonomy: application to the genus Arthrobacter. *Bioscience, Biotechnology, and Biochemistry, 59,* 1679–1687.

Petti, C. A., Polage, C. R., & Schreckenberger, P. (2006). The role of 16S rRNA gene sequencing in identification of microorganisms misidentified by conventional methods. *Journal of Clinical Microbiology, 44,* 3469–3470.

Rathgeber, C., Beatty, J. T., & Yurkov, V. (2004). Aerobic phototrophic bacteria: new evidence for the diversity, ecological importance and applied potential of this previously overlooked group. *Photosynthesis Research, 81,* 113–128.

Rossman, A. Y., & Palm, M. E. (2006). Why are Phytophthora and other Oomycota not true fungi? *Outlooks on Pest Management, 17,* 217–219. Also see www.pestoutlook.com.

Saksena, D. N. (2006). Rotifers as indicators of water quality. *Acta Hydrochimica et Hydrobiologica, 15*, 481–485.

Samson, R., & Pitt, J. I. (1989). *Modern concepts in Penicillium and Aspergillus classification*. New York and London: Plenum Press.

Sander, D. (2007). The big picture book of viruses. http://www.virology.net/Big_Virology/BVFamilyGroup.html. Accessed 2 Sept 2010.

Sze, P. (1986). *A biology of the algae*. Dubuque: Brown.

Trainor, F. R. (1978). *Introductory phycology*. New York: Wiley.

Van Regenmortel, M. H. V. (1999). How to write the names of virus species. *Archives of Virology, 144*, 1041–1042.

Van Regenmortel, M. H. V., Bishop, D. H. L., & Fauquet, C. M. (Eds.). (2005). *Virus taxonomy: Eighth report of the International Committee on Taxonomy of Viruses*. London: Elsevier.

Woese, C. R. (1987). Bacterial evolution. *Microbiological Reviews, 51*, 221–271.

Woese, C. R. (2000). Interpreting the universal phylogenetic tree. *Proceedings of the National Academy of Sciences of the United States of America, 97*, 8392–8396.

Woese, C. R. (2002). On the evolution of cells. *Proceedings of the National Academy of Sciences of the United States of America, 99*, 8742–8747.

Yamanaka, T. (2008). *Chemolithoautotrophic bacteria: Biochemistry and environmental biology*. Tokyo: Springer.

Part III

The Ecology of Microorganisms in Natural Waters

Ecology of Microorganisms in Freshwater

5

Abstract

Freshwaters are defined as natural waters containing less than 1,000 mg per liter of dissolved solids, most often salt. Globally, freshwaters are scarce commodities and make up only 0.009% of the earth's total water. Although they generate only about 3% of the earth's total primary biological productivity, they contain about 40% of the world's known fish species. Natural freshwaters are classifiable into atmospheric, surface, and underground waters each type having a unique microbial ecology. Atmospheric waters lose their microorganisms as they fall as rain or snow. Surface freshwaters are found in rivers and lakes, and contain large and diverse groups of microorganisms. Using molecular methods such as 16S rRNA analysis, which enables the study of unculturable microorganisms, new information regarding freshwater microbial ecology has emerged in recent times: There are more phylogenetic groups of bacteria than are observed by cultural methods; there is a unique and distinct bacterial group, which can be termed "typical freshwater bacteria"; contrary to previous knowledge when aquatic bacteria were thought to be mostly Gram-negative bacteria, Gram-positive bacteria are in fact abundant in freshwaters; finally, marine-freshwater transitional populations exist in coastal waters. Ground waters suffer contamination from chemicals and less from microorganisms; the deeper the groundwater, the less likely it is to contain microorganisms, which are filtered away by soil.

Keywords

Underground waters • Pollution of underground waters • Typical freshwater bacteria • Ingoldian fungi • Abundance of Gram-positive bacteria • Transitional microbial population in estuaries

5.1 Microbial Ecology of Atmospheric Waters

Atmospheric water occurs in the form of rain, snow, or hail; all three forms occur in temperate countries. In tropical countries, however, snow is unknown, except perhaps at the peaks of the high mountains which are found, in East Africa, the Cameroon, etc. Hail only falls occasionally. The major source of atmospheric water in tropical countries is therefore rain.

As it falls from clouds (which are themselves condensates of water vapor), rain water collects with it dissolved gases, dust particles, and microorganisms.

N. Okafor, *Environmental Microbiology of Aquatic and Waste Systems*,
DOI 10.1007/978-94-007-1460-1_5,

The result is that if rain falls for a sufficiently long period, rain water would have brought down dust particles and any microorganisms circulating in the atmosphere. Water resulting from such heavy and continuous rain in a rural area free of industrial gases, should be expected to be virtually sterile and almost of the quality of the distilled water prepared in the laboratory. It should be possible to collect such water when heavy rain has continued uninterrupted for upwards of 90 min, if proper aseptic conditions are followed. Indeed in parts of the rural areas of the developing world where water is scarce, harvested rain water often directly forms the only source of drinking water.

It should not be taken for granted, however, that microorganisms from the atmosphere make significant contributions (Jones et al. 2008) to aquatic systems as can be seen from Fig. 5.1.

5.2 Microbial Ecology of Surface Waters

Surface waters include rivers, streams, lakes, ponds, and wetlands. Surface waters may consist of either freshwater or saline water. Freshwater is water with a salt content (salinity) of less than $1 \, g \, L^{-1}$, while saline waters (seas and oceans) are characterized by salinities $>1 \, g \, L^{-1}$. This chapter will examine the ecology of microorganisms in fresh water (Hahn 2006). The next chapter will look at marine microbiology.

5.2.1 Rivers and Streams

Fresh water systems can be divided into lotic systems composed of running water (rivers, streams, creeks, springs) and lentic systems which comprise still water (ponds and lakes).

5.2.2 Lakes and Ponds

Lakes have been defined as large ponds, the defining size varying according to authors. Thus, the size of defining lakes as opposed to ponds have varied from 5 acres (2 ha) through 12 acres (5 ha) to 20 acres (8 ha). Most lakes have a natural outflow in the form of a river or stream, but some do not, and lose water solely by evaporation or underground seepage or both. Many lakes are artificial

and are constructed for hydro-electric power generation, recreational purposes, industrial use, agricultural use, or domestic water supply. Table 1.4 shows the world's largest lakes. Most of them are fresh water.

5.2.3 Wetlands

Wetlands are land pieces either temporarily or permanently submerged or permeated by water. They are characterized by plants adapted to soil conditions saturated by water. Wetlands include fresh and salt water marshes, wooded swamps, bogs, seasonally flooded forest, and sloughs.

Wetland vegetation is typically found in distinct zones that are related mainly to water depth and salinity. Since wetlands are partly defined by the vegetation found in them, the typical vegetation found in wetlands are described briefly below:

1. *Shoreline*: Plants that grow in wet soil on raised hummocks or along the shorelines of streams, ponds, bogs, marshes, and lakes. These plants grow at or above the level of standing water; some may be rooted in shallow water. Examples of plants are temperate climate shoreline plants, e.g., western coneflower (*Rudbeckia occidentalis* Nutt.) and buttonbush (*Cephalanthus occidentalis*). In tropical and semi-tropca climates, for example, the Florida everglades mangroves are common. These are salt tolerant plants climates, two examples of which are red mangrove (*Rhizophora mangle*) and black mangrove (*Avicennia germinans*).

2. *Emergent*: Plants that are rooted in soil that is underwater most of the time. These plants grow up through the water, so that stems, leaves and flowers emerge in air above water level. Some of the commonest plants in the emergent zone are arrowheads (*Sagittaria* spp.). Nearly all parts of the plants, including the roots are used by wild life as food. In Japan and China, they are used as human food. Among the most common wetland plants worldwide are cattails (*Typha* spp.).

3. *Floating*: Plants whose leaves mainly float on the water surface. Much of the plant body is underwater and may or may not be rooted in the substrate. Only small portions, namely flowers, rise above water level. Examples are duckweeds (family Lemnaceae), which are used for animal and human food and pondweeds (*Potamogeton* spp.).

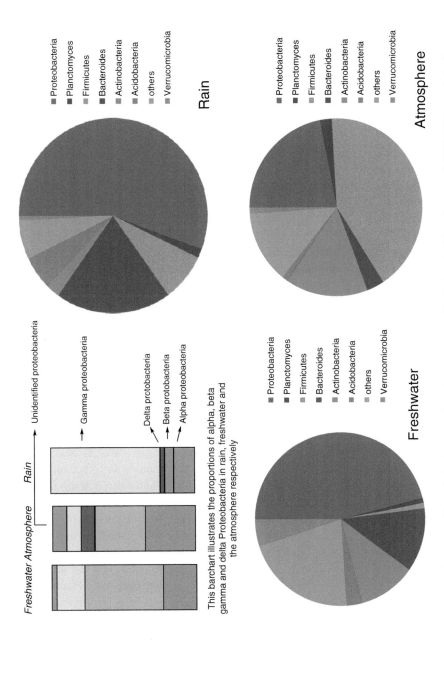

Fig. 5.1 Proportions of taxonomic groups of bacteria in the atmosphere, in rain water and in freshwater (From Jones et al. 2008. With permission)

4. *Submerged*: Plants that are largely underwater with few floating or emergent leaves. Flowers may emerge briefly in some cases for pollination. Examples are water milfoil (*Myriophyllum heterophyllum*) and watercress (*Rorippa nasturium-aquaticum*), a member of the mustard family (Cruciferae), used for food and medicinal purposes.

Wetlands are the only ecosystem designated for conservation by international convention. They have been recognized as particularly useful for the following reasons:

(a) They act as buffers and absorb the impact of water-related traumatic occurrences such as large waves or floods.
(b) They filter off sediments and toxic substances.
(c) They supply food and essential habitat for many species of fish, shellfish, shorebirds, waterfowl, and furbearing mammals.
(d) They provide products for food (wild rice, cranberries, fish, wildfowl), energy (peat, wood, charcoal), and building material (lumber).
(e) They provide avenues for recreation such as hunting, fishing, and bird watching.

Previously wetlands were considered wasteland, and in some countries, they were drained or filled in, so that they could be farmed or built upon. In recent times, however, the value of wetlands has been recognized and efforts have been made to protect them. However, they are still disappearing under the pressure of human activity, and are being threatened by air pollution and climate change.

Some of the steps taken in recent times to preserve wetlands are as follows:

1. Adding sediment to coastal wetlands to keep up with rising sea levels
2. Planting grass to protect coastal sands from erosion
3. Building dikes or barrier islands to protect the wetlands
4. Controlling water levels artificially to ensure the wetland is flooded

5.3   Ground Waters

Groundwater flows slowly through water-bearing formations (aquifers) at different rates. In some places, where groundwater has dissolved limestone to form caverns and large openings, its rate of flow can be relatively fast but this is exceptional.

Many terms are used to describe the nature and extent of the groundwater resource. The level below which all the spaces are filled with water is called the *water table*. Above the water table lies the *unsaturated zone*. Here, the spaces in the rock and soil contain both air and water. Water in this zone is called *soil moisture*. The entire region below the water table is called the *saturated zone*, and water in this saturated zone is called *groundwater*.

Groundwater is not confined to only a few channels or depressions in the same way that surface water is concentrated in streams and lakes. Rather, it exists almost everywhere underground. It is found underground in the spaces between particles of rock and soil, or in crevices and cracks in rock.

The water filling these openings is usually within 100 m of the surface. Much of the earth's fresh water is found in these spaces. At greater depths, because of the weight of overlying rock, these openings are much smaller, and therefore hold considerably smaller quantities of water.

Although groundwater exists everywhere under the ground, some parts of the saturated zone contain more water than others. An *aquifer* is an underground formation of permeable rock or loose material which can produce useful quantities of water when tapped by a well. Aquifer sizes vary. They may be small, only a few hectares in area, or very large, underlying thousands of square kilometers of the earth's surface. They may be only a few meters thick, or they may measure hundreds of meters from top to bottom.

Groundwater circulates as part of the hydrologic cycle. As precipitation and other surface water sources recharge the groundwater, it drains steadily, and sometimes very slowly, toward its discharge point.

When precipitation falls on the land surface, part of the water runs off into the lakes and rivers. Some of the water from melting snow and from rainfall seeps into the soil and percolates into the saturated zone. This process is called *recharge*. Places where recharge occurs are referred to as *recharge areas*.

Eventually, this water reappears above the ground. This is called *discharge*. Groundwater may flow into streams, rivers, marshes, lakes, and oceans, or it may discharge in the form of springs and, when tapped, wells.

The *residence time* of groundwater, i.e., the length of time water spends in the groundwater portion of the

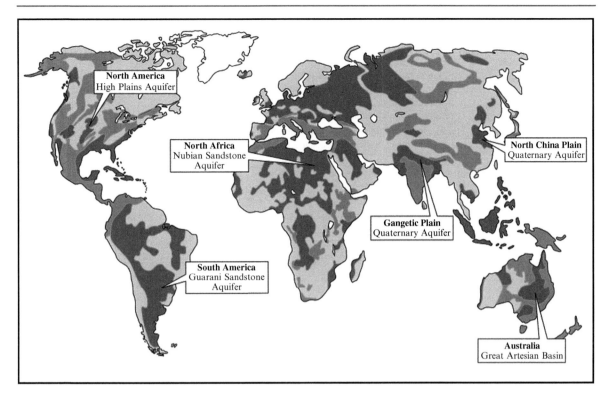

Fig. 5.2 Distribution of underground water world-wide (From Foster and Chilton 2003, with permission). Note: Hydrogeological map of the world showing widespread occurrence of geological formations containing useful ground water and the locations of some of the world's largest aquifers with vast storage reserves. The *dark gray* areas are major regional aquifers, while the midgrey areas contain some important but complex aquifers. The *light gray* areas contain only small aquifers of local importance

hydrologic cycle, is highly variable. Water may spend as little as days or weeks underground, or as much as 10,000 or more years.

Groundwater systems constitute the predominant reservoir and strategic reserve of freshwater storage on earth; about 30% of the global total and as much as 98% if that bound up in the polar ice caps and glaciers is discounted. As seen in Fig. 5.2, some aquifers extend quite uniformly over very large land areas and have much more storage than all of the world's surface reservoirs and lakes. In sharp contrast to surface water bodies, they lose very little of their stored water by direct evaporation.

Although there are country to country differences and local variations, globally, groundwater is estimated to provide at least 50% of current potable water supplies; 40% of the demand from those industries that do not use mains water, and 20% for water use in irrigated agriculture. Compared with surface water, groundwater use often brings large economic benefits per unit volume, because of ready local availability, high drought reliability and generally good quality requiring minimal treatment.

Groundwater contaminants come from two categories of sources: Point sources and distributed or non-point sources. Landfills, leaking gasoline storage tanks, leaking septic tanks, and accidental spills are examples of point sources. Infiltration from farm land treated with pesticides and fertilizers is an example of a non-point source.

Microorganisms are trapped by the particles of soil which act as components of a vast filter. The result of this is that the deeper the ground water, the less likely it is to have microorganisms. Waters collected from deep wells are therefore usually found to contain few bacteria. Artesian springs which erupt from the ground under great pressure have been found to contain less than 300 bacteria per ml. and it is doubtful whether such bacteria have not arisen from the side of the well or introduced during sampling. In very deep bores dug for the exploration of oil, sulfate-reducing bacteria e.g., *Desulphovibrio* have been reported, especially in

Table 5.1 Sources of deterioration of underground waters (Modified from Foster and Chilton 2003. With permission)

Type of problem	Cause	Parameters of concern
Salinization processes	Mobilization and/or fractionation of salinity due to inadequate management of ground water irrigation, mine reservoir exploitation drainage or petroleum; extensive and prolonged surface water irrigation without adequate drainage	Na, Cl and sometimes F, Br SO4
Anthropogenic pollution	Inadequate protection of vulnerable aquifers against man-made discharges/leachates from urban activities	Pathogens, NO_3, NH_4, Cl, SO_4, B, heavy metals, DOC, Aromatic and halogenated hydrocarbons, etc. NO3, Cl2, some pesticides and derivatives
Well head contamination	Inadequate well construction and completion, allowing direct ingress of polluted surface water or shallow ground water	Mainly pathogens, NO3, Cl
Naturally occurring contamination	Related to pH-Eh evolution of ground water and dissolution of minerals from aquifer matrix (can be aggravated by anthropogenic pollution and/or uncontrolled exploitation)	Mainly Fe, F, and sometimes As, I, Mn, Al, Mg, SO_4, Se and NO_3 (from paleo recharge): F and As in particular represent a serious public health hazard for potable supplies

the zones with oil-water mixtures. Some of such findings were made in bores as deep as 2,000 m.

In many parts of the world, drinking water is obtained from underground water. In the US, for example, over 50% of the United States population utilizes groundwater as its drinking water source. Approximately 96% of the groundwater users live in small rural areas often utilizing small individual wells, where the resources for treatment and monitoring are limited. While microorganisms can be filtered out by soil, chemicals are not; ground water is therefore frequently contaminated by chemicals.

Although ground waters are generally free of contaminants, the quality of ground waters can be compromised by both man-made and other sources (see Table 5.1).

5.4   Some Microorganisms Usually Encountered in Fresh Water

5.4.1   Bacteria

Early work (in the 1980s and earlier) on freshwater bacteria was based on bacteria which were culturable. Freshwater bacteria were merely grouped fresh water into:

(a) The flourescent group
(b) The chromogenic bacteria including violet, red and yellow forms
(c) The coliform group
(d) The Proteus group

(e) Non-gas-forming, non-chromogenic, non-spore-forming rods which do not produce Proteus-like colonies and may not acidify milk and liquify gelatin
(f) Aerobic spore-forming rods
(g) White, yellow, and pink cocci

It was believed that there was no clear separation between soil bacteria and aquatic bacteria and that fresh water bacteria did not have a unique population of its own. However, it soon became known that bacteria cultivated from the environment including the soil, and fresh and marine waters represented only a fraction of the actual total, as the bulk of such bacteria were not culturable (Oren 2004).

The advent of molecular techniques and especially the polymerase chain reaction (PCR) has made it possible to obtain information on microbial community composition directly, without cultivation. With new molecular techniques of the 1990s, it became possible to assess the microbial population of a natural environment through culture-independent techniques by isolating the nucleic acid in it, followed by the amplification and sequencing of bacterial 16S rRNA genes (Miskin et al. 1999).

5.4.1.1 New Data Regarding Freshwater Bacteria

Assessing the freshwater environment as described above through the work of Zwart and others (Zwart et al. 2002) led to new conclusions regarding the bacterial population of that environment. After analyzing available database of 16S rRNA sequences from freshwater plankton, including rivers and lakes in the USA, South Korea, and the Netherlands the following conclusions were reached (Table 5.2):

1. *More phylogenetic groups of bacteria observed by molecular methods*

   Most bacterial sequences retrieved from freshwater habitats were neither affiliated with known bacterial species nor represented those phylogenetic groups previously obtained from freshwater habitats using cultivation methods. Bacteria in the freshwater environment which could not be cultivated were observed by direct nucleic acid assessment.

2. *Existence of a unique and distinct bacterial group, "typical freshwater bacteria"*

   Sequences retrieved from freshwater habitats were most closely related to other freshwater clones, whereas relatively few were most closely related to sequences recovered from soil or marine habitats. Thus, there appears to be a specific group of bacteria which are typical of, and thus indigenous in, freshwater. Thirty-four phylogenetic clusters of bacteria representing typical inhabitants of freshwater systems were initially recorded but more have been since added.

3. *The abundance of Gram-positive bacteria in freshwater*

   Prior to the advent of molecular methods, it was shown, because of the limitations of the cultural methods, that most freshwater bacteria were Gram negative and only a few Gram positive bacteria were found therein. With the new techniques, it is now known that Gram positive bacteria, especially *Actinobacteria* are plentiful in freshwater.

4. *A marine-freshwater transitional population exists in coastal waters*

   Distinct bacterial populations exist in both marine and the freshwater environments. However in coastal areas, while there is a preponderance of marine bacteria, there are also some bacteria regarded as typically of the freshwater niche. This shows that waters in the coastal areas of the seas and oceans can be regarded as being transitional (Rappe et al. 2000).

5. *Some factors affecting the distribution of bacteria in freshwater*

   The factors which affect the distribution of bacteria in freshwater are complex. At present, not much information is available, but some factors affecting the nature of the bacterial community in a body of fresh water include the following: Water chemistry (including pH), water temperature, metazooplankton predation, protistan predation, phytoplankton composition, organic matter supply, intensity of ultraviolet radiation, habitat size, and water retention time (Lindstrom et al. 2005).

Table 5.2 Examples of "typical freshwater bacteria" (Modified from Zwart 2002. With permission)

S/No	Bacterial cluster name	Division
1	LD12	α-Proteobacteria
2	*Brevundimonas intermedia*	α-Proteobacteria
3	CR-FL11	α-Proteobacteria
4	GOBB3-C201	α-Proteobacteria
5	*Novosphingobium subarctica*	β-Proteobacteria
6	*Polynucleobacter necessaries*	β-Proteobacteria
7	LD28	β-Proteobacteria
8	GKS98	β-Proteobacteria
9	*Ralstonia picketti*	β-Proteobacteria
10	*Rhodoferax* sp. Bal47	β-Proteobacteria
11	GKS16	γ-Proteobacteria
12	*Methylobacter psychrophilus*	Verrucomicrobia
13	CL120-10	*Verrucomicrobia*
14	CL0-14	*Verrucomicrobia*
15	FukuN18	*Verrucomicrobia*
16	Sta2-35	*Verrucomicrobia*
17	LD19	*Actinobacteria*
18	ACK-m1	*Actinobacteria*
19	STA2-30	*Actinobacteria*
20	MED0-06	*Actinobacteria*
21	URK0-14	*Actinobacteria*
22	CL500-29	*Cytophaga-Flavobacterium-Bacteroides*
23	LD2	*Cytophaga-Flavobacterium-Bacteroides*
24	FukuN47	*Cytophaga-Flavobacterium-Bacteroides*
25	PRD01a001B	*Cytophaga-Flavobacterium-Bacteroides*
26	CL500-6	*Cytophaga-Flavobacterium-Bacteroides*
27	*Synechococcus* 6b	*Cyanobacteria*
28	*Planktothrix agardhii*	*Cyanobacteria*
29	*Aphanizomenon flos aquae*	*Cyanobacteria*
30	*Microcystis*	*Cyanobacteria*
31	CL500-11	Green non-sulfur bacteria
32	CLO-84	OP10
33	CL500-15	*Planctomyces*

5.4.1.2 Some Bacteria in Freshwater

More recent publications list the organisms, but it is clear from these that much work remains to be done in fresh waters of various categories, especially in tropical countries for a rounded picture to be built.

The nature of the bacteria flora of a fresh water varies. The bulk of such bacteria are heterotrophic while a small proportion are photo- or chemo-autotrophic.

In oligotrophic surface waters, e. g., springs recently emerging from their source, the bacterial population consists among the nonbenthic population largely of Gram-negative, non-spore- forming rods, especially *Achromobacter* and *Flavobacterium*. With increasing eutrophication, *Pseudomonas, Proteus, Bacillus,* and *Enterobactetriaceae* become more important. Other bacteria usually encountered include *Vibrio* and *Actinomycetales*. When eutrophication is heavy, and due to organic material. *Zooglea* tends to occur; if the water is lotic or fast flowing, then the "sewage fungus" *Sphaerotilus natans* may also develop. Most of the above-mentioned bacteria are Gram-negative; in general, Gram-negative bacteria are planktonic or tectonic. Among the "Aufwuchs" are actinomycetes, which are usually epiphytic. Most benthic bacteria are generally Gram-positive and include spore-formers, and Gram-positive cocci, *Clostridium,* and pleomorphic forms (e.g., *Arthrobacter, Nocardia*).

Chemo-autotrophic bacteria, e. g., *Nitrosomonas* and *Nitrobacter* are found in some lake waters while photosynthetic bacteria are found in some rivers. Certain aquatic bacteria are considered as nuisance bacteria in drinking water. These include iron, sulfur, and sulfate-reducing bacteria. These nuisance bacteria may cause odor, taste or turbidity in water as well as destroy water pipes. "Iron" bacteria withdraw iron which is present in the environment and deposit it in the form of hydrated ferric hydroxide in mucilaginous secretions· and this imparts a reddish tinge to water and may stain clothes. Some well-known iron bacteria genera include *Sphaerotilus, Leptothrix, Toxothrix, Crenothrix, Callionella, Siderobacter,* and *Ferrobacillus*.

The sulfur bacteria include the green photosynthetic and nonphotosynthetic members. Many of the photosynthetic bacteria produce H_2S which may impart odor to drinking water. Among some well-known colorless (non-photosynthetic) sulfur bacteria are *Thiobacterium* (short rods which deposit sulfur within or outside the cells), *Macromonas* (large slow-moving organisms

which contain $CaCO_3$ as, well as sulfur), and *Thiovulum* (spherical cells up to 20 μ). A well-known color-less sulpur bacterium is *Thiobacillus,* an autotrophic organism, the best known of which is *T. thioxidans*. It oxidizes thiosulphate first to sulfur and then to sulphuric acid. The H_2SO_4 may then corrode pipes and concrete sewers. All the above usually occur singly.

Beggiatoa and *Thiothrix* are usually filamentous. *Beggiatoa* move by gliding and contain sulfur granules. Some species may be up to 50 μ in diameter. *Thiothrix* is non-motile.

5.4.2   Fungi

Fungi are primarily terrestrial, but some are aquatic. Most of the water-dwelling fungi are *Phycomycetes,* although representatives of the other groups (*Ascomycetes, Basidiomycetes*, and especially Fungi imperfecti (*Hyphomycetes*)) contain some aquatic counterparts. Aquatic *Hyphomycetes* have been well described.

It is perhaps not surprising that it is only among the *Phycomycetes* that fungi with motile zoospores are to be found. Among the Phycomycetes, the following orders are aquatic: *Chytriditlles, Blastocladiales, Monoblepharidales, Hyphochytriales, Leptomitales* and *Lagenidiales*. These fungi are saprophytes or parasites on various plants and animals or their parts in water. In other words, they are mostly, "Aufwuchs," that is attached and may be found on any aquatic plant or animal, algae, fish or even other fungi.

Among the more common genera of fungi encountered in water are: *Allomyces, Achyla, Sapromyces,* and various chytrids. *Fusarium* (not a phycomycete but in the Fungi Imperfecti) is also common in water. Aquatic yeasts have been discovered in large numbers in recent times particularly in waters with high organic matter contents and many *Hyphomycetes* (Fungi Imperfecti) have also been recorded. There are more than 600 species of freshwater fungi with a greater number known from temperate, as compared to tropical, regions.

The Fungi Imperfecti (called mitosporic fungi by some authors) are classified into two main classes, namely hyphomycetes, and coelomycetes. The hyphomycetes produce conidia directly from vegetative structures (hyphae) or on distinct conidiophores (a specialized hypha that bears conidiogenous cells and conidia, for example in *Aspergillus* or *Penicillium*) whereas, the coelomycetes produce conidia within asexual fruit bodies called pycnidia.

The freshwater Fungi Imperfecti (mitosporic fungi) are classified into three groups: The Ingoldian hyphomycetes (also called the aquatic hyphomycetes), the aeroaquatic hyphomycetes, and the dematiaceous (dark colored) and hyaline (light colored) hyphomycetes and coelomycetes (see Fig. 2.2).

(a) *The Ingoldian hyphomycetes*

The aquatic or fresh water hyphomycetes are also known as the Ingoldian hyphomycetes, after Ingold who first described them in 1942. They are Fungi Imperfecti (i.e., Ascomycetes whose perfect or sexual stages have not been described) as well as some Basidiomycetes. Ingoldian hyphomycetes produce conidia that are mostly unpigmented and branched or long and narrow, and are adapted for life in running water. The Ingoldian hyphomycetes most commonly occur on dead leaves, and wood immersed in water.

(b) *Aeroaquatic hyphomycetes*

The aeroaquatic hyphomycetes produce purely vegetative mycelium in substrates under water, but produce conidia with special flotation devices, only when the substrates on which the fungus is growing are exposed to a moist aerial environment. They are found in stagnant ponds, ditches, or slow flowing freshwater. Their vegetative hyphae grow on submerged leaves and woody substrates under semi-anaerobic freshwater conditions and are found around the world.

(c) *Dematiaceous and hyaline fungi imperfecti*

The dematiaceous and hyaline hyphomycetes and coelomycetes are distinct from Ingoldian hyphomycetes, because the conidia is not specifically adapted for aquatic existence. The fungi occur mainly on decaying herbaceous plant material and woody debris in aquatic and semi aquatic habitats worldwide.

They are classified into two main groups: indwellers and immigrants. Indwellers have been reported only from freshwater habitats, whilst immigrants have been reported from terrestrial as well as freshwater habitats.

Freshwater fungi are thought to have evolved from terrestrial ancestors. Many species are clearly adapted to life in freshwater as their propagules have specialized aquatic dispersal abilities. Freshwater fungi are involved in the decay of wood and leafy material and also cause diseases of plants and animals (see Fig. 4.21).

5.4.3 Algae

Algae are primarily aquatic organisms and hence are to be found in large numbers in water, including freshwater. Some algae commonly encountered in drinking water include the blue-green algae: *Microcystis aeroginesa* (which yields a material toxic to man and animals), *Aphanozomenon flos-aquae,* and *Anabaena circinalis,* all of which hamper filteration processes in water purification. The chrysophyte *Sunura uvella* imparts to drinking water a cucumber taste.

The following diatoms also cause filteration problems:

Asterionella formosa, Nitzschia acicularis, Stephanodisus astrea, Melosira spp. as well as the Xanthophyte, *Tribonema bombycinum.*

5.4.4 Protozoa

The Protozoa also show a pattern of succession in their use of bacteria as food. The earliest occurring Protozoa during the eutrophication of a fresh body of water are Sarcodina and Mastigophora which are found in large numbers sonly in freshly contaminated waters.

Like the bacteria which are induced to develop by eutrophication, they also absorb soluble nutrients (Fig. 5.3).

5.5 Succession of Organisms in the Breakdown Materials Added to Aquatic Systems

5.5.1 The Physiological Basis of Aquatic Microbial Ecology

In water, as in most other natural environments, microorganisms do not usually exist as pure cultures, rather they exist as mixed cultures. The abundance of any particular organism or groups of organisms depends on the conditions operating at a given time in the given environment. Some of such conditions are discussed below, although it ought to be noted that the situation is complex in that several of these conditions may operate at the same time.

1. *Nutrient availability*: The quality as well as the quantity of nutrients entering water is important in the microbial ecology of aquatic systems. Any

Fig. 5.3 Succession
of organisms in the
decomposition of materials
added to aquatic systems

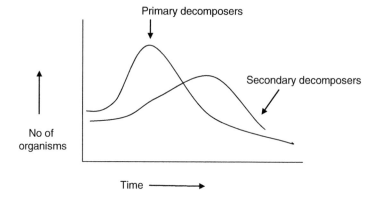

increase in the concentration of available nutrients in water is known as *eutrophication,* which may be man-made as in sewage discharge into a stream, or natural as with rain water washings.

In the mixed populations of freshwater systems, the addition of the easily digested carbohydrates such as starches and sugars ·stimulate the general development of bacteria and fungi. Among the bacteria, the versatile *Pseudomonas* is usually present and proteins when added to water stimulate the development of bacteria such as *Alcaligines* and *Flavobacterium.*

As a consequence of eutrophication, the organisms which can utilize any particular added materials grow rapidly. On the exhaustion of the substrate, these primary invaders die, releasing cellular materials which are predominantly protein. This release of proteins encourages a secondary development of proteinaceous and other bacteria.

Among the Protozoa, the tectonic cliates soon over run their predecessors, Sarcodina and Mastigophora which then die off. The bacterial population, the source of nourishment for the tectonic ciliates, soon decline in population to the extent that they no longer satisfy the needs for these active organisms. At this stage, the stalked ciliates *(Suctoria),* needing less nourishment because of lesser activity suck in the remaining bacteria; but sooner or later, these remaining bacteria may not be enough to sustain even these sessile ciliates. It is at this stage that rotifers and crustaceans move in, to scavenge off any living microorganisms available such as the insoluble remains of the dead ones (Hugenholtz et al. 1998). This relationship can be represented as shown in Fig. 5.4.

2. *pH*: At pH values of 4.0–5.0, the fungi predominate and bacteria are completely excluded. At pH values

of 7 and above, the predominant organisms are bacteria. Thus *Vibrio* will predominate at pH values of over 8.0.

3. *The oxygen tension*: When the system is aerated, fungi, bacteria, and protozoa grow rapidly, and organic materials are broken down ultimately into CO_2 and water and new cells are produced.

Under anaerobic conditions, anaerobic bacteria develop and fungi and protozoa are generally absent. If sulfate is present, the sulfate-reducing bacteria e.g., *Desulfovibrio* (small Gram-negative anaerobic curved rods) occur in large numbers, forming sulfide in the process. Some clostridia, e.g., *C nigrificans* also reduce sulfate to H_2S.

If sulfate is absent methane-producing bacteria occur. These are able to use CO_2 as electron transport acceptor for anaerobic respiration, resulting in its reduction to the gas methane (CH_4). Three genera are involved: *Methanobacterium, Methanococcus,* and *Methanosarcina* (see Chap. 10).

Anaerobic bacteria are found in muds where their activities lead, as shown above, to the formation of foul gases such as methane and hydrogen sulfide.

4. *Temperature*: High temperatures (about 30–37°C) encourage enteric organisms, while extremely high temperatures (higher than 45°C) select thermophiles. Thus in hot springs, thermophilic bacteria and bluegreen algae are encountered. Low temperatures such as the ones found in oceans (4–10°C) encourage the growth of psychrophyls.

5. *Depth of the water*: More bacteria are generally found in the upper portion of the water than the lower. Furthermore, the type of organisms developing at the various regions of a stream or lake differ one from the other. Cocci, for example, are generally benthic whereas the Gram-negative mobile rods are generally tectonic.

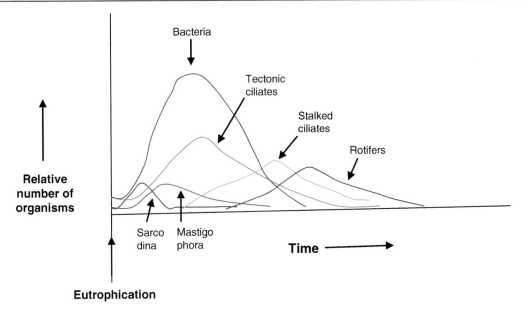

Fig. 5.4 Succession of microorganisms in an aquatic system

6. *Flow velocity in moving waters*: The velocity of the water affects the type of organisms developing in water. It has been established for example that some aquatic algae and bacteria (e.g., *Sphaerotilus natans* and *Leptomitus lacteus)* grow better in flowing waters under natural conditions than in stagnant water. The protozoan *Vorticella* (attached ciliate) and *Nitzschia paleae* (diatoms) also grow better in slow-moving waters.

7. *Light*: Light tends to inhibit the growth of bacteria, but encourages the development of the algae. When nutrients are plentiful under aerobic and light conditions such as in an oxidation pond (see Chap. 10), the bacteria break down the organic matter releasing CO_2 in the process. The CO_2 is then used up, by the algae, for photosynthesis. The latter process releases oxygen which is required by the bacteria.

5.6 Microbial Loop and the Food Web in a Freshwater System

The succession of organisms in an aquatic system is intimately related to the food web (see Fig. 5.5). With better techniques for assessing the load of viruses in aquatic system, we now know that viruses are not only very abundant in aquatic systems, freshwater and saline, but that they also play a major role in the food web of

the system. They attack all classes of microorganisms in the system, bacteria, fungi algae, and protozoa, and the carcasses of these organisms contribute to the dissolved organic matter which are absorbed bacteria in the microbial loop shown in Fig. 5.5. The microbial loop is a new concept in the food chain of aquatic systems.

In Fig. 5.5, the regular food chain (shown in unbroken lines) begins with primary produces such as algae which capture the energy of the sun in photosynthesis. These are eaten by the smaller aquatic life such as small fish and krils. The fish are eaten by bigger fish which are eventually eaten by humans. Small fish, big fish, and humans all produce wastes which are broken down by aerobic bacteria in water. When the algae and the aquatic animals die, they are also broken down by aerobic bacteria. The bacteria themselves are eaten by protozoa; the protozoa are eaten by krils and small fish and the cycle goes on. The release of breakdown products from wastes and the carcasses of aquatic organisms contributes to the dissolved organic matter of aquatic systems and provides nourishment for aquatic organisms. The dissolved organic matter of aquatic systems is also increased by eutrophication, i.e., the addition of organic matter.

The new concept is the "microbial loop." The concept has come into being because of the recent recognition of the place of viruses in aquatic microbial ecology. As seen from Fig. 5.5, viruses attack and kill bacteria,

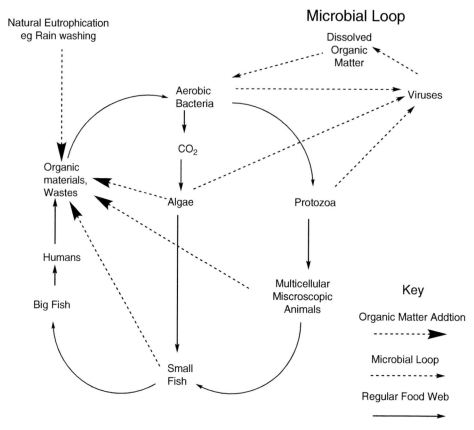

Fig. 5.5 The microbial loop in the food web in a freshwater system

fungi, protozoa, and algae and release materials which contribute to the dissolved organic matter. The microbial loop is that aspect of the food chain which depends on the activities and contribution of microorganisms.

References

Foster, S. S. D., & Chilton, P. J. (2003). Groundwater: The processes and global significance of aquifer degradation. *Philosophical Transactions of the Royal Society of London, B, 358,* 1957–1972.

Hahn, M. W. (2006). The microbial diversity of inland waters. *Current Opinion in Biotechnology, 17,* 256–261.

Hugenholtz, P., Goebel, B. M., & Pace, N. R. (1998). Minireview: Impact of culture-independent studies on the emerging phylogenetic diversity. *Journal of Bacteriology, 180,* 4765–4774.

Jones, S. E., Newton, R. J., & McMahon, K. D. (2008). Potential for atmospheric deposition of bacteria to influence bacterioplankton communities. *FEMS Microbial Ecology, 64,* 384–394.

Lindstrom, E. S., Agterveld, M. P. K., & Zwart, G. (2005). Distribution of typical freshwater bacterial groups is associated with pH, temperature, and lake water retention time. *Applied and Environmental Microbiology, 71,* 8201–8206.

Miskin, I. P., Farrimond, P., & Head, I. M. (1999). Identification of novel bacterial lineages as active members of microbial populations in a freshwater sediment using a rapid RNA extraction procedure and RT-PCR. *Microbiology, 145,* 1977–1987.

Oren, A. (2004). Prokaryote diversity and taxonomy: Current status and future challenges. *Philosophical Transactions of the Royal Society of London, B, 359,* 623–638.

Rappe, M. S., Vergin, K., & Giovannoni, S. J. (2000). Phylogenetic comparisons of a coastal bacterioplankton community with its counterparts in open ocean and freshwater systems. *FEMS Microbiology Ecology, 33,* 219–232.

Zwart, G., Crump, B. C., Agterveld, M. P. K., Hagen, F., & Han, S.-K. (2002). Typical freshwater bacteria: An analysis of available 16S rRNA gene sequences from plankton of lakes and rivers. *Aquatic Microbial Ecology, 28,* 141–155.

Ecology of Microorganisms in Saline Waters (Seas and Oceans)

6

Abstract

The waters of the seas and oceans of the world contain large amounts of solutes; mainly, salt of about 3.5 g/L, occupy about 71% of the earth's surface and have an average depth of 3.8 km. The photic zone of seas and oceans, about 200 m deep, is the region permeated by sunlight where photosynthesis can take place. It has the greatest biodiversity, and all food for the marine population arises from the photic zone; such food includes marine snow which consists of globules of mucopolys-sacharides containing dead and living microorganisms floating downward toward the deep ocean. Marine organisms are adapted to the unique conditions found in the marine open sea (pelagic zone) environment: high salinity (3.5 g/L), low temperature (about 4°C), and high barometric pressure of up to 500 bar depending on the depth. Thermophilic organisms grow near the occasional hot thermal vents where hot magma spews out onto the ocean floor.

Using the technique of 16S rRNA, it has been found that over 70% of marine bacteria have not been cultured and hence have no counterparts among known bacteria. Microscopic cyanobacteria (picophytoplankton) make up 15% of all the bacteria. Among them, *Synechoccus* and *Prochlorococcus,* predominate and constitute the most abundant photosynthetic microbes on earth, contributing more than 50% of the total marine photosynthesis. Of the cultivated bacteria, *Roseobacter* spp. form about 15% of the total bacteria, while green non sulfur bacteria make up about 6%.

Keywords

Marine environment • Hydrothermal vents • Pelagic zone • Marine snow • Microbial loop • Redfield ratio • Nitrogen transformations in the ocean • *Pelagibacter ubique* • Marine viruses • Global marine nutrient recycling • Marine organisms and global climate change • Dimethyl sulfide (DMS) • Albedoes • Ammonnox

6.1 The Ocean Environment

Saline waters are waters with high concentrations of dissolved solutes, mostly salt, NaCl. The US Geological Survey classifies saline waters into three: Slightly saline, 1,000–3,000 ppm (1–3 g/L), moderately saline, 3,000–10,000 ppm (3–10 g/L) and highly saline water 10,000–35,000 ppm of solutes (10–35 g/L). Seawater has a salinity of about 35 g/L (Table 6.1).

The seas and oceans constitute planet earth's principal component of the hydrosphere: A major body of saline water that, in totality, covers about 71% of the

Table 6.1 Concentrations of the 11 most abundant constituents in sea water (From Allaby and Allaby 1990. With permission)

Constituent	Ion symbol	Parts per thousand by weight (g/kg)	Percentage of dissolved material
Chloride	Cl⁻	18.980	55.05
Sodium	Na⁺	10.556	30.61
Sulfate	SO_4^{2-}	2.649	7.68
Magnesium	Mg^{2+}	1.272	3.69
Calcium	Ca^{2+}	0.400	1.16
Potassium	K⁺	0.380	1.10
Bicarbonate	HCO_3^-	0.140	0.41
Bromide	Br⁻	0.065	0.19
Borate	$H_3BO_3^-$	0.026	0.07
Strontium	Sr^{2+}	0.008	0.03
Fluoride	F⁻	0.001	0.00
Total		34.447	99.99

earth's surface (or an area of some 361 million square kilometers). The average depth of the oceans is 3.8 km, but a number of deep sea trenches exist. The deepest sea trench is Marianas Trench, 11 km deep, in the Pacific Ocean. Though somewhat arbitrarily divided into several "separate" oceans, these waters comprise one global, interconnected body of salt water often referred to as the World Ocean or global ocean. The major oceanic divisions are defined in part by the continents, various archipelagos, and a number of other criteria; these divisions are (in descending order of size) the Pacific Ocean, the Atlantic Ocean, the Indian Ocean, the Southern Ocean (which is sometimes subsumed as the southern portions of the Pacific, Atlantic, and Indian Oceans), and the Arctic Ocean (which is sometimes considered a sea of the Atlantic). Smaller regions of the oceans are called seas, gulfs, bays, and other names (see Fig. 1.6).

There are also some smaller bodies of salt water that are inland and not interconnected with the World Ocean; e.g., the Caspian Sea, the Aral Sea, and the Great Salt Lake. These are not considered to be oceans or parts of oceans, though some are called "seas."

The ocean floor contains large mountain ridges and most, and about 90%, of the earth's volcanic activity take place in these undersea mountains.

The total mass of the hydrosphere is about 1.4×10^{21} kg, which is about 0.023% of the earth's total mass. Less than 2% of the earth's waters or hydrosphere is freshwater; the rest is saltwater, mostly found in the oceans. Some features peculiar to the ocean environment are described below.

1. *Hydrothermal vents* are hot water fountains which occur on the sea floor. They continuously gush out super-hot, mineral-rich water that supports a diverse community of organisms. Although most of the deep sea is sparsely populated, vent sites teem with a wide array of life, including bacteria. Hydrothermal vents were discovered in 1977 in the Pacific Ocean, and have since been found in the Atlantic, the Indian, and the Arctic Oceans. They occur at depths of about 2,100 m in areas of seafloor spreading along the Mid-Ocean Ridge system – the underwater mountain chain that occurs around the globe. They form when the huge plates that form the earth's crust move apart, causing deep cracks in the ocean floor. Seawater seeps into these openings and is heated by the molten rock, or magma, beneath the crust. When the hot springs gush out into the ocean, their temperature may be as high as 360°C, but the water does not boil because it is under so much pressure from the tremendous weight of the ocean above. Hydrothermal vents support the growth of many organisms which live in complete darkness, including many thermophilic bacteria and Archae which grow at temperatures as high as 112°C.

2. *Cold seeps* (also called *cold vents*) are areas of the ocean floor where hydrogen sulfide, methane, and other hydrocarbon-rich fluid seepage occur. Cold seeps are distinct from hydrothermal vents in that their temperature is same as the surrounding sea water. Chemoautotrophic Archae and bacteria, utilize sulfides and methane therein for energy and

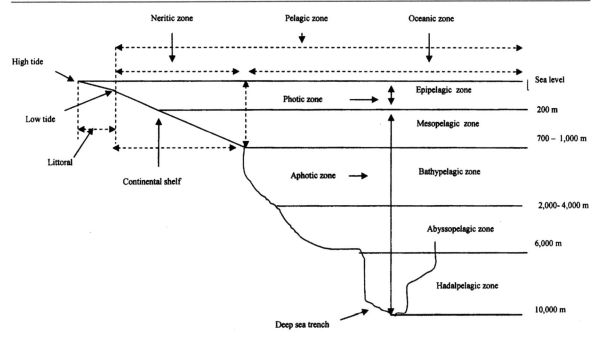

Fig. 6.1 Schematic illustration of ecological zones of the oceans, showing marine microbial habitats (Not drawn to scale)

growth. A unique biological community subsisting in the dark develops through the use of chemosynthesis to produce chemical energy. Higher organisms, namely vesicomyid clams and siboglinid tube worms, which harbor these bacteria, use this energy to power their own life processes, and in exchange provide both safety and a reliable source of food for the bacteria. Other bacteria form mats, blanketing sizable areas in the process.

Unlike hydrothermal vents, which are volatile and ephemeral environments, cold seeps emit at a slow and dependable rate. Perhaps owing to the differing temperatures and stability, cold seep organisms are much longer-lived than those inhabiting hydrothermal vents. Indeed, the seep tube-worm *Lamellibrachia luymesi* is believed to be the longest living noncolonial invertebrate known, with a minimum lifespan of between 170 and 250 years.

Cold seeps were first discovered in 1984 in the Gulf of Mexico at a depth of 3,200 m. Since then, seeps have been discovered in other parts of the world. The deepest seep community known is found in the Japan trench at a depth of 7,326 m.

3. *The continental shelf* surrounds the continents and is a shallow extension of the landmass of a continent. This shelf is relatively shallow, tens of meters

deep compared to the thousands of meters of depth in the open ocean, and extends outward to the continental slope where the deep ocean truly begins. Sediment from the erosion of land surfaces, washed into the sea by rivers and waves, nourishes microscopic plants and animals. Larger animals then feed upon them. These larger animals include the great schools of fish, such as tuna, menhaden, cod, and mackerel, which humans catch for food. The continental shelf regions also contain the highest amount of benthic life (plants and animals that live on the ocean floor). Combined with the sunlight available in shallow waters, the continental shelves teem with life compared to the biotic desert of the oceans' abyssal plain. The pelagic (water column) environment of the continental shelf constitutes the neritic zone, and the benthic (sea floor) province of the shelf is the sublittoral zone (see Fig. 6.1).

The continental slope connects the continental shelf and the oceanic crust. It begins at the continental shelf break (see Fig. 6.1), or where the bottom sharply drops off into a steep slope. It usually begins at 430 ft (130 m) depth and can be up to 20 km wide. The continental slope, which is still considered part of the continent, together with the continental shelf is called the continental margin.

Beyond the continental slope is the continental rise. As currents flow along the continental shelf and down the continental slope, they pick up and carry sediments along and deposit them just below the continental slope. These sediments accumulate to form the large, gentle slope of the continental rise.

Most commercial exploitation of the sea, such as oil and gas extraction, takes place on the continental shelf. Sovereign rights over their continental shelves were claimed by the marine nations that signed the Convention on the Continental Shelf drawn up by the UN's International Law Commission in 1958 partly superseded by the 1982 United Nations Convention on the Law of the Sea.

4. The *zones of the seas and oceans* are as follows: The *pelagic zone* (see Fig. 6.1) is the part of the open sea or ocean that is not near the coast or sea floor. In contrast, the demersal zone comprises the water that is near to (and is significantly affected by) the coast or the sea floor. The pelagic zone (also known as the open-ocean zone) is further subdivided, creating a number of subzones. These subzones are based on their different ecological characteristics, which roughly depend on their depth and the abundance of light. The subzones of the pelagic zone are as follows:

(a) *Epipelagic* (from the surface down to around 200 m) – the illuminated surface zone where there is enough light for photosynthesis. Due to this, marine plants and animals are largely concentrated in this zone. Here, one will typically encounter fish such as tuna and many sharks. This zone is also known as the *photic (sunlight) zone.* This is the region where the photosynthesis most commonly occurs and therefore contains the largest biodiversity in the ocean, including bacteria. Any life found in regions of the sea lower than the photic zone must either rely on material floating down from above known as *marine snow* (see below). Zones of the seas and oceans lower than the photic, or 200 m, are the *aphotic zone.*

(b) *Mesopelagic* (from 200 m down to around 1,000 m) – the twilight zone. Although some light penetrates this deep, it is insufficient for photosynthesis. The name stems from Greek for *middle.* Its lowermost boundary has a temperature of about10°C, and, in the tropics generally lies between 700 and 1,000 m.

(c) *Bathypelagic,* from the Greek for *deep* (from 1,000 m down to around 4,000 m) – by this depth, the ocean is almost entirely dark (with only the occasional bioluminescent organism). There are no living plants, and most animals survive by consuming the "snow" of detritus falling from the zones above, or (like the marine hatchetfish) by preying upon others. Giant squid live at this depth, and here they are hunted by deep-diving sperm whales. The temperature lies between 10°C and 4°C.

(d) *Abyssopelagic* (from 4,000 m down to above the ocean floor) – no light whatsoever penetrates to this depth, and most creatures are blind and colorless. The name is derived from the Greek for *abyss,* meaning bottomless (because the deep ocean was once believed to be bottomless).

(e) *Hadopelagic* (the deep water in ocean trenches) – the name is derived from *Hades,* the classical Greek underworld. This lies between 6,000 m and 10,000 m and is the deepest oceanic zone. This zone is relatively unknown and very few species are known to live here (in the open areas). However, many organisms live in hydrothermal vents in this and other zones.

The bathypelagic, abyssopelagic, and hadopelagic zones are very similar in character, and some marine biologists put them into a single zone or consider the latter two to be the same. Some define the hadopelagic as waters below 6,000 m, whether in a trench or not.

The pelagic (open sea) zone can also be split into two subregions, the *neritic* zone and the *oceanic* zone. The neritic encompasses the water mass directly above the continental shelves, while the oceanic zone includes all the completely open water. In contrast, the *littoral* zone covers the region between low and high tide and represents the transitional area between marine and terrestrial conditions. It is also known as the *intertidal* zone because it is the area where tide level affects the conditions of the region.

5. *Marine snow* is found in the deep ocean. It is a continuous shower of mostly organic detritus falling from the upper layers of the water column. This continuous shower appeared to deep sea divers like flakes of snow, hence the name. Marine snow is a mucopolysaccharide matrix, extracellular product released marine organisms, especially bacteria and phytoplankton, in which living and dead organisms and their parts are embedded. The origin of marine snow lies in activities within the productive photic (epipelagic) zone. Consequently, the prevalence of

marine snow changes with seasonal fluctuations in photosynthetic activity and ocean currents. Thus marine snow is heavier in the spring, and the reproductive cycles of some deep-sea animals are synchronized to take advantage of this (Anonymous 2010a).

Many marine snow "flakes" are sticky and fibrous like a crumbled spider net, and particles easily adhere to them, forming aggregates. Marine snow is composed of tiny leftovers of animals, plants (plankton), and non-living matter in the ocean's sun-suffused upper zones. Among these particles are chains of diatoms, shreds of protozoan mucous food traps, soot, fecal pellets from upper ocean zooplankton, dust motes, radioactive fallout, sand grains, pollen, and microorganisms which live inside and on top of the flakes. An aggregate begins to sink when it attracts fecal pellets, foraminifera (coats) tests, airborne dust, and other heavier particles. Many zooplankton fecal pellets are covered with a thin coating material. Although individual particles sink very slowly or are even buoyant, when they are bundled into a tight package and ballasted with particles of calcite – one of the densest materials produced in the ocean – they sink as rapidly as 100–200 m a day. As it descends, more suspended particles are added, making the aggregate even heavier and thus faster moving. An aggregate may break apart, spilling its contents into the water, but soon the spilled particles are picked up or "scavenged" by other falling aggregates. Thus aggregates are reorganized constantly with individual particles jumping on and off them before they arrive on the ocean floor. Meanwhile, a large portion of the organic matter in marine snow is recycled by microorganisms and upper and middle water column animals which again generate fecal pellets. Local concentrations of protozoa, principally bacterivorous flagellates and *ciliates*, are found associated with "marine snow."

The "snowflakes" (which are more like clumps or strings) are aggregates of smaller particles held together by a sugary mucus, transparent exopolymer particles (TEPs); natural polymers exuded as waste products by bacteria and phytoplankton. These aggregates grow over time and may reach several centimeters in diameter, traveling for weeks before reaching the ocean floor. Marine snow is everywhere in the ocean, and sometimes, it reaches blizzard proportions, and divers cannot see beyond a few feet

Most organic components of marine snow are consumed by microbes, zooplankton, and other filter-feeding animals within the first 1,000 m of their journey. In this

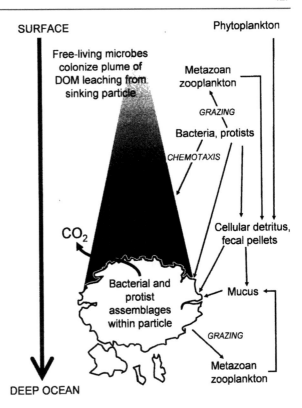

Fig. 6.2 Formation of and fate of marine snow (From Munn 2004. With permission)

way, marine snow may be considered the foundation of deep-sea mesopelagic and benthic ecosystems. Because sunlight cannot reach them, deep-sea organisms rely heavily on marine snow as an energy source. The small percentage of material not consumed in shallower waters becomes incorporated into the muddy ocean floor, where it is further decomposed through biological activity. Bacteria transported within the flakes may exchange genes with what were previously thought to be isolated populations of bacteria inhabiting the breadth of the ocean floor. The ocean bacteria are also now being exploited as sources of new pharmaceutical bioactive products (Fig. 6.2).

6.2   Some Properties of Sea Water

6.2.1   Salinity

Sea water is slightly alkaline (pH 7.5–8.4), with numerous chemicals, organic and inorganic, and gases dissolved in it. The concentration of these varies according to geographic and physical factors and is generally referred to as salinity. Salinity is defined as the mass of materials dissolved in 1 kg of sea water (denoted as salinity per thousand, 0/00). The global average salinity of ocean waters is about 35 g/kg. Oceans in subtropical regions have higher salinity due to higher evaporation, while those in tropical areas are lower due to dilution by higher rainfall. In coastal areas, salinity is diluted by runoffs and rivers. The major ions of sea water are sodium, chloride, magnesium, calcium, and potassium (see Table 6.1).

Salinity, along with temperature, affects the density and thus stability of the water column. In turn, this greatly affects many biological processes in the upper ocean. Saltier water is more dense and thus tends to sink below fresher water. The source of a water mass can be determined from its salinity and temperature. Ocean salinity is measured, either in situ by measuring conductivity or sea water through an instrument lowered from a ship, or by chemical analysis when water is brought to the laboratory.

6.2.2   Temperature

The temperature of deep sea water at 35/00 is −1.9°C. Oxygen and CO_2 are more soluble in cold water and are more abundant at 10–20 m of the sea and this affects the living things in the ocean. The concentration of these gases increases with depth until about 1,000 m when anaerobic conditions set in.

The surface temperatures of sea water in tropical regions can be as high as 25–30°C giving rise to temperature differences in density between surface waters at deep sea waters; the temperature dropping to about 10°C at 150–200 m. In Arctic regions, the water remains cold for most of the year; temperate regions show great variations in the temperature, being warmer in summer and cold in winter

The temperature is fairly uniform at the top layers of the sea, about 20–30°C, depending on the part of the world; this top layer is known as the *mixed layer*. In the deep sea, the temperature is low and fairly uniform. The transition zone between the mixed layer and the deep sea is the thermocline. In the thermocline, the temperature decreases rapidly from the mixed layer temperature to the much colder deep water temperatures, which vary from 0°C to about 3°C. The thermocline varies with latitude and season; it is permanent in the tropics, variable in the temperate climates (strongest during the summer), and weak to nonexistent in the polar regions, where the water column is cold from the surface to the bottom. In the earth's oceans, 90% of the water is below the thermocline (Anonymous 2003) (Fig. 6.3).

As will be seen below, temperature, along with salinity, affects the density and thus the stability of the

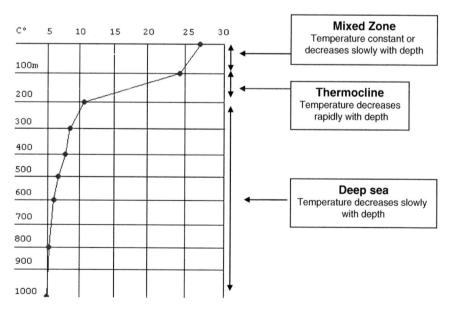

Fig. 6.3   The oceanic temperature profile showing temperatures in the mixed zone, the thermocline and the deep sea (Modified from http://www.windows2universe.org/earth/Water/temp.html; Anonymous 2010b)

water column. Warmer water is less dense and thus tends to stay on top of colder water. During winter, storm winds mix the water column, and the temperature is more uniform in the top – several 100 m. As spring approaches, increasing solar radiation warms the surface waters and this warmer, buoyant water stays on top. This increases the stability of the water column, preventing deeper, nutrient-rich water from being mixed into the surface from below. The stable surface layer keeps the planktonic cyanobacteria and algae near the surface where there is plenty of light. Under this situation, nutrients brought to the surface by winter mixing encourages the rapid growth of the organisms and spring bloom may occur.

Marine organisms are also adapted to live at different temperatures, which can thus determine the diversity or the numbers of organisms at different levels of the water column. As temperature changes with season

and location, the diversity and numbers of organisms also change.

Temperature data are recorded in situ with instrumentation lowered from a ship; the temperature is instantaneously recorded at various depths of the water column (Anonymous 2010b).

6.2.3   Light

Light has a major influence on the distribution of photosynthetic organisms. Light is generally limited to the upper 150–200 m or the photic region. Blue light has the deepest penetration in sea water and photosynthetic organisms at the lowest level of the photic region have mechanisms with which they are able to collect blue light. The relationship between light in the ocean with temperature and depth are depicted in Fig. 6.4.

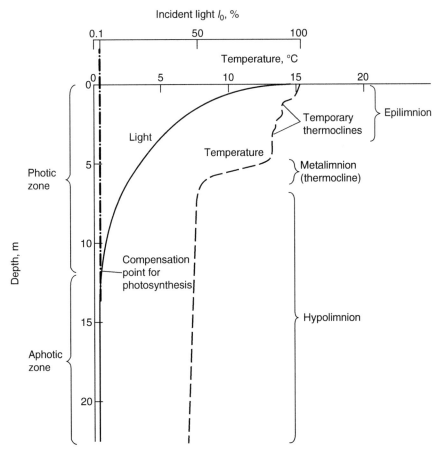

Fig. 6.4 Relationship between light, temperature and increasing depth in the ocean (Modified from http://www.windows2universe.org/earth/Water/temp.html; Anonymous 2010b)

6.2.4   Nutrients

Dissolved organic matter (DOC) is an important source of nutrition for marine bacteria; most of the organic carbon in the sea is in a dissolved form. As can be seen from the microbial loop concept, much of the DOC results from materials released when viruses attack microorganisms (Pomeroy 1974).

In addition to DOC, sea water contains many floating and sinking particles. Collectively, these tiny particles contain large amounts of carbon (particulate organic carbon, POC), and nitrogen (particulate organic nitrogen, PON), which supply nutrients to marine organisms.

Because light only penetrates a few hundred meters into the sea, no photosynthetic organisms occur in the ocean's permanently dark depths. Thus there are no organisms in these regions to remove nutrients from the water, and the ocean's deeper waters tend to be enriched in nutrients compared to its surface waters. Upwelling brings nutrients from the ocean depths to the sunlit surface waters where they can be used by photosynthesis. In some areas of the world's oceans for example, the Sargasso Sea, nutrients are not replenished continually, and phytoplankton can often use them up. These regions become nutrient-poor "ocean deserts" at certain seasons of the year.

6.2.5   Oxygen and CO$_2$ in the Marine Environment

Oxygen is needed by aerobic living things in the seas and oceans, from bacteria to fish to whales. Oxygen enters the ocean in two main ways: It diffuses into the ocean surface from the atmosphere, and it is produced by photosynthetic marine organisms. The amount of oxygen in the water is controlled by the temperature, as well as by the quantity produced photosynthetically.

Oxygen data are collected by chemical sensors, which work by binding to oxygen molecules. The bound oxygen can be determined by titration.

Carbon dioxide is a greenhouse gas that diffuses into the ocean surface from the atmosphere, and is taken up by marine photosynthetic organisms. The magnitude of the greenhouse effect depends on the amount of carbon dioxide in the atmosphere and how much carbon dioxide the oceans can take up from the atmosphere. Uptake of carbon dioxide by the oceans is

affected by the temperature of the water because colder water holds more gas, and by the abundance of photosynthetic organisms.

6.2.6   Sea Sediments

Billions of tons of sediment accumulate in the ocean basins every year. The nature of such sediments may be indicative of climate conditions near the ocean surface or on the adjacent continents. Sediments are composed of both organic and inorganic materials.

The organic component of sea sediment includes the remnants of sea-dwelling microscopic plankton, which provide a record of past climate and oceanic circulation. For example, by studying the chemical composition of plankton shells, we can reveal information about past seawater temperatures, salinity, and nutrient availability. Indeed, such techniques have been used to reconstruct ocean temperatures over the last 100 Ma, and have confirmed continental drift theories of climate change that a long term global cooling has taken place since the extinction of the dinosaurs.

Most inorganic material comes from adjacent landmasses, eroded from rocks and washed down to the coast by river channels, or blown from soils, dusty plains, and deserts. The nature and abundance of inorganic materials provides information about how wet or dry the nearby continents were, and the strengths and directions of winds.

6.3   Microbial Ecology of the Seas and Oceans

In the open ocean, far from the influences of coastal human habitation, sea water still contains huge numbers of microbes: bacteria, archae, protozoa, algae, fungi, and viruses. The presence and significance of these organisms will be discussed below. Coastal areas can contain even greater concentrations.

6.3.1   Bacteria

Using the technique of 16S rRNA it has been found that over 70% of marine bacteria have not been cultured and hence have no counterparts among known bacteria. As shown in Fig. 6.5, SAR11, SAR116,

Fig. 6.5 Frequency of the most common marine bacteria based on 16 S rRNA (From Giovannoni and Rappe 2000. With permission)

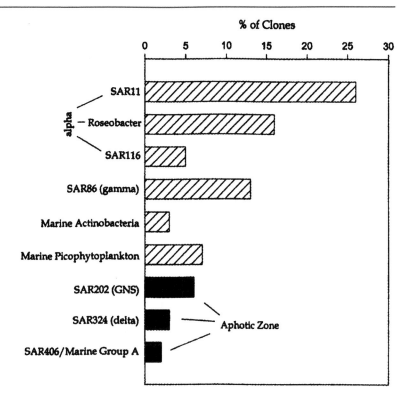

SAR86, SAR324, and SAR406, all of which do not have known cultivated counterparts make up about 70% of the bacteria. Microscopic cyanobacteria (picophytoplankton) make up 15% of all the bacteria. Of the cultivated bacteria, *Roseobacter* spp. form about 15% of the total bacteria, while green non sulfur bacteria make up about 6%. About 90% of the bacteria are Gram negative; Gram positive Actinobacteria form 3% of the total. Most of the sea bacteria belong to the Proteobacteria: SAR11, Roseobacter, and SAR116 belong to the "α" sub group, while SAR86 and SAR324 belong to "γ" and "δ" sub groups respectively (Irenewagner-D Obler and Biebl 2006).

SAR11 is ubiquitous and widely distributed at all levels of the pelagic zone from the shallow coastal waters to depths of about 3,000 m. SAR116 appears to be confined to the upper portion of the oceans.

SAR-11 was isolated from the Sargasso Sea in 1990 using the r RNA gene. It is said to be so dominant in the sea (about 10^{28}/ml) that its combined weight is more than that of the fishes put together. It was cultivated in 2002 and tentatively designated as a single species, *Pelagibacter ubique*. It belongs to the α-Proteobacteria, and the Order Rickettsiales. It is one of the smallest cells known, being only 0.37–0.89 μm long and 0.12–0.20 μm in diameter. It has very few genes (1,354) while humans have 18,000–25,000. It has no superfluous genes and uses base pairs with less nitrogen, since nitrogen is relatively difficult to obtain by biological objects. All these make the organism very efficient (Fig. 6.5).

Most of the bacteria are in the photic zone, while SAR202 (green non sulfur bacteria), SAR324 and SAR406 are confined to the aphotic zone.

Of the cyanobacteria found in marine environments, two genera, *Synechoccus* and *Prochlorococcus,* predominate and constitute the most abundant photosynthetic microbes on earth, contributing more than 50% of the total marine photosynthesis. *Prochlorococcus* occurs ubiquitously in surface waters between latitudes 40°N and 40°S. *Synechococcus* occurs more widely, but it decreases in abundance beyond 14°C; *prochlorococcus* is about ten times more abundant than *Synechococcus*. Some cyanobacteria encountered in the marine environment are depicted in Fig. 6.6.

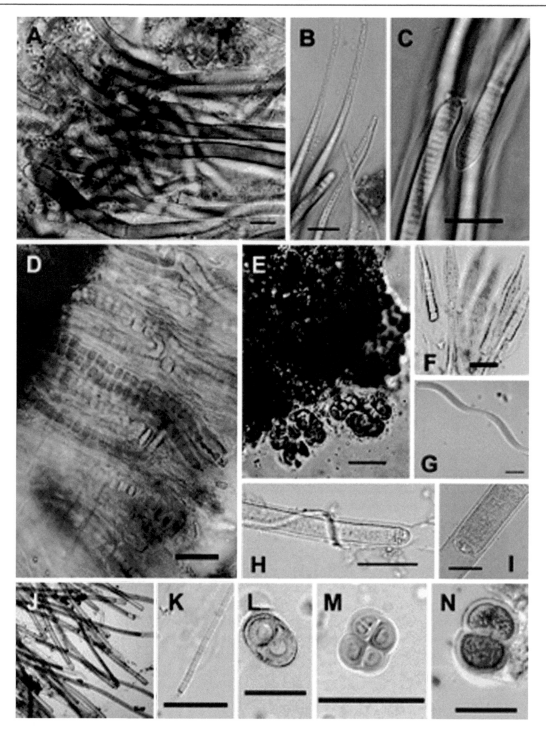

Fig. 6.6 Cyanobacteria found in a tropical marine environment. (**a**) *Calothrix* species 1, (**b**) *Calothrix* species 1, (**c**) *Blennothrix* sp., (**d**) *Kyrtuthrix* sp., (**e**) *Enthophysalis* species 1, (**f**) *Calothrix* species 2, (**g**) *Arthrospira* sp., (**h**) *Lyngbya* species 2, (**i**) *Lyngbya* species 1, (**j**) *Lyngbya* species, (**k**) *Pseudanabaenaceae* family (?), (**l**) *Gloeocapsa* sp., (**m**) *Chroococcidiopsis* sp (?), (**n**) *Chroococcus* sp. Scale bar = 20 μm. (j) Scale bar = 50 μm (From Díez et al. 2007. With permission)

6.3.2 Archae

Archae are divided into, Euryacheota and Crenarcheota. Several groups of Archae have been found in the sea.

Euryacheota contains methanogens, and hyperthermophilic and hyperhalophilic members.

Methanogens are strict anaerobes and produce methane. Thermophilic methanogens are found in thermophilic vents in the deep sea. *Methanococcus jannaschi* and *Methanococcus pyrus* are found in hydrothermal vents; the latter are among the most thermophilic organisms known, being able to grow at 110°C.

Hyperthermophilc Archae have optimal temperatures of growth of 100°C. They include *Thermococcus celer* and *Pyrococcus furiosus*.

Hyperhalophilic Archae can grow in salt concentrations of more than 9%. Examples are *Halobacterium, Halococcus*, and *Halomegaterium*.

The Crenarcheota are also found in thermophilic vents. *Desulfurococcus* is found in the upper layers of thermophilic vents, where the temperature is highest and is the most thermophilic organism known, being able to grow at 113°C.

6.3.3 Fungi

Fungi of all classes have been encountered in the marine environment, from Phycomycetes through Ascomycetes and Deuteromycetes to Basidiomycetes. In nearly all the cases, they are found attached to dead matter and in some cases, living matter, occasionally as parasites. Thus, the Phycomycete *Atkinsiella dubia,* has been found parasitizing eggs of crabs, while a strain of the plant pathogenic Phycomycete, *Pythium* sp, has been found growing on the marine brown red alga, *Porphyra*. When cultivated in the lab, many marine phycomycetes fail to complete their life history unless sea water or a high (4%) salt concentration is used.

The other three fungal groups Ascomycetes, Fungi Imperfecti, and Basidiomycetes occur in the marine environment in the above order of abundance on live plants or inanimate debris. In the mangrove swamp of *Rhizophora apiculata,* there is vertical distribution of different fungi. Thus some fungi are limited to upper zones of tidal flow, such as *Pyrenographa xylographoides, Julella avicennia,e* and *Aigialus grandis*, while others are found at lower reaches of the tidal ebb such as *Trichocladium achrasporum* and *T. alopallonellum*.

Around the world in both temperate and tropical regions, numerous fungi in the three groups have been found in the order given above on detritus in the intertidal regions of coastal areas on leaves, seaweeds, seagrass, chitinous substrates, even on sand (the *arenicolous* or sand-dwelling fungi), but most frequently on decaying wood. Some of the fungi encountered include *Torpedospora radiate, Antennospora quadricornuta, Clavatospora bulbosa, Crinigera maritima, Periconia prolifica*, and *Torpedospora radiata*.

As has been seen, the most abundant filamentous marine fungi are Ascomycetes. Marine Ascomycetes are peculiar in that their spores show adaptation to the marine ecosystem in the production of appendages, which facilitate buoyancy in water, entrapment, and adherence to substrates. The filamentous ascomycete *Halosphaeria mediosetigera* and the deuteromycete *Culcitalna achraspora* are designated marine and are able to grow in natural and artificial seawater media. Marine fungi are generally able to grow on woody materials in the ocean (see Fig. 6.7).

Fungi are the principal degraders of biomass in most terrestrial ecosystems. Fungi are usually found on drifting wood in the oceans or in the interstitial regions. They decompose the wood making it more available to other inhabitants of the marine ecosystem.

In contrast to surface environments, however, the deep-sea environment (1,500–4,000 m; 146–388 atm) has been shown using fungal-specific 18S rRNA gene analysis to contain very few fungi, which occur mainly as yeasts.

Culturable fungi recovered from sediments and bottom of the deep sea have been found to require salt concentrations of up to 4% and barometric pressures of up to atm 500 bar hydrostatic pressures at 5°C. Among them are strains of the Deuteromycete *Aspergillus sydowii* and the phycomycete *Thraustochytrium globosum*.

6.3.4 Algae

Marine algae vary from tiny microscopic unicellular forms of 3–10 μm (microns) to large macroscopic multicellular forms up to 70 m long and growing at up to 50 cm per day, kwon as seaweeds. Seaweeds include Green algae (Chlorophyta), Brown algae (Phaeophyta), Red algae (Rhodophyta) and some filamentous Blue-green algae (Cyanobacteria). Most of the seaweeds are red (6,000 species) and the rest known are brown (2,000 species) or green (1,200 species). Seaweeds are used in

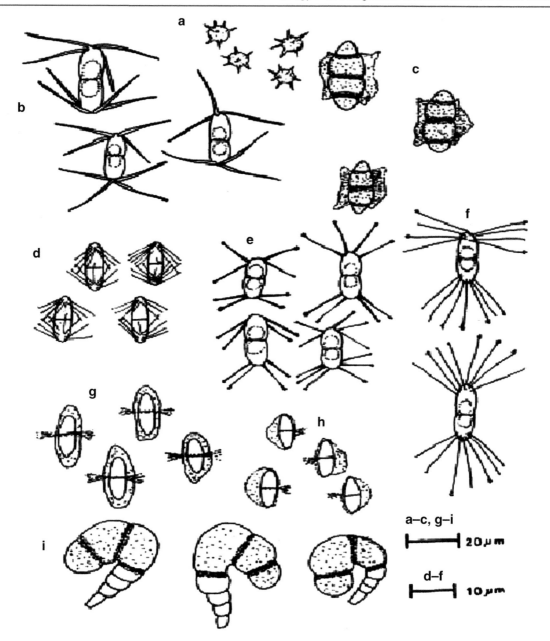

Fig. 6.7 Ascospores of fungi growing on wood in the marine environment. Note the structures for floatation and/or attachment. Ascospores of (**a**) *Amylocarpus encephaloides*, (**b**) *Arenariomyces majusculus*, (**c**) *Carbosphaerella leptosphaerioides*, (**d**) *Crinigera maritima*, (**e**) *Dryosphaera navigans*, (**f**) *Dryosphaera tropicalis*, (**g**) *Nimbospora bipolaris* and (**h**) *Nimbospora effusa*; (**i**) conidia of *Cirrenalia pseudomacrocephala* (From Prasannarai and Sridhar 2001. With permission)

many maritime countries as a source of food, for industrial applications and as a fertilizer. Nori (*Porphyra* spp.), a Japanese red seaweed, is very popular in the Japanese diet, has a high protein content (25–35% of dry weight), vitamins (e.g., vitamin C), and mineral salts, especially iodine. Industrial utilization is at present largely confined to extraction of phycocolloids, industrial gums classified as agars, carrageenans, and

alginates. Agars, extracted from red seaweeds such as *Gracilaria* , are used in the food industry and in laboratory media culture. In addition, some tuft-forming blue-green algae (e.g., the poisonous *Lyngbya majusculac*) are sometimes considered as seaweeds.

Seaweeds grow in marine environments, where there is sunlight to enable them carry out photosynthesis. Many of them contain air vacuoles to aid them in

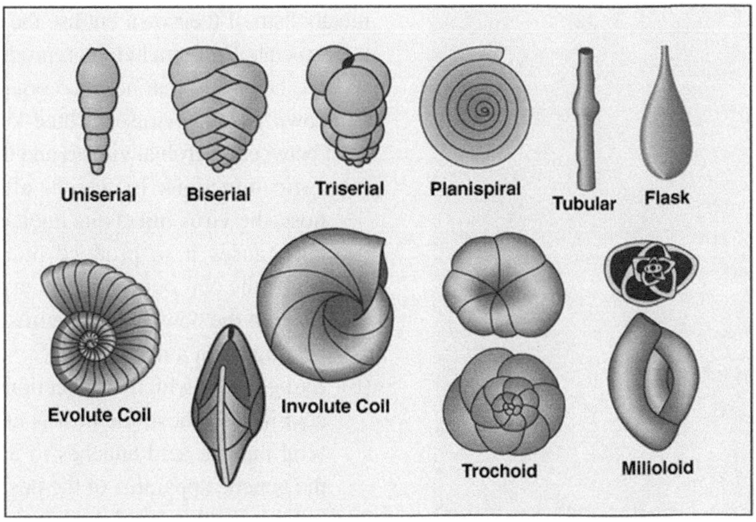

Fig. 6.8 Forminfera testa. (**1**, **2**) *Discorinopsis aguayoi* (Bermudez); (**3**, **4**) *Helenina anderseni* (Warren); (**5**, **6**) *Haplophragmoides wilberti* Andersen; (**7**) *Miliammina* Fusca; (**8**) *Polysaccammina ipohaiina* Scott (From Javaux and Scott 2003. With permission) Note: The illustrations above are the commoner of the patterns found in foraminifera testa. Some organisms follow a single test construction pattern throughout their life; others can change patterns during their life cycle, switching, for example, from uniserial to biserial chambering or from evolute to involute coiling

floatation. Often, they require a point to which they are attached; some, however, are free floating.

Diatoms are golden brown algae of the group Chrysophyta. They contain chlorophylls "a" and "c," and the carotenoid fucoxanthin. They have a distinctive cell wall with a top and bottom section as in a Petri dish. They are an important part of the primary producers of the colder parts of the world oceans (see Chap. 4).

Coccolithophorids are unicellular flagellated golden brown algae with chlorophylls "a" and "c" and the carotenoids diadinoxanthin and fucoxanthin. They are mostly marine and are found in tropical waters. They sometimes form heavy growths, blooms, during which they may clog the gills of fish. They also produce dimethyl sulfide (DMS), a foul-smelling compound which sometimes turns fish away from their normal migratory routes.

The *phytoflagellates* are algae which are motile with flagella, such as *Euglena*. *Dinoflagellates* are distinguished by having two flagella, one of which is a transverse flagellum that encircles the body in a groove; the other flagellum is longitudinal and extends to the rear. They also have vesicles under their cell membrane. They are classified as *Pyrrophyta*.

6.3.5   Protozoa

All groups of protozoa, except Sporozoa, (i.e., Sarcodina, Mastigophora, Cilophora, and Suctoria) are found in the marine environment; Sporozoa which are exclusively parasitic are not found (Fig. 6.8).

Sarcodina

Foraminifera: In the deep ocean, a group of Sarcodina which form shells (tests) (Fig. 6.8) and have fine radiating pseudopodia and known as foraminifera, are abound. These shells are made of calcium carbonate ($CaCO_3$). They are usually less than 1 mm in size, but some are much larger. Some have algae as endosymbionts. Foramanifera typically live for about a month.

Flagellates

Flagellates are protozoa which move with flagella and are classified as Mastigophora. Many flagellates are marine (see Fig. 6.9).

Suctoria

The suctoria are exclusively marine. The juvenile stage is a ciliate and moves about. The adult stage is sessile and catches food with tentacles.

Ciliates

Ciliates are protozoa with cilia, and are classified as *Ciliophora* (see Chap. 4). They posses two nuclei. Marine ciliates are large, about 20–80 μm with some as large as 200 μm. Ciliates are important in the marine food web because they ingest (graze) bacteria and other smaller organisms in the marine environment.

Fig. 6.9 Different flagella types among marine flagellates. A variety of marine flagellates from the various genera (*left* to *right*): *Cryptaulax, Abollifer, Bodo, Rhynchomonas, Kittoksia, Allas*, and *Metromonas* (From http://www.tolweb.org/notes/?note_id=50. With permission) Note: Marine phytoplankton Cryptomonad (Rhodomonas salina). Cryptomonads are flagellate protists, most of which have chloroplasts,and live in marine, brackish water, and freshwater. Cells are ovoid and flattened in shape with an anterior groove and two flagella used for locomotion. Because some contain chlorophyll, botanists treat them as a division of the Plant Kingdom, Cryptophyta, while zoologists place them in the Animal Kingdom as Cryptomonadida. (Image by Robert Folz, Visuals unlimited Inc. www.visualsunlimited.com reproduced with permission)

Some ciliates contain photosynthetic organisms as endosymbionts; they are able to obtain food by photosynthesis as well as by grazing and are said to be mixotrophic. Some ciliates do selective grazing, ingesting some organisms and leaving others. Freshwater protozoa regulate the water content of their bodies by the expansion and periodic collapse of their contractile vacuole, a vesicle which increases in size as it extracts water from the interior of the cell and collapses to nothing as it expels the extracted water. Because of the high osmotic pressure of the marine environment, contractile vacuoles are not observed in marine protozoa.

6.3.6 Viruses

Viruses are small particles, 20–30 nm long and made of either DNA or RNA, covered with a protein coat and sometimes also with lipid (see Chap. 4). Viruses have no

metabolism of their own but use the host mechanism for their metabolism, including reproduction. They attack specific hosts and marine microorganisms seem to have their own peculiar viruses. Three kinds of relationships exist between microbial viruses and their microbial hosts:

(a) *Lytic* infections in which, after attaching to its host, the virus injects its nucleic acid into the host and causes it to produce numerous viruses like itself; the cell bursts and releases the new viruses. Each of the new daughter viruses can start the process again in a new host.

(b) *Lysogeny*, in which on injection of the viral nucleic acid into the host, the host is not lysed. Rather, the viral nucleic acid attaches to and becomes part of the genetic apparatus of the host. It may be induced to become lytic by ultra violet and some chemicals.

(c) *Chronic relationship*, in which the new viruses do not lyse the host, but are released by budding over many generations.

Marine viruses can be both detrimental and beneficial to the ocean's health. Some viruses attack and kill plankton, eliminating the base of the ocean food chain in a particular area. At the same time, the dead plankton can become a source of carbon that is not otherwise readily available to other sea life. It is estimated that up to 25% of all living carbon in the oceans is made available through the action of viruses. When these aspects remain in proper balance, the ocean functions normally.

Until the 1990s, it was believed that since the oceans were then believed to deserts in terms of microorganisms, few viruses would be in the sea. Since then, using the transmission electron microscope and epiflorescence microscopy (see Chap. 2) coupled with uranium stains, viruses have been shown to occur up to 10^{10} per ml in sea water. The distribution of viruses follows the relative abundance of microorganisms along the water column in the ocean. Some viruses affecting marine organisms are shown in Table 6.2.

Viruses are important in the food economy of marine organisms because the materials released when they lyse their hosts contribute to the dissolved organic matter (DOM) and the particulate organic matter (POM) of the oceans. They also contribute to gene transfer among marine organisms (Fuhrman 2000).

6.3.7 Plankton

Plankton are drifting organisms that inhabit the water column of oceans and seas; they also occur in fresh

Table 6.2 Some viruses infecting marine organisms (From Munn 2004. With permission)

Virus family	Nucleic acid	Shape	Size (nm)	Host
Myoviridae	dsDNA	Polygonal with contractile tail	80–200	Bacteria
Podovirrdae, Siphoviridae	dsDNA	Icosahedral with noncontractile tail	60	Bacteria
Microviridae	ssDNA	Icosahedral with spikes	23–30	Bacteria
Leviviridae	ssRNA	Icosahedral	24	Bacteria
Corticoviridae, Tectiviridae	dsDNA	Icosahedral with spikes	60–75	Bacteria
Cystoviridae	dsRNA	Icosahedral with lipid coat	60–75	Bacteria
Lipothrixviridae	dsDNA	Thick rod with lipid coat	400	Archaea
SSV1 group	dsDNA	Lemon shaped with spikes	60–100	Archaea
Parvoviridae	ssDNA	Icosahedral	20	Crustacea
Caliciviridae	ssRNA	Spherical	35–40	Fish, marine mammals
Totiviridae	dsRNN	Icosahedral	30–45	Protozoa
Reoviridae	dsRNA	Icosahedral with spikes	50–80	Crustacea, fish
Birnaviridae	dsRNA	Icosahedral	60	Molluscs, fish
Adenoviridae	dsDNA	Icosahedral with spikes	60–90	Fungi
Orthomyxoviridae	ssRNA	Various, mainly filamentous	20–120	Marine mammals
Baculoviridae	dsDNA	Rods, some with tails	100–400	Crustacea
Phycodnaviridae	dsDNA	Icosahedral	130–200	Algae

Table 6.3 Plankton classification by size and their abundance in the marine environment (Modified from Munn 2004. With permission)

Size category	Size range (µm)	Microbial group
Femtoplankton	01–0.2	Viruses
Picoplankton	0.2–2	Bacteria, archae, some flagellates
Nanoplankton	2–20	Flagellates, diatoms, dinoflagellates
Microplankton	20–200	Ciliates, diatoms, dinoflagellates, other algae

Size (decreasing arrow at left), *Abundance* (increasing arrow at right)

water. They are important in the food webs of aquatic systems because they provide food for the biotic communities. Organisms which spend their entire life cycle free floating as part of the plankton such as most algae, copepods, salps, and jellyfish are *holoplankton*. Those that are only plankton for part of their lives, usually the larval stage, and later move either to the nekton (free swimming) or a benthic (sea floor) life, are *meroplankton*. Fish, marine crustaceans, starfish, sea urchins belong to this latter group.

The organisms discussed above: Viruses, bacteria, archae, protozoa, algae constitute plankton. Plankton are small and are usually classified by size rather than by their taxonomic composition. Based on size, plankton are grouped into femtoplankton, picoplankton, nanoplankton, and microplankton (see Table 6.3). Some plankton engulf others of about their size. As seen from Table 6.3, the most abundant plankton are the smallest in size, while the largest in size are the fewest.

6.4 Unique Aspects of the Existence of Microorganisms in the Marine Environment

The marine environment has the following peculiarities: The temperature is low, except in the thermophilic vents; the pressure is high; and nutrients are sparse. Microorganisms existing under these conditions have adapted to the conditions

6.4.1 Low Temperature

The differences in temperature between the photic (or sunlit) zones nearer to the surface and the deep sea are dramatic. Temperatures vary more in the waters of the mixed zone and thermocline than the deep sea as shown in Fig 6.3. In most parts of the deep sea, the water temperature is more uniform and constant. With the exception of hydrothermal vent communities where hot water is emitted into the cold waters, the deep sea temperature remains between 2°C and 4°C.

The decreasing temperature of sea water with depth is shown in Fig. 6.3. The temperature in the deep sea is about −1°C. Psychrophiles (bacteria with optimum temperature of 0–5°C, but maximum growth temperature of 15°C) and psychrotolerant (bacteria which can grow at 0°C, but have maximum growth at about 20–25°C) abound in the sea. To enable them to survive and grow in cold environments, psychrophilic bacteria have evolved a complex range of adaptations to all of their cellular components, including their membranes, energy-generating systems, protein synthesis machinery, biodegradative enzymes, and the components responsible for nutrient uptake.

6.4.2 High Pressure

Considering the volume of water above the deepest parts of the ocean, it is not surprising that pressure is one of the most important environmental factors affecting deep sea life. Pressure increases 1 atmosphere (atm) for each 10 m in depth. The deep sea varies in depth from 700 meters to more than 10,000 m; therefore, pressure ranges from 20 atm to more than 1,000 atm. On average, pressure ranges between 200 and 600 atm. Advances in deep sea technology now make it possible for samples under pressure to be collected in such a way that they reach the surface without much damage. Without this technology, the animals would die shortly after being collected and the absence of pressure would cause their organs to expand and possibly explode. Deep sea creatures have adapted to pressure by developing bodies with no excess cavities, such as swim bladders, that would collapse under intense pressure. The flesh and bones of deep sea marine creatures are soft and flabby, which also helps them withstand the pressure. Recent results indicate that the deep-sea strain bacteria express different DNA-binding factors under different pressure conditions.

6.4.3 Oxygen

The dark, cold waters of the deep sea are also oxygen-poor environments. Consequently, deep sea life requires little oxygen. Oxygen is transported to the deep sea from the surface where it sinks to the bottom when surface temperatures decrease. Most of this water comes from arctic regions. Surprisingly, the deep sea is not the most oxygen-poor zone in the ocean. The most oxygen-deficient zone lies between 500 and 1,000 m, where there are more species that require oxygen depleting the oxygen in this zone during respiration. In addition, the bacteria that feed on decaying food particles descending through the water column also require oxygen. Oxygen is never depleted in the deepest parts of the ocean because there are fewer animals to deplete the available oxygen.

6.4.4 Food/Nutrients

Deep sea creatures have developed special feeding mechanisms because of the lack of light and because food is scarce in these zones. Some food comes from the detritus, of decaying plants and animals from the upper zones of the ocean. The corpses of large animals that sink to the bottom provide infrequent feasts for deep sea animals and are consumed rapidly by a variety of species. The deep sea is home to jawless fish such as the lamprey and hagfish, which burrow into the carcass quickly consuming it from the inside out. Deep sea fish also have large and expandable stomachs to hold large quantities of scarce food. They do not expend energy swimming in search of food; rather, they remain in one place and ambush their prey.

6.4.5 Light

The deep sea ocean waters are as black as night. The deep is also known as the twilight zone. The only light is produced by bioluminescence, a chemical reaction in the creature's body that creates a low level light; so, deep sea life must rely on alternatives to sight. Many deep sea fish have adapted large eyes to capture what little light exists. Most often, this light is blue-green, but some creatures have also developed the ability to produce red light to lure curious prey. Lack of light also creates a barrier to reproduction. Bioluminescent light is also used to signal potential mates with a specific light pattern. Deep sea creatures are also often equipped with a powerful sense of smell so that chemicals released into the water can attract potential mates.

6.4.6   High Temperature

Thermophilic bacteria and archae are found near thermal vents, and active underwater volcanoes. The temperature in deep sea vents can be as high as 350°C. A high temperature is created as the hot water mixes with sea water and a variety of bacteria and archae develop in these gradients. Among the hyperthermophilic bacteria found in these regions are *Aquiflex pyrophilus* and *Thermotoga maritama* (bacteria) and *Pyrococcus furiosus* and *Pyrolobus fumarii* (Archae).

6.4.7   Size in Marine Microorganisms

Marine microorganisms are generally small because the marine environment is oligotrophic (low in nutrients). Small cells can absorb nutrients more efficiently (see Table 6.3). The surface area /volume ratio is important. Many microorganisms are 0.6 µm at their widest dimension, and most are less than 0.3 µm, a genetic adaptation due to starvation on account of the scarcity of food in the environment. The spherical shape is inefficient in the absorption of nutrients in the sparse environment of the seas. On account of this, most marine bacteria are elongated. In addition, many marine bacteria have invaginations or buds which increase surface areas leading to greater absorption. There are exceptions and some marine bacteria are large. It is thought that these large bacteria have invaginations in their cell membranes which help increase their surface areas.

6.5   The Place of Microorganisms in the Food Chains of the Oceans and Seas

Until recently, it was thought that microorganisms did not play any role in the ecological hierarchies of the biotic life of the sea and oceans. It was thought that microorganisms were few especially because the open seas and oceans were oligotrophic (i.e., they contain little nutrients); if they played any part at all, it was to breakdown detritus (Azam et al. 1983).

The earlier oceanic (see Fig. 6.10) food chain formulated for such conditions in the photic section of the pelagic region excluded microorganisms and was based on primary producers being photosynthetic plankton, including the following: Diatoms (yellow-brown), single-celled algae typically about 30 µm in diameter and dominant in temperate and polar waters; dinoflagellates, dominant in subtropical and tropical seas and oceans and ranging from 30 to 200 µm (2 mm); coccolithophores dominant in tropical waters, typically 5–10 µm, and cyanobacteria.

The phytoplanktons are eaten by very small floating animals, zooplankton, such as copepods. The zooplankton are eaten by larger zooplankton such as shrimps, fish larvae, and jelly-fish.

The zooplankton are eaten by small fish such as sardines and herrings; these fish are eaten by larger fish. Some of these are birds, fish, and marine mammals.

At the top of the marine food web are the large predators including tuna, seals, and some species of whales.

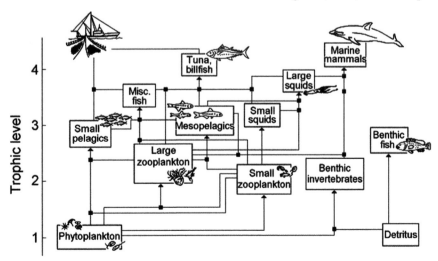

Fig. 6.10 Diagrammatic illustration of the marine food chain without microbial interaction (From Pauly 2007. With permission)

In more recent times (see Fig. 6.11) and following the use of newer techniques including epifluorescence light, confocal laser scanning microscopy, flow cytomtery, but especially molecular methods, it has been found that there are far more bacteria than previously known. Furthermore, it is also known that bacteria are central to the biotic activities of oceans and seas.

Modern studies, especially the use of controlled pore-size filters and fluorescent DNA stains, showed that the oceans are rich in very small (picoplanktons, <2 μm) cyanobacteria, to the order of about 10^7 per ml. These methods revealed that two cyanobacterial genera, *Synechoccus* and *Prochlorococcus* constitute the most abundant photosynthetic microbes on earth, contributing more than 50% of the total marine photosynthesis. *Prochlorococcus* occurs ubiquitously in surface waters between latitudes 40°N and 40°S. *Synechococcus* occurs more widely, but it decreases in abundance beyond 14 °C; *Prochlorococcus* is about ten times more abundant than *Synechococcus*.

These methods also show that viruses are more important in the food economy of the oceans than originally recognized in the traditional food chain. Dissolved and particulate organic matter derived from the feces and excreta of fishes and breakdown materials from dead are consumed by bacteria.

Figure 6.11 shows the modern food web of oceans and seas in which the primary producers through photosynthesis are diatoms, dinoflagellates, and cyanobacteria. These are consumed by protozoa (bacterivorous flagellates, ciliates), which are in turn eaten by larger zooplankton (e.g., squids) which are then eaten by fish. Viruses play important parts in the modern food web; they lyse heterotrophic bacteria and Archae, cyanobacteria, and diatoms and flagellates. Dissolved and particulate organic matter which are so important in the nutrition of deep sea organisms are feces of fish and breakdown products from all the members of the community and occur in the sea snow already described (Azam et al. 1983; Stewart 2005).

In particular, modern food chains emphasize the *microbial loop* which illustrates the crucial role of microorganisms in recycling food in the marine environment (Figs. 6.11 and 8.12). The microbial loop is a term coined to describe a trophic pathway in aquatic environments where dissolved organic carbon (DOC) is reintroduced to the food web through the incorporation into bacteria. Bacteria are consumed mostly by

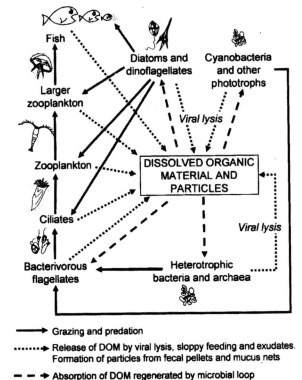

—→ Grazing and predation

·······▶ Release of DOM by viral lysis, sloppy feeding and exudates. Formation of particles from fecal pellets and mucus nets

− − ➔ Absorption of DOM regenerated by microbial loop

Fig. 6.11 Modern marine food webs. *Arrows* show transfer of organic matter (From Munn 2004. With permission)

flagellates and ciliates. They in turn, are consumed by larger aquatic organisms (for example small crustaceans like copepods).

The dissolved organic carbon (DOC) matter is introduced into aquatic environments from several sources, such as the leakage of fixed carbon from algal cells or the excretion of waste products by aquatic animals and microbes. DOC is also produced by the breakdown and dissolution of organic particles. In inland waters and coastal environments, DOC can originate from terrestrial plants and soils. For the most part, this dissolved carbon is unavailable to aquatic organisms other than bacteria. Thus, the reclamation of this organic carbon into food web results in additional energy available to higher tropic levels (e.g., fish). Because microbes are the base of the food web in most aquatic environments, the tropic efficiency of the microbial loop has a profound impact on important aquatic processes. Such processes include the productivity of fisheries and the amount of carbon exported to the ocean floor (Fig. 6.12).

Fig. 6.12 The microbial loop (From Schulz 2006. With permission)

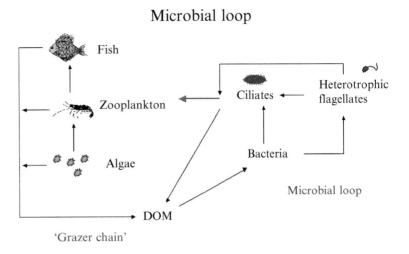

Microbial loop

'Grazer chain'

6.6   Marine Microorganisms and Their Influence on Global Climate and Global Nutrient Recycling

The sea and oceans occupy 71% of the earth's surface and 97% of all the waters on earth (i.e., the biosphere). The marine environment contains the greatest biological diversity up to 11 m in ocean waters and up to 400 m in the sediments. At the same time, it has been estimated that the biomass of microorganisms in seas and oceans is more than gigatons, many times more than other marine lives put together. On account of their capacity for rapid growth, and their diverse ability to bring about biochemical transformations, marine microorganisms are major movers of global nutrient cycles. It is not surprising that the activities of marine microorganisms have global effects in the area of global climate as well as the recycling of nutrients on a global scale. This section will discuss these important consequences of marine microbial activities on planet earth (Arrigo 2005).

6.6.1   The Influence of Marine Microorganisms on Global Climate and Global Nutrient Recycling

In microorganisms living in environments such as sea water where the osmotic pressure of the surrounding liquid is higher than that of the cell, the osmoregulation of the organisms is achieved by K^+, especially KCl, as well as one or more of a few low molecular weight organic compounds known as "compatible solutes" due to their compatibility in the cells where they are found. A list of selected compatible solutes is given in Table 6.4.

Apart from their osmoprotectant functions, they also have other uses in the cell. For instance, they act as cell reserves of carbon and nitrogen and are utilized when necessary. They also act to protect the cell against other forms of stress such as high or low temperature,

Table 6.4 Some compatible solutes (Modified from Welsh 2000. With permission)

Microbial group	Compatible solute
Algae	Sucrose
	Glycerol
	Mannitol
	Proline
	Glycine betaine
	Dimethylsulfoniopropionate
Cyanobacteria	Sucrose/trehalose
	Glycine betaine
	Glucosylglycerol
Phototrophic bacteria	Sucrose/trehalose
	Glycine betaine
Sulfate reducing bacteria	Trhalose
	Glycine betaine
Archaebacteria	Glycine betaine
	β-Glutamate

or desiccation. One compound which has been suggested as having general protective capabilities against many types of stress is trehalose. Another important compatible solute is dimethylsulfoniopropionate (DMSP), which is produced by some algae. DMSP is broken down to dimethyl sulfide and acrylate by the enzyme DMSP-lyase which is produced by the algae. It has been suggested that blooms formed by planktonic algae producing DMSP are not grazed by certain protozoa because of the DMS which the protozoa find unpalatable, and that the acrylate inhibits bacterial growth.

6.6.1.1 Global Marine Algal Sulfur Recycling, Dimethylsulfoniopropionate, Dimethyl Sulfide and Climate Change

DMSP is an osmoprotectant in some planktonic algae. One group of planktons, which contains DMSP as an osmoprotectant are the coccolithophorids, which belong to algal group, the Prymnesiophyta which are also known as Haptophyta. Coccolithophores are exclusively marine and are found in large numbers throughout the surface euphotic zone of the ocean. An example of a globally significant coccolithophore is *Emiliania huxleyi*. It is responsible for the release of significant amounts of dimethyl sulfide (DMS) into the atmosphere (Fig. 6.13).

release of DMS. Among bacteria which can breakdown DMSP into DMS are members of the *Roseobacter* group and those of the SAR-11 (*Pelagibacter ubique*) clade. The rate of DMS flux from the ocean to the atmosphere depends on its concentration in sea water (Welsh 2000).

DMS is the most important biologically produced sulfur compound in the marine atmosphere and it is essential in the global sulfur cycle. Gaseous DMS is photo-oxidized to sulfated aerosols in the atmosphere and a relationship has been established between DMS, sulfate aerosols, and cloud condensation nuclei.

The sulfate aerosols function as cloud condensation nuclei and on account of this, DMS has a significant impact on the earth's climate. Plankton production of DMS and its escape to the atmosphere is believed to be one of the mechanisms by which the microorganisms can regulate the climate.

The radiation balance has a fundamental effect on earth's climate. Approximately 33% of the solar radiation that reaches the earth is reflected back into space by clouds and from earth surfaces, such as ice and snow. The atmosphere absorbs some solar energy, but most of the other two thirds is absorbed by the land and oceans, which are warmed by the sunlight. The sun's energy is converted into heat, and the land and oceans then radiate a portion of this energy back as outgoing long-wave radiation (infrared), also known as terrestrial radiation. As this energy is radiated back

$$(CH_3)_2S^+CH_2CH_2COO^- \xrightarrow{\quad DMSP\text{-lyase}\quad} (CH_3)_2S + CH_2 = CHCOO^- + H^+$$

DMSP   DMS   Acrylate

DMSP is synthesized as a compatible solute by many algae as well as by aquatic angiosperms. It is released from plankton by damaged phytoplankton cells due to physical stress (e.g., turbulence, zooplankton grazing, or viral-lysis). It is subsequently transformed by phytoplankton and bacterial enzymes to DMS (see formula above). Many bacteria have DMSP-lyase and are thought to play a significant part in converting the algal DMSP to DMS; some bacteria actually consume the DMSP. Photochemical reactions and ultraviolet radiation can degrade DMS to further break down products, removing DMS. Under aerobic conditions, DMS can be oxidized by chemolithotrophic bacteria such as *Thiobacillus* sp. to CO_2 and sulfate. The various pathways by which DMSP may be metabolized in water are shown in Fig. 6.14, with ultimate

out, it warms the atmosphere and continues on into space. The amount of solar energy received by the earth, the planetary albedo (the amount reflected back) and the emitted terrestrial radiation, makes up the earth's radiation balance. If the earth receives more energy than it loses, the result is global warming, and if it loses more energy than it receives, the result is global cooling.

The oxidation of DMS by hydroxyl (Fig. 6.14) and nitrate radicals results in the formation of sulfate aerosols, which on advection into water saturated air cause cloud formation. Both increased cloud formation and dry sulfate aerosols increase planetary albedo resulting in a relative cooling effect. Dry deposition of sulfate aerosols and precipitation of sulfate enriched rainwater over the continents couples the marine and terrestrial

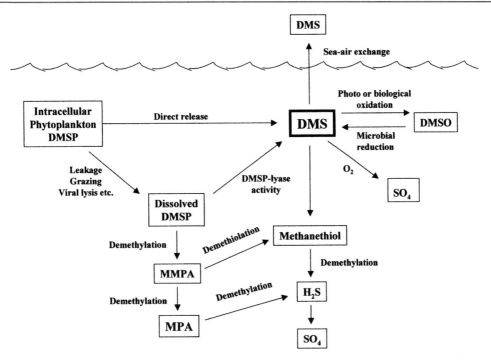

Fig. 6.13 Breakdown of DMSP in water. Microbial transformations involved in the turnover of DMSP and DMS in marine surface waters. Abbreviations: *DMSP* dimethylsulfoniopropionate, *DMS* dimethyl sulfide, *DMSO* dimethylsulfoxide, *MMPA* methylmercaptopropionate, *MPA* mercaptopropionate. DMS is the most important biologically produced sulfur compound in the marine atmosphere (From Welsh 2000. With permission)

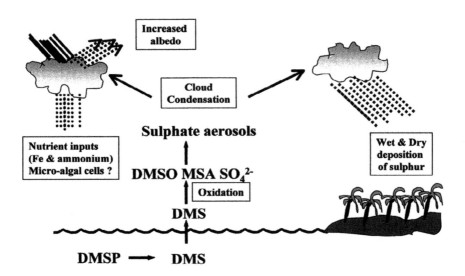

Fig. 6.14 Schematic representation of the processes involved in DMS oxidation, climate regulation, and coupling of the oceanic and terrestrial sulfur cycles. The oxidation of DMS by hydroxyl and nitrate radicals results in the formation of sulfate aerosols, which on advection into water saturated air cause cloud formation. Both increased cloud formation and dry sulfate aerosols increase planetary albedo resulting in a relative cooling effect. Dry deposition of sulfate aerosols and precipitation of sulfate enriched rainwater over the continents couples the marine and terrestrial sulfur cycles. Rainfall over the oceans may provide a feedback between DMS production by micro-algae and their productivity due to increased inputs of dissolved nutrients or by providing a dilute inoculum of micro-algal cells transported within the clouds. (From Welsh 2000. With permission)

Table 6.5 Reflectivity values of various surfaces (From Encyclopedia of Earth; http://www.eoearth.org/article/albedo)

Surface	Description	Albedo
Soil	Dark and wet	0.05–0.40
	Light and dry	0.15–0.45
Grass	Long	0.16
	Short	0.26
Agricultural crops		0.18–0.25
Tundra		0.18–0.25
Fresh asphalt		0.04
Forest	Deciduous	0.15–0.20
	Coniferous	0.05–0.15
Water	Small zenith angle	0.03–0.10
	Large zenith angle	0.10–1.0
Snow	Old	0.40
	Fresh	0.95
Ice	Sea	0.30–0.45
	Glazier	0.20–0.40
Clouds	Thick	0.60–0.90
	Thin	0.30–0.5

sulfur cycles. Rainfall over the oceans may provide a feedback between DMS production by micro-algae and their productivity due to increased inputs of dissolved nutrients or by providing a dilute inoculum of micro-algal cells transported within the clouds.

The word *albedo* comes from the Latin for white. The albedo of an object is the extent to which it reflects light. Typical albedos are given in Table 6.5.

Albedo is an important concept in climatology and astronomy. In climatology, it is sometimes expressed as a percentage. Albedo is an important factor in the radiation balance and clouds have major effect on albedo. The optical properties of a cloud are a key issue to understanding and therefore predicting global climate change. A cloud's optical properties are related to the size distribution and number of its droplets. The more cloud condensation nuclei, the smaller the size of its water droplets and the higher the density of water droplets since the same amount of water vapor is distributed among a greater number of CCN. This affects properties (reflectance, transmittance, and absorbance) of the cloud (Budikova 2010).

Clouds affect both incoming solar and outgoing thermal infrared fluxes; low thick clouds act as shields, blocking and reflecting solar radiation back into space which cools the planet, but high clouds can also trap outgoing heat (longwave radiation), warming the planet. Data indicate that clouds have an overall net cooling effect. The smaller droplet size will likely decrease precipitation, resulting in a longer lifetime for a cloud. Because the models had a poor ability to reproduce the effects of clouds, a priority was set to observe, measure, and learn about clouds' physical properties and radioactive fluxes. DMS may influence both the hydrologic cycle and the global heat budget through its part in cloud formation, and may alter rainfall patterns and temperatures

6.6.1.2 Carbon Recycling by Marine Algae and Reduction of Global Warming

Green house gases are those which stop the heat absorbed by the earth's surface from returning to outer space thus causing the earth's temperature to rise in the "green house effect." Photosynthetic microorganisms in the marine environment, including cyanobacteria fix atmospheric carbon converting it to carbohydrates at the starting point of the food chain, and thus reduce the CO_2 of the atmosphere and its contribution to global warming.

Some of the planktonic organisms, notably microalgae, such as members of the Prymnesiophyta (Haptophyta) including coccolithophores, convert some of the carbon to calcium carbonate in their shells. This formation of calcareous skeletons by marine planktonic organisms and their subsequent sinking to depth generates a continuous rain of calcium carbonate to the deep ocean and underlying sediments. This is important in regulating marine carbon cycling and ocean-atmosphere CO_2 exchange. A rise in the atmospheric CO_2 levels causes significant changes in surface ocean pH and carbonate chemistry. Such changes have been shown to slow down calcification in corals and coralline macroalgae, but the majority of marine calcification occurs in planktonic organisms.

Another way of reducing the CO_2 of the atmosphere and hence its green house effect is by increasing the photosynthetic activity of marine organisms through increasing their supply of nutrients. Seeding the oceans with iron appears a viable way to permanently lock carbon away from the atmosphere and potentially tackle climate change. In this process, several tons of iron are put into the ocean. Although there is some concern about the long term effect of the action, it is seen as a possible way to slow down global warming. Marine algae and other phytoplankton capture vast quantities of carbon dioxide from the atmosphere as they grow, but this growth is often limited by a lack of essential nutrients such as iron. Artificially adding these nutrients would make algae bloom and, as the

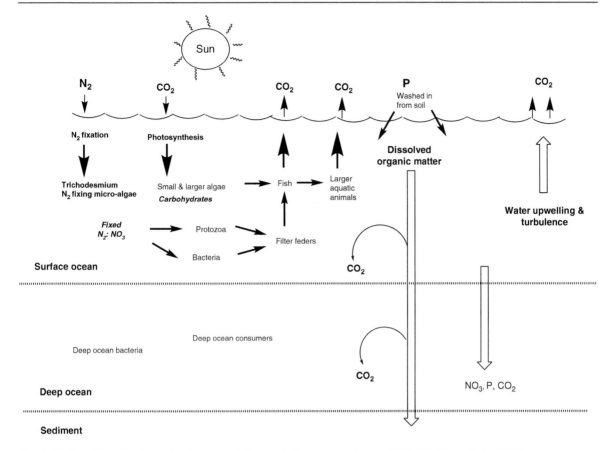

Fig. 6.15 Recycling of carbon, phosphorous and nitrogen in the ocean environment (Modified from Arrigo 2005)

organisms grow, they take up more CO_2. When they die, some of the organisms sink to the bottom of the ocean, taking their carbon with them. Experimentally iron fertilization has been shown to triple the growth of photosynthetic algae and the amount which sink to the ocean's bottom.

Some photosynthetic microorganisms in the marine environment also fix nitrogen. The link between carbon recycling, nitrogen fixation, and phosphorous recycling is further discussed below and shown in Fig. 6.15.

6.6.1.3   Marine Microorganisms and the Nitrogen Economy of Seas and Oceans

Nitrogen Fixation in the Ocean

Oligotrophic oceanic waters of the central ocean sites typically have extremely low dissolved fixed inorganic nitrogen concentrations, and few nitrogen-fixing microorganisms from the oceanic environment have been cultivated. The picture has changed in recent times. Nitrogenase is the enzyme involved in nitrogen fixation. Nitrogenase gene (*nifH*) sequences amplified directly from oceanic waters showed that the open ocean contains more diverse diazotrophic microbial populations and more diverse habitats for nitrogen fixers than previously observed by classical microbiological techniques. Nitrogenase genes derived from unicellular and filamentous cyanobacteria, as well as from the "α" and "γ" subdivisions of the class *Proteobacteria*, have been found in both the Atlantic and Pacific oceans (Zehr et al. 1998).

The current view is that the abundance of the N_2 fixers such as the cyanobacterium, *Trichodesmium*, the most common representative, has been severely underestimated and that N_2 fixation is a more important component of the marine N cycle than previously realized. It was once thought that nitrogen fixation would be more in coastal waters where Fe necessary for full function of nitrogenase would be more abundant. But it is now known that the amount required for the proper functioning of nitrogenase is much less than previously thought and the amount is present in the oceans away from the coast. Apart from *Trichodesmium,* other high N_2 fixers include the diatom genera *Rhizosolenia* and *Hemiaulus* which harbor the endosymbiotic high N_2-fixing cyanobacterium *Richelia intracellularis.* Other nitrogen-fixing cyanobacteria are species of *Synechococcus, Prochlorococcus, Trichodesmium,* and *Crocosphaera* (Ward 2005).

Unlike other nitrogen fixing bacteria, *Trichodesmium* also called sea sawdust, does not have heterocysts (spore-like structures in which nitrogen is synthesized in many cyanobacteria), nor any other specialized cells for this task. Furthermore, nitrogen fixation peaks at mid-day, i.e., occurs during the same time as photosynthesis.

Photosynthetic fixation of CO_2 in the oceans accounts for approximately half of total global primary production. Cyanobacteria, including species of *Synechococcus, Prochlorococcus, Trichodesmium,* and *Crocosphaera,* are prominent constituents of the marine biosphere that contribute significantly to this "biological carbon pump." The factors that control the growth of these cyanobacteria directly impact not only the carbon pump, but also the global nitrogen cycles through the activity of nitrogen-fixing organisms (i.e., *Trichodesmium* and *Crocosphaera*). The introduction of this new nitrogen to the euphotic zone is significant since it allows further CO_2 fixation by the remaining non-diazotrophic (i.e., non-nitrogen fixing) phytoplankton community.

Trichodesmium spp. are the most significant cyanobacterial primary producers in tropical and subtropical North Atlantic as well as in the tropical Pacific Ocean. In these regions, *Trichodesmium* introduces the largest fraction of new nitrogen to the euphotic zone, even exceeding the estimated flux of nitrate across the thermocline. In short, cyanobacteria are significant primary producers at the center of the marine food chain, with the genus *Trichodesmium* being particularly important in the tropical and subtropical oceans due to both its high abundance and high N_2-fixation rates.

Anaerobic Oxidations of Ammonium and of Methane

Interest in marine anaerobic ammonium oxidation (anammox), i.e., the microbiological conversion of ammonium and nitrite to dinitrogen gas, is a very recent addition to our understanding of the biological nitrogen cycle. It was discovered in 1986, and so far is the most unexplored part of the cycle. Given its basic features, the anammox process is a viable option for biological wastewater treatment. Very recently, it was discovered

that anammox makes a significant (up to 70%) contribution to nitrogen cycling in the World's oceans.

Anaerobic methane oxidation (AMO) and anaerobic ammonium oxidation (Anammox) are two different processes catalyzed by completely unrelated microorganisms. Still, the two processes do have many aspects in common. First, both of them were once deemed biochemically impossible and nonexistent in nature, but have now been identified as major factors in global carbon and nitrogen cycling. Second, the microorganisms responsible for both processes cannot yet be grown in pure culture; their detection and identification were based on molecular ecology, tracer studies, use of lipid biomarkers, and enrichment cultures. Third, these microorganisms grow extremely slowly (doubling time varies from weeks to months). Fourth, both processes have a good potential for application in biotechnology (Strous and Jetten 2004).

The processes can be represented thus:

$$CH_4 + SO_4^{2-} + H^+ = CO_2 + HS^- + 2H_2O \quad (6.1)$$

$$NH_4^+ + NO_2^- = N_2 + 2H_2O \quad (6.2)$$

The anammox and amo bacteria form a monophyletic cluster branching off deep in the order.

Planctomycetales. With 16S rDNA probes based on the sequences derived from enrichment cultures, anammox bacteria were detected in various ecosystems such as the suboxic zone of the water of the Black Sea. Three anammox bacteria has been tentatively named "*Kuenenia stuttgartiensis,*" "*Scalindua sorokinii,*" and "*Brocadia.*" All three genera share the same metabolism. Anammox was found to contribute up to 50% to marine N_2 production.

With regard to the roles of the two processes, current information is that AMO and anammox are responsible for more than 75% of marine methane oxidation and 30–50% of marine ammonium oxidation. Since the marine biosphere is strongly coupled to global climate, AMO and anammox play important parts in this regard.

Anaerobic methane and ammonium oxidation have two properties in common: Slow microbial growth and mutualism. AMO is mediated by syntrophic reversed methanogenic archaea and sulfate-reducing bacteria. The two are always found in close proximity to one another.

The figure shows loss of ammonium and nitrites respectively due to anaerobic ammonium oxidation and anaerobic methane oxidation (AMO). particulate organic nitrogen (PON), including plankton; DON, dissolved organic matter; DNRA, dissimilatory nitrate

Fig. 6.16 The nitrogen cycle in the marine environment (Modified from Arrigo 2005) The figure shows loss of ammonium and nitrites respectively due to anaerobic ammonium oxidation (AMMONOX) and anaerobic methane oxidation (AMO). *PON* particulate organic nitrogen, including planton; *DON* dissolved organic matter; *DNRA* dissimilatory nitrate reductase to ammonium

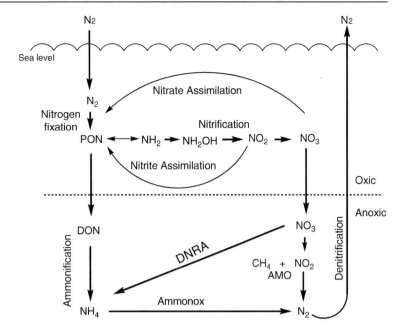

reductase to ammonium unknown if these archaea are simply inactive, are capable of AMO without a sulfate-reducing partner, or are doing something completely different. For anammox, the anaerobic ammonium oxidizers always depend on a nearby source of nitrite. Denitrifiers reduce nitrate to nitrite, which is then used by anammox bacteria. The doubling time of anammox bacteria is 2–3 weeks on average, rivaled in their slow growth only by *Mycobacterium leprae* grown in the nine-banded armadillo as a surrogate host.

A possible application is wastewater treatment. The introduction of anammox to nitrogen removal would lead to a reduction of operational costs of up to 90%. Anammox would replace the conventional denitrification step completely and would also save half of the nitrification aeration costs. In feasibility studies with sludge digestor effluents on laboratory scale, the effluents did not negatively affect the anammox activity and anammox biomass could be enriched from activated sludge within 100 days.

6.6.1.4 The Global C:N:P Marine Ratio and Its Maintenance Through Microbial Activity: The Redfield Ratio

The Redfield ratio (named after Alfred Redfield, its discoverer) is that the marine nitrate:phosphate ($NO_3:PO_4$) ratio of 16:1 in the oceans is controlled by the requirements of phytoplankton, which subsequently release nitrogen (N) and phosphorus (P) to the environment at this ratio as they die, are broken down, and remineralized. Redfield (1934) had analyzed thousands of samples of marine biomass from all ocean regions. He found that globally the elemental composition of marine organic matter (dead and living) was constant. The ratios of carbon to nitrogen to phosphorus remained the same from coastal to open ocean regions. Redfield's initial observations has been extended to include other elements, most notably carbon (C), and it now links these three major biogeochemical cycles through the activities of marine phytoplankton in the ratio of 106:16:1 for phytoplankton and the deep ocean (Redfield 1934).

Under certain conditions, the phytoplankton chemical ratio diverges from the expected Redfield ratio. The change in the ocean ratio could be changes in exogenous nutrient delivery and microbial metabolism (e.g., nitrogen fixation, denitrification and anammox). These changes are reflected as N deficit or excess relative to P for a given water mass.

These biologically mediated cycles modulate, and are themselves modulated by, processes operating at scales ranging from algal photosynthesis to the global climate. The amount of atmospheric carbon dioxide removed by the ocean is very sensitive to the ratio relationships between phytoplankton and nutrients, including inputs by humans of ordinarily limiting elements such as nitrogen, through, for example, fertilizer eutrophication, or by natural nitrogen fixation and losses due to AMO and anammox (see Fig. 6.16).

References

Allaby, A., & Allaby, M. (1990). *Sea water. The concise Oxford dictionary of earth sciences*. Oxford: Oxford University Press.

Anonymous (2003). The three-layered ocean. http://www.villasmunta.it/oceanografia/the_three.htm. Accessed 4 Sept 2010.

Anonymous (2010a). Marine snow. http://www.sos.bangor.ac.uk/marine/mb/O3B14/Marine_snow.html. Accessed 12 May 2010.

Anonymous (2010b). Temperature of ocean water. http://www.windows2universe.org/earth/Water/temp.html. Accessed 31 Aug 2010.

Arrigo, K. R. (2005). Marine microorganisms and global nutrient cycles. *Nature, 437*, 349–355.

Azam, F., Fenchel, T., Field, J. G., Gray, J. S., Meyer-Reil, L. A., & Thingstad, F. (1983). The ecological role of water-column microbes in the sea. *Marine Ecology Progress Series, 10*, 257–263.

Budikova, D. (2010). Albedo. *Encylopaedia of Earth*. http://www.eoearth.org/article/albedo. Accessed 26 Sept 2010.

Díez, B., Bauer, K., & Bergman, B. (2007). Epilithic cyanobacterial communities of a marine tropical beach rock (Heron Island, Great Barrier Reef): Diversity and Diazotrophy. *Applied and Environmental Microbiology, 73*, 3656–3668.

Fuhrman, J. (2000). Impact of viruses on bacterial processes. In D. L. Kirchman (Ed.), *Microbial ecology of the sea* (pp. 327–350). New York: Wiley-Liss.

Giovannoni, S. J., & Rappe, M. (2000). Evolution, diversity and molecular ecology of marine prokaryotes. In D. L. Kirchman (Ed.), *Microbial ecology of the sea* (pp. 47–84). New York: Wiley-Liss.

Irenewagner-DObler, I. D., & Biebl, A. H. (2006). Environmental biology of the marine *Roseobacter* lineage. *Annual Review of Microbiology, 60*, 255–280.

Javaux, E. J., & Scott, D. B. (2003). Illustration of modern benthic foraminifera from Bermuda and remarks on distribution in other subtropical/tropical areas. *Palaeontologia Electronica, 6*, 29 pp.. http://palaeo-electronica.org.

Munn, C. B. (2004). *Marine microbiology*. London: Bios Scientific.

Pauly, D. (2007). 'Fishing down marine food webs' as an integrative concept. ACP – EU fisheries research report number 5/page 8. http://cordis.europa.eu/inco/fp5/acprep8_en.html. Accessed 10 Sept 2010.

Pomeroy, L. R. (1974). Oceans food web, a changing paradigm. *Bioscience, 24*, 499–504.

Prasannarai, K., & Sridhar, K. R. (2001). Diversity and abundance of higher marine fungi on woody substrates along the west coast of India. *Current Science, 81*, 308–316.

Redfield, A. C. (1934). On the proportions of organic derivations in sea water and their relation to the composition of plankton. In R. J. Daniel (Ed.), *James Johnstone memorial volume* (pp. 177–192). Liverpool: University Press of Liverpool.

Schulz, K. (2006). Bacteria and the microbial loop. http://www.esf.edu/efb/schulz/Limnology/Bacteria.html. Accessed 3 Feb 2007.

Stewart, R.R. (2005). Marine fisheries food webs. http://oceanworld.tamu.edu/resources/oceanography-book/marinefoodwebs.htm. Accessed 21 Sept 2010.

Strous, M., & Jetten, M. S. M. (2004). Anaerobic oxidation of methane and ammonium. *Annual Review of Microbiology, 58*, 99–117.

Ward, B. B. (2005). Molecular approaches to marine microbial ecology and the marine nitrogen cycle. *Annual Review of Earth and Planetary Science, 33*, 301–333.

Welsh, D. T. (2000). Ecological significance of compatible solute accumulation by micro-organisms: from single cells to global climate. *FEMS Microbiology Reviews, 24*, 263–290.

Zehr, J. P., Mellon, M. T., & Zani, S. (1998). New nitrogen-fixing microorganisms detected in oligotrophic oceans by amplification of nitrogenase (*nifH*) genes. *Applied and Environmental Microbiology, 64*, 3444–3450.

Part IV

Pollution and Purification of, and Disease Transmission in, Water

Pollution of Aquatic Systems: Pollution Through Eutrophication, Fecal Materials, and Oil Spills

7

Abstract

Natural bodies of water will purify themselves and remove added materials given sufficient time. Pollution therefore occurs when the self-purifying powers of a body of water have not had enough time to remove the pollutant and return itself to its original state. Pollutants in water include bacteria, chemicals, heat and by the addition of organic or inorganic nutrients or eutrophication. Eutrophication causes excesses growth of cyanobacteria or blooms.

Pollution by fecal matter is determined principally by the identification of *E. coli* in water. Water bodies are expected to meet standards of the maximum content of microbial and chemical pollutants set by governments all over the world. The Total Maximum Daily Load (TDML) is a calculation of the maximum amount of a pollutant that a water body can contain and still meet water quality standards set by the regulating authority for a particular water use. Since not all *E coli* is necessarily of human origin, its source is determined through methods of microbial source tracking.

Oil spills are major sources of marine pollution. They are remediated physically by skimmers which collect the oil, adsorption onto suitable materials, the use of dispersants or in-situ burning. Biological methods include addition of nutrients to stimulate the growth of oil-degrading microorganisms, the use of surfactants to emulsify the oil and increase contact between oil and microorganisms, and the introduction of organisms specially adapted to growth on oil.

Keywords

Self-purification • Eutrophication • "Algal blooms" • Fecal pollution • Indicator organisms in pollution • Oil spills • Remediation of oil spills • Standard water analysis • Total Daily Maximum load (TDML) • Microbial source tracking (MST)

7.1 Nature of Pollution

Pollution of water may be defined as the introduction into a body of water of any material or condition which may be injurious to human health or offend aesthetic sensibilities and thus limit the use of water for drinking, recreation, fishing, and other activities to which water is normally put.

It should be borne in mind that the idea of pollution in the wider sense is anthropocentric (i.e., centered around man's needs). Seen in this light, pollution should be understood as a relative term, which depends

N. Okafor, *Environmental Microbiology of Aquatic and Waste Systems*,
DOI 10.1007/978-94-007-1460-1_7, © Springer Science+Business Media B.V. 2011

on the specific use to which a particular body of water is put. A good example relates to the bacterium, *Sphaerotilus natans*. This bacterium grows profusely into strands when there is eutrophication (i.e., the availability of more nutrients than is usual), especially by the introduction of sewage. On account of the bacterium's abundant growth accompanied by its formation of strands when eutrophication occurs through the introduction of sewage, the trivial name of the bacterium is "sewage fungus." The sewage fungus is objectionable from the point of view of man's use of water, especially as it is reminiscent of sewage, although the excess nutrient causing its growth may not in fact come from sewage. Despite being ordinarily objectionable, *S. natans* is sometimes deliberately cultivated in the rivers of some countries. This is because the larvae of certain insects (*Chironomidae*) are encouraged to develop by the bacterium. The larvae in turn are fed upon by fish (trout), which is then farmed. When it is associated with fish farming, the sewage fungus is certainly not a pollutant!

7.1.1 The Concept of the Self-purification of Water as Basis for the Understanding of Pollution

When a large body of water either flowing, such as a river, or fairly static, such as a reservoir or lake, suffers eutrophication (the addition of nutrients and/or microorganisms), the organic materials and microorganisms introduced thereby will gradually disappear and the water will return to the state it was before the eutrophication. The process, which brings about this return to the original state is known as *self-purification*. Self-purification takes place in all bodies of water, including seas and oceans. The processes underlying self-purification can be grouped into physicochemical and biological. These will be discussed below to provide a basis for a better understanding of pollution.

1. *Physicochemical mechanisms in the self-purification of water*
    (a) Flocculation
        The materials introduced into water sediment at rates which depend on their weight; they remove themselves from general circulation as they sediment. Equally important is that, during this sedimentation, microorganisms, which bind to heavier organic materials as well as clay particles

are dragged down as they sediment, and are removed from circulation.
    (b) Light
        Sunlight contains ultraviolet light, which has germicidal effects. Aquatic bacteria exposed to light are therefore killed, although the effect of light is limited to the uppermost portion of water.
    (c) Aeration
        Aeration is brought about by both the physical diffusion of air and the release of oxygen by the photosynthetic activity of algae and cyanobacteria. Aeration encourages the growth of aerobic bacteria, which participate in the self-purification of water by breaking down organic matter introduced into water. Additionally, certain compounds, for example, those of manganese and iron, are oxidized in the presence of oxygen and hence are precipitated, contributing to the self-purification of water.
    (d) Dilution
        Materials added to natural bodies of water are usually less in volume than the water body. As a consequence, the added materials are necessarily diluted when introduced. The dilution of added organic material helps reduce its concentration at the point of entry in the case of nonflowing waters such as lakes. In the case of flowing waters such as streams, the concentration of added organic matter is reduced both at the point of introduction and downstream as the water flow moves the added material downstream. Provided that the load of added organic matter is not so heavy that it cannot be sufficiently diluted, the process of dilution helps reduce the onset of pollution.
2. *Biological factors*
    Biological factors are by far the most important in the self-purification of water. Biological factors act through the breakdown (stabilization) of added materials by microorganisms, mainly by bacteria and occasionally by fungi. By far, the bulk of the breakdown is brought about by bacteria and among these, aerobic bacteria are most important. The activities of anaerobic bacteria lead to only partial breakdown of organic matter, and take much longer time. The factors, which affect the breakdown of materials added to water are discussed below.
    Factors affecting the biological breakdown of materials added to water

(a) The chemical nature of the added material
    The nature of the material is crucial in deciding whether or not, and how fast, the added material is degraded or stabilized. For instance, sugars, starch, amino acids, and proteins are easily broken down, whereas cellulose is not as easily degraded. Organic chemicals from industrial activity are often only slowly degraded. In some cases they are not broken down at all unless microorganisms specially developed by enriching the materials to be degraded, are introduced.

(b) Predation by Protozoa
    The bacteria, which breakdown organic materials, are themselves consumed by Protozoa, particularly the ciliates. Most of the bacteria, which are introduced in the process of eutrophication, for example, with sewage, are removed in this way. Among the ciliates, which are encountered in eutrophic waters are *Paramecium*, *Colpoda*, and *Carchesium*.

(c) Bacteriophages
    Because bacteriophages attack and destroy specific bacteria and are so commonly found in polluted waters, some authors have suggested that they should be used as indicators of the presence of fecal bacteria.

(d) *Bdellovibrio* spp.
    Bacteria belonging to this group are small (0.3–0.4-μm wide), Gram-negative, polarly flagellated, and comma-shaped rods. They are found in high numbers in eutrophic waters, especially those polluted with sewage effluent. They attack Gram-negative, but not Gram-positive, bacteria including *Escherichia coli*, *Pseudomonas*, *Erwinia*, *Salmonella*, and *Shigella*.

(e) Antibiotic activity
    Antagonistic activity between some bacteria against other bacteria and between some algae against some bacteria has been reported. Thus, *Pseudomonas aeruginosa* and other bacteria have been reported to produce anti-coliform factors against *E. coli*. The production of similar materials has also been reported from *Bacillus coagulans* and *B. licheniformis*. Algae are well known as producers of antibacterial substances. The green alga *Scendemus obliquus*, for example, has been shown to be active against *Salmonella*, while the diatom *Skelotonoma costatum* has been shown to be active against *Vibrio* and *Pseudomonas* (Okafor 1985).

7.1.2   Definition of Pollution

As described above, organic materials and microorganisms added to water are removed by the self-purifying power of water acting through physicochemical and biological processes. The self-purifying power of water requires time to be effective. The definition of pollution to be given here relies on the idea that *given sufficient time* the self-purifying powers of any body of water will restore that body of water to the state it was before the added material entered the water, even with the addition of the most recalcitrant items.

Given sufficient time, even the most recalcitrant among degradable added materials will be removed because microorganisms able to degrade it will develop. In the process of the degradation through the development of the appropriate organisms, sufficient time will have occurred for the component parts of self-purification to occur.

Based on this idea, pollution occurs when the self-purifying powers of a body of water are not able to remove the materials added to a body of water because the component parts of the self-purifying process, including sufficient time, have not been able to achieve the removal. When the added materials have not been removed by the processes of self-purification, they may render the body of water unsuitable for normal uses to which a body of water is put, such as drinking, washing, irrigation, or for recreation such as swimming.

A 1971 United Nations report defined ocean pollution as: "The introduction by man, directly or indirectly, of substances or energy into the marine environment (including estuaries) resulting in such deleterious effects as harm to living resources, hazards to human health, hindrance to marine activities, including fishing, impairment of quality for use of sea water and reduction of amenities." The definition given in this book tallies well with the United Nations definition as *given sufficient time*, "substances or energy (introduced into the) marine (or any aquatic environment) will be removed and will have no chance to "result in such deleterious effects as harm to living resources, hazards to human health, hindrance to marine activities, including fishing, impairment of quality for use of sea water and reduction of amenities" (Okafor 1985).

## 7.1.3	Kinds of Pollutants

In general terms, a pollutant is anything, which when added to a body of water, directly or indirectly limits man's legitimate use or appreciation of that body of water because sufficient time has not occurred for the self-purifying powers of the water to remove the added materials or conditions. As mentioned earlier, the use of water can be found in the domestic, agricultural, transportation, and recreational arenas. A material, of whatever nature, which when added to water does violence to the observer's aesthetic appreciation would also be regarded as a pollutant. When this broader view of definition is adopted, one may classify pollutants into the following four categories: pollution by eutrophication, pollution by the addition of harmful or undesirable organisms, pollution by toxic materials and industrial effluents, and thermal pollution. The bulk of the discussion will be on pollution by eutrophication; the other three will be discussed very briefly before the comparatively extensive discussion on eutrophication.

1. *Pollution by the addition of harmful or undesirable organisms*

   This type of pollution typically occurs when sewage is the cause of the eutrophication. It can also occur when effluents rich in easily degradable organic materials are added to water, such from a food processing factory. When sewage derives from a population harboring enteric pathogens, water pollution caused by such sewage will naturally contain the pathogens of that community.

2. *Toxic materials and industrial effluents*

   These include industrial heavy metals, radioactive materials, and pesticides, which can be harmful to man in one way or the other. In this group also would be placed effluents from the chemical industry and pollution by crude oil when oil-tankers run aground. Pollution by crude oil during spills and methods of remediating the situation will be discussed later in this chapter.

3. *Thermal pollution*

   Industrialization uses power. When such power is derived from a nuclear plant, water is needed for cooling. Such cooling water becomes hot in the process and is usually discharged into rivers or estuaries. Cooling is also done in other industries outside nuclear establishments, for example, in metal smelting. As a result of the addition of hot water to a natural body of water the balance of the population in the ecosystem may alter, giving rise to a preponderance of thermophilic organisms. Heated water also renders bodies of water unusable for some of man's legitimate undertakings such as fishing.

4. *Pollution by eutrophication*

   Pollution by eutrophication is a regular occurrence in rivers and lakes and will be discussed in the following section.

   Oil spills important in oil producing countries and when accidents occur in the oceans with oil tankers will be examined at the end of this chapter.

## 7.1.4	Pollution by Eutrophication

Eutrophication, also known as nutrient pollution, is a process whereby water bodies, such as lakes, estuaries, or slow-moving streams, receive excess nutrients that stimulate excessive growth of microorganisms including algae and cyanobacteria. This enhanced growth often results in a bloom (see below).

Point and Nonpoint Eutrophication
Pollution by eutrophication may be *point* or *nonpoint*. *Point* pollution occurs when pollutants enter directly into a body of water from identifiable sources. The well-known Exxon Valdez oil spill best illustrates a point source water pollution. *Nonpoint* sources deliver pollutants indirectly through environmental changes, and cannot always be indentified directly. An example of this type of water pollution is when fertilizer from a field is carried into a stream by rain, in the form of runoff, which in turn affects aquatic life. Pollution arising from nonpoint sources accounts for a majority of the contaminants in streams and lakes.

The Role of Oxygen in Eutrophication
Dissolved oxygen (DO) is a most important factor in aquatic pollution by eutrophication. The excessive growth of aerobic bacteria (and to a lesser extent fungi), which are the primary stabilizers of added material results in greater demand for oxygen. Under normal conditions, the nutrient load of a stream is low and hence the bacterial and fungal populations are also low. With heavy eutrophication and the consequent increased microbial growth, the depletion of oxygen may be so severe that the dissolved oxygen is finally exhausted and anaerobiosis sets in; consequently, fish, which also depend on dissolved O_2, may die, followed by deaths among crustaceans, rotifers, and protozoa.

Under anoxic conditions (i.e., conditions of oxygen shortage), anaerobic bacteria develop. Their metabolism gives rise to the foul odors of H_2S and methane, as well as the black color of the mud, due to sulfide production by the anaerobic bacteria.

Furthermore, under anaerobic conditions the silting up of rivers and lakes occurs because the organic materials are not completely broken down, and hence the partially decomposed organic matter accumulates.

Reoxygenation of Water

Oxygenation of water helps reverse the effects of eutrophication by encouraging the growth of aerobic microorganisms. The reoxygenation of water bodies is brought about in two ways: (a) by the activities of aquatic photosynthetic cyanobacteria and algae and (b) by physical reaeration.

(a) *Photosynthetic organisms* in water use CO_2 released by the aerobic bacteria to produce carbohydrates and release O_2 as a by-product. Reoxygenation by photosynthetic organisms may be prevented or reduced if the water contains large amounts of suspended particles, which prevent light penetration.

(b) *Physical reaeration*: The temperature of water and the speed of its flow are two factors affecting the rate at which oxygen is physically reintroduced into water. The higher the temperature, the less the amount of oxygen or any other gas dissolved in water.

Regarding the speed of water, in lotic (swift-moving) waters there is turbulent movement of water; hence, there is regular change of water at the air/water interface where O_2 exchange actually takes place, leading to rapid reoxygenation of water. In slow-moving (or limnetic) waters, oxygen permeates into water from the fixed air–liquid interfaces by simple diffusion. Since this is slow, the lower layers of the water column will be more oxygenated in river under lotic conditions than when the same river is under limnetic conditions. Because the breakdown of organic materials in water is brought about by aerobic bacteria, whose growth is encouraged by the availability of oxygen, pollution by eutrophication (i.e., nondisappearance of added materials) is more likely to occur in limnetic waters, for example, lakes where the slower movement of water leads to less available oxygen for the activity of aerobic bacteria, than in rapidly moving or lotic waters (Okafor 1985).

7.1.4.1 "Algal Blooms" and Eutrophication

"Algal bloom" is a term used to describe an abundance of blue-green algae (Cyanobacteria) (and occasionally, proper algae) at the surface of lakes or reservoirs, which abundance is such as to confer on the water the color of the predominant organism (s) in the bloom. Blooms are also known as "water blooms," "flowering of waters," or "flos-aquae." Blooms can form and disperse again within a matter of hours and are found the world over but in temperate countries they form mainly in summer while in the tropics they may form at any time of the year (e.g., in Lake George, Uganda and Lake Kilotes, Ethiopia). As a rule, blooms form only in waters rich in dissolved nutrients, especially phophorus.

Blooms are formed mainly by cyanobacteria and those in which blooms have been found are listed in Table 7.1.

Blooms are sometimes regarded as useful since they provide additional food for fish and hence may increase the productivity of freshwater fisheries. Cyanobacteria of the genus *Spirulina* form blooms, which are consumed directly as food in Chad Republic in central Africa and probably other countries. It is used as a food supplement in Europe and the USA, because it is believed to impact positively on the health of the consumers (see Anonymous 2010a).

Blooms are, however, regarded as nuisance in many parts of the world, for the following reasons:

(a) In water-treatment plants they sometimes clog the filters.

Table 7.1 Cyanobacteria which form blooms (From http://www.cdc.gov/hab/cyanobacteria/table1.htm; Credit: Center for Disease Control [CDC]; Anonymous 2010a)

Order	Family	Genus
Chrococales	Chrococcaceai	*Coelosphaerium*
		Gomphosphaeria
		Microsystis
Nostocales	Oscilatoriaceae	*Oscillaloria*
		Spirulina
		Trichodesmium
	Nostocaceae	*Anabaena*
		Anabaenopsis
		Aphanizomenon
	Rivulariacea	*Gloetrichia*

(b) Many species produce unpleasant odors and tastes in drinking water.

(c) Some blooms produce toxic materials, which kill fish and domestic animals.

(d) Blooms make recreational lakes unavailable for fishing and other water sports.

Blooms are restricted mainly to blue-green algae (cyanobacteria) because most of them are able to float at some stage in their lives, whereas other organisms are always heavier than water and hence sink to the bottom; not all blue-green algae can, however, form blooms. Buoyancy is conferred on those, which form blooms by gas vacuoles present in the protoplasm of these bacteria.

Factors Encouraging Bloom Formation

The following factors affect the development of algal blooms:

(a) Availability of inoculum: In tropical lakes, which regularly form blooms, the collapsed cells and resistant structures sink to the bottom and may be lodged in the mud from where they develop when conditions are favorable.

(b) Sudden influx of organic nutrients: One of the major predisposing factors of bloom formation is a sudden influx of organic material such as is found in sewage. Since cyanobacteria are photo-autotrophic and can manufacture their own food, they would appear to absorb the added nutrients mostly in the dark. It is also possible that the addition of sewage introduces trace elements, which are required by the bacteria. Furthermore, eutrophication with organic matter induces the growth of aerobic bacteria, which not only use up the oxygen and create anaerobiosis, but also release CO_2. The CO_2 would encourage greater photosynthesis by the cyanobacteria, while anaerobiosis is known to favor nitrogen fixation in them, producing nitrogenous compounds, which encourage the growth of aerobic bacteria; the net result of these activities is the growth of the cyanobacteria and the aerobic decomposers.

(c) Availability of light and a suitable temperature: Light is required for photosynthesis by cyanobacteria. Suitably warm temperatures dictate that blooms are formed in the summer and autumn in the temperate countries and year-round in the tropics.

(d) Availability of suitable pH: Blooms tend to form more in hard waters (which contains Mg^{2+} and Ca^{2+} carbonates) than in soft waters. Furthermore, they will not form in water with a pH value less than 6. Indeed, most cyanobacteria grow best in the pH range 7.9–9.0. The Mg^{2+} and Ca^{2+} carbonates would appear to act as buffers, which prevent the pH from changing drastically.

(e) The availability of phosphorous and nitrogen: These two nutrients, especially phosphorous, are critical and must be available in suitable quantities before blooms can form.

Factors Adversely Affecting Blooms

Blooms may be suppressed by a number of biological agents within a body of water. These include fungi of the order *Chytridiales* and *Blastocladiales*, which attack cyanobacteria. Some viruses ("cyanophages") attack cyanobacteria, while some protozoa, for example, *Pelomyxa* (amoeboid) and *Ophryoglena* (ciliate) graze them; finally, fish, for example, *Tilapia* and *Haplochromis* consume large quantities of cyanobacteria.

When blooms are a definite nuisance such as in reservoirs, they may also be destroyed by vigorous shaking.

7.1.5   Biological Indicators of Pollution by Eutrophication

Pollution by eutrophication is difficult to measure by the chemical analysis of the components of water. This is because it would involve an unthinkable amount of work if the various organic and inorganic compounds, which cause pollution were to be determined by this means, even if they were all identifiable. For this reason, the method used for determining the decomposable organic matter added to water is the biochemical oxygen demand (BOD), which determines the amount of oxygen consumed by aerobic bacteria in the decomposition of organic matter in water (see Chap. 11).

However, a method simpler than the BOD technique can be employed. This method is based on knowledge of the change in the flora and fauna of natural waters as a natural water changes from a eutrophic to postpollution status. It consists of examining the various organisms found in water. Although not favored by all aquatic biologists the method recommends itself to the environment of developing countries where equipment and reagents for the most simple analyses are often lacking. Among biological indicators of water pollution, the following have been used: fish, algae, sponges, and benthic invertebrates. However, bacteria and protozoa are the principal organisms used.

(a) *Use of bacteria*

The use of bacteria to detect pollution is based on the principle that when there is a high nutrient concentration the biological population is high and diverse (i.e., a large number of a wide range of bacteria are present) unless toxic materials are present in the water or there is a shortage of some key nutrients. This high bacterial population (apart from using up the nutrients) uses up the oxygen and hence creates anaerobic or near-anaerobic conditions. The result is that many bacteria die out, and only a few types will remain. This is even accentuated by the grazing activity of protozoa.

In order to detect pollution changes it is necessary to make observations at least from two different points along the stream or to do so at some point at two different occasions. By taking plate counts on a variety of media and both under aerobic and anaerobic conditions, it is possible to follow the progress of pollution.

(b) *Use of protozoa and other organisms*

The types of organisms present besides the bacteria will indicate whether the water is just receiving pollutants or is just recovering from them by natural self-purification. If bacteria, protozoa of group Sarcodina, flagellates, and free-swimming ciliates are seen in that order of abundance, then the pollution is fresh. On the other hand, if stalked ciliates, free-swimming ciliates, and bacteria are seen in that order of abundance, then the water is recovering from pollution. A predominance of insects and rotifers indicates that the pollution is over as these organisms are encountered in bodies of water low in organic matter.

7.2   Pollution of Water with Reference to Human Health: Bacterial Indicators of Fecal Pollution

Many parameters are used by different authorities as indicators of water pollution. These parameters include biological oxygen demand (BOD), dissolved oxygen (DO), pH, specific conductance, water temperature, and chemical constituents including nitrates and phosphates. BOD and DO will be discussed in Chap. 11.

The most widely used parameters are, however, bacterial indicators of fecal pollution. The indicators have been undertaken in order to protect human health.

They have been in use for over a 100 years and the procedures involved have become highly standardized.

Bacterial indicators are used to indicate if:

(a) Drinking water sources are fit for drinking.

(b) Treatment of drinking water has been adequate.

(c) Drinking water in the distribution system continues to be protected.

(d) Recreational and shellfish waters are microbiologically safe.

7.2.1   Microbiological Examination of Water for Fecal Contamination

The purpose of examining water microbiologically is to help in the determination of its sanitary quality and its suitability for general use. The sanitary quality of water may be defined as the relative extent of the absence of suspended matter, color, taste, unwanted dissolved chemical, bacteria indicative of fecal presence, and other "aesthetically offensive" objects or properties. In short, the sanitary quality of water depends on its acceptability for internal consumption and other uses in which water comes directly or indirectly in contact with man.

The current bacterial indicator approaches have become standardized, are relatively easy and inexpensive to use, and constitute the cornerstone of many monitoring and regulatory programs. Due to an increased understanding of the diversity of waterborne pathogens, including their sources, physiology, and ecology, a growing understanding has occurred that the use of bacterial indicators may not be as universally protective as was once thought. Thus, the greater environmental survival of pathogenic viruses and protozoa, when compared with indicator bacteria (coliforms) has raised serious questions about the suitability of relying on relatively short-lived organisms as an indicator of the microbiological quality of water (Anonymous 2004, 2005).

The implication of this situation is that while the presence of coliforms could still be taken as a sign of fecal contamination, the absence of coliforms can no longer be taken as assurance that the water is uncontaminated. Thus, existing bacterial indicators and indicator approaches do not, in all circumstances, identify all potential waterborne pathogens. Furthermore, recent advances in microbiology, molecular biology, and analytical chemistry make it necessary to reassess the current reliance on traditional bacterial indicators for waterborne pathogens.

Nevertheless, indicator approaches will still be required for the near future for two reasons. Firstly, it will be impracticable to monitor all known waterborne pathogens directly. Secondly, pathogens usually occur in very low numbers and even where it is possible to culture them, this may present difficulties on account of their low numbers.

Organisms to be used as indicators of pollution should have certain attributes. Although the attributes of the ideal bacterial have been refined, they are still based on those described by Bonde (1966) (see Tables 7.2 and 7.3).

7.2.1.1 Principle of Indicator Organisms

The greatest hazard associated with drinking water is that it may recently have been contaminated with sewage or by human (or even animal) excrement. Water recently contaminated by feces from patients or carriers of waterborne pathogens, for example, cholera, *Salmonella*, and *Shigella* may carry the live pathogens and thus be a source of fresh outbreaks. It is, however, not practicable to isolate and identify these pathogenic organisms as a routine practice. When pathogens are present in sewage or feces, they are, however, usually out-numbered by bacteria normally present in the alimentary canal and hence in the feces. These normal inhabitants are easier to detect. If they are not found in water it can be inferred in general that the water is free of pathogens, but it should be borne in mind that viruses may well be present.

Table 7.2 Criteria for an ideal indicator (Bonde 1966. With permission)

An ideal indicator should
1. Be present whenever the pathogens are present
2. Be present only when the presence of the pathogens is an imminent danger (i.e., they must not proliferate to any greater extent in the aqueous medium)
3. Occur in much greater numbers than the pathogens
4. Be more resistant to disinfection and to the aqueous environment than the pathogen.
5. Grow readily on simple media
6. Yield characteristic and strong reactions enabling as far as possible an unambiguous identification of the group
7. Be randomly distributed in the sample to be examined, or it should be possible to obtain a uniform distribution by simple homogenization procedures, and
8. Grow widely independent of other organisms present when inoculated into artificial media (i.e., indicator bacteria should not be seriously inhibited in their growth by the presence of other bacteria)

Table 7.3 Suggested refinements of biological attributes of the fecal indicator and its detection methods (Bonde 1966. With permission)

Desirable biological attributes of indicators
1. Correlated to health risk: the indicator should be present whenever the pathogen is present
2. Indicator should have similar or (greater) survival than pathogen to adverse conditions:
• Ultraviolet exposure
• Temperature
• Salinity
• Predation by indigenous flora
• Desiccation
• Freezing
• Response to disinfectants
• Biologic survival mechanisms (where available) would advantageous
Sporulation
Cyst and other latency mechanisms
Arrested metabolism (viable, but nonculturable)
Shock proteins and other survival strategies
3. Similar or greater transport than pathogens during
• Filtration
• Sedimentation
• Adsorption to particles
4. Indicator should always be present in greater numbers than pathogen
5. Indicator should be specific to fecal source or identifiable as to source of origin
Desirable attributes of detection methods for fecal indicators
1. Specificity to desired target organism which is
• Independent of matrix effects
2. Broad applicability
3. Precision: method should be precise
4. Rapidity of results
5. Quantifiable
6. Measures viability or infectivity: method should measure only living organisms
7. Logistical feasibility
• Training and personnel requirements should be easy and accessible
• Utility in field
• Cost
• Volume requirements

The organisms which are used as indicators of fecal contamination are the following:

E. coli
Streptococcus fecalis
Clostridium welchii (*C. perfringens*)
Bifidobacteria

Among these the most commonly used indicators are *E. coli* and the coliform group as a whole. *E. coli* is of fecal origin and *E. coli* Type I (Eijkmann positive), which is of human (and warm-blooded animal) origin grows at 44.5°C. Hence, the presence of fecal *E. coli* is a definite indication of fecal pollution. All coliforms in general are not necessarily of fecal origin; nevertheless, since they are not *indigenous* to water their presence in drinking water should cast suspicion on the water and should indicate pollution in the widest sense.

Coliform bacteria are defined as Gram-negative, nonspore-forming rods capable of fermenting lactose, aerobically or facultatively, with production of acid and gas at 35°C in less than 48 h.

S. fecalis occurs regularly in feces in numbers that usually probably disappear at the same rate as *E. coli*, but quicker than other coliforms. When, therefore, it is found in water in which *E. coli* cannot be detected, this is an important confirmatory evidence of fecal pollution.

C. welchii is also regularly present in human feces but fewer than *E. coli*. Spores survive longer in water and usually resist chlorination. Its presence suggests fecal contamination and the absence of *E. coli* suggests contamination in the distant past.

7.2.1.2 Procedure for the Determination of Fecal Contamination

1. *Collection of samples*

   Samples of water for bacteriological examination should be collected in clean sterile bottles, and should not be less than 100 ml. If the water to be examined is chlorinated, then a dechlorination agent should be added to the bottle before sterilizing. Sodium thiosulfate is usually used for the purpose, the amount employed depending on the amount of residual chlorine in the water (0.1 ml of 10% solution of sodium thiosulfate will neutralize a 250-ml sample of water containing 15 mg of residual chlorine per liter). For samples containing metals such as tin or copper, a chorinating agent such as EDTA should be added. If the water is collected from a tap it should be allowed to run for 2–3 min before the bottle is filled. If from a river, stream, lake, reservoir, etc. the aim should be to collect the water in the same way as for consumers. In samples suspected to be highly contaminated, dilutions of up to 10^4 may be required. The colonies should, as in routine plate counting, lie between 30 and 300 (Anonymous 2006).

2. *Media used for enumerating indicator organisms of fecal pollution*

   Various media are used as indicator organisms. In effect they are selective media, which eliminate other organisms while encouraging the development of the indicators. The active components of the more common media are discussed briefly.

   (a) Coliform media

   Among the short Gram-negative rods the ability to ferment lactose appears to be limited to the family *Enterobacteriaceae* within which the coliforms are found. Anaerobic spore formers notably *C. welchii*, which are also of fecal origin also ferment lactose, as do some aerobic spore formers. In order to inhibit these Gram-positive organisms during the initial (presumptive) tests for coliforms, a number of inhibitors are employed; some of the Gram-positive inhibitors' activities promote the growth of coliforms. The inhibitors act by lowering the surface tension of the medium thereby making the medium more favorable for the growth of intestinal organisms, which are already adapted to the low surface tension (due to bile salts) of the lower alimentary canal. Sometimes dyes and other chemicals, which selectively inhibit Gram-positive bacteria are used. Some surface tension-reducing salt compounds appear to act by both methods.

   The surface tension-reducing agents, which have been used include the following: Ox bile, bile salt (or sodium taurocholate), lauryl sulfate, and sodium ricinoleate (i.e., soap made from castor oil). The dyes, which have been used are the triphenyl methane dyes which in low concentrations inhibit Gram-positive organisms. In much higher concentrations they inhibit even Gram-positives. Thus, brilliant green inhibits *C. welchii* at 1:30,000, but coliforms grow at 1:350. Gentian (crystal) violet inhibits the anaerobe at 1:9,000, but at 1:100,000 begins to inhibit the coliforms. Rosolic acid, a dye in this family, has also been used.

   After the initial (presumptive and confirmatory) tests, which are described latter in this chapter, the media used in the subsequent completed tests do not need to contain Gram-positive inhibitors. The media used on both sides of the Atlantic vary slightly, but the principles dictating which medium is used at which stage are

Table 7.4 Some bacteriological media used for the presumptive and confirmed tests in water examination (Compiled from *Standard Methods for the Examination of Water and Wastewater*; Anonymous 2006)

Medium	Dye for inhibiting Gram-positive bacteria	Reducing agent	Buffer	Nitrogen source	pH indicator	Remarks
Media for presumptive test						
Brilliant green lactose bile (BGLB)	Brilliant green	Ox bile	Peptone	Peptone	Brilliant green	–
Lauryl sulfate tryptose broth (LST)	–	Sodium lauryl sulfate/bile salts	Na/K$_2$HPO$_4$, KH$_2$PO$_4$	Tryptose	–	–
McConkey broth	–	Bile salts	Peptone	Peptone	Neutral red	
Media for confirmed test						
Eosin methylene blue (EMB) agar	–	–	Peptone, K$_2$HPO$_4$	Peptone	Eosin	Metallic sheen produced by *Escherichia coli*
Endo agar	–	–	Peptone	Peptone	Basic fuchsin	

the same. Table 7.4 lists some of the media, which have been used. They all contain lactose, a buffer, a nitrogen source, while those used for the presumptive test contain besides the above, a pH indicator, Gram-positive inhibitor, and/or a surface-tension reducer.

One of these media, E.M.B. agar, characteristically shows *E. coli* colonies as having a metallic sheen, whereas other coliforms, particularly *Enterobacter* and *Klebsiella,* are not only larger on this agar, but do not show this metallic sheen The metallic sheen typical of *E. coli* depends on the action of the eosin and methylene blue to form a methylene blue eosinate, which acts as an acidic dye. Absorption of the eosinate by the *E. coli* cells is facilitated by the lowering of the pH by the acid produced by the fermentation of the lactose.

(b) Streptococcus medium

Several media are available for selectively growing *Streptococcus*, including azide dextrose broth, Ethyl violet azide broth, and KF streptococcus, the compositions of which are found in *Standard Methods.* They all have in common the presence of sodium azide, which inhibits Gram-negative bacteria. They also contain buffers, nitrogen source, and other requirements of a bacteriological medium.

(c) *C. perfringens* (welchii) medium

C. perfringens produces a large number of gas bubbles in litmus milk. The clot of protein produced by the acid from the fermentation of the milk lactose is broken up by the gas bubbles to produce the so-called stormy clot.

7.2.1.3 Methods Used in the Enumeration of Indicator Organisms in Water

Two general methods are used for enumerating fecal contaminants in water (and indeed any liquid): the multiple tube method and the membrane-filter method. With either method it is customary to relate the numbers to 100 ml of water. Both methods may be used with all three indicator organisms.

1. *The multiple-tube fermentation method*

The multiple tube method is basically the dilution count (or *dilution to extinction*), which is made more accurate by the use of several tubes. In the ordinary dilution count, a serial tenfold dilution of the liquid whose bacterial load is to be counted is made. A loopful is introduced from each dilution to a suitable liquid medium in a tube, which contains a Durham tube to trap any gas produced. This method can only be used if no growth occurs in one of the dilutions introduced into the broth (hence the alternate name of "dilution to extinction" (see Fig. 2.2). In the ordinary dilution, the number of organisms in the original liquid is assumed to be the reciprocal of the dilution just before the negative one, that is, the highest dilution. Thus, if of a series of ten tubes in serial tenfold dilution, the highest dilution in which growth occurred is 10^3, the number of organisms in the original tube is taken as 10^3 or 100 (or more accurately more than 10^3 but less than 10^4. The rationale behind this is illustrated in Fig. 2.2.

The basic reason for the development of the most probable number (MPN) method can be seen from Fig. 2.2. Suppose several replicates were taken of the 10^3 dilution, then growth might occur in one at least if the number was less than ten. The MPN tables were developed to take care of this possibility by using statistical tables.

In the multiple tube method several replicates are used in accordance with probability tables to determine the most probable numbers (for details of the procedures, *Standard Methods* should be consulted). Table 7.5 shows a table of MPN numbers. The number of organisms is recorded in terms of most probable numbers (MPN). Figures obtained in MPN determinations are indices of numbers which, more probably than any others, would give the results shown by laboratory examination. It should be emphasized that they are not actual numbers, which, when determined with the plate count are usually much lower.

Membrane filtration method
This method consists in filtering a measured volume of the sample through a membrane composed of cellulose esters and other materials. All the bacteria present are retained on the surface of the membrane and, by incubating the membrane face upward on a suitable medium and temperature, colonies develop on the membrane, which can then be counted. The volume of liquid chosen will depend on the expected bacterial density of the water and should be such that colonies developing on the membrane lie between 10 and 100.

The great advantage of the filtration method over the multiple-tube method is its rapidity. Thus, it is possible to obtain direct counts of coliforms and *E. coli* in 18 h without the use of probability tables. Secondly, labor, media, and glassware are also saved. Thirdly, neither spore-bearing anaerobes nor mixtures of organisms, which may give false presumptive reactions in McConkey broth cause false-positive results on membranes. Finally, a sample may be filtered on the spot or in a nearby laboratory with limited facilities instead of taking the liquid to the laboratory.

However, membranes have the following disadvantages:
(a) They are unsuitable for waters of high turbidities in which the required indicator organisms are low in number, since they are blocked before enough organisms have been collected.
(b) When non-coliforms predominate over coliforms, the former may overgrow the membrane and make counting of coliforms difficult.
(c) If nongas-producing lactose fermenters predominate in the water, false results will be obtained.

Both methods do not give the same results. Parallel tests should, therefore, be used to determine corresponding results but using the multiple-tube method as the standard.

2. *Examination of water for general coliforms*
Coliforms may be enumerated by two methods: the multiple-tube and the membrane filtration methods.
(a) Multiple-tube method
In the examination of water for coliforms the procedure is carried out as the *Presumptive Test*, the *Confirmed Test*, and the *Completed Test*. The schematic outline indicating the appropriate media and conditions for these tests is set out in Table 7.6. The details of the various methods are available in *Standard Methods* (21st edition) (Anonymous 2006) and only an outline will be given here. Not all the tests need to be employed all the time. What is used depends on the nature of water and the use to which it is to be put. Thus, the presumptive test alone is used for a polluted water, whose consideration for direct potability without further treatment is not in question. The Confirmed Test alone is applied to potable water in the process of purification, chlorinated sewage effluents and bathing waters, and the finished water samples. The Completed Test is used for drinking water.

In computing the MPN numbers for presumptive, confirmed, or completed test, in the multiple-tube method of determining numbers, the combination of sample sizes to be used should be obtained by a knowledge of the quality of the water as indicated below. Furthermore, the appropriate combination of water sample sizes to be used should be chosen by examining the MPN. Different sets of tables are available in *Standard Methods for the Examination of Water and Waste Water*. One such set of tables is given in Table 7.5. The combination of positive and negative tubes is read off the table against the sample sizes to determine the MPN.

The combination of sample sizes used in Table 7.5 are 5 of 10 ml, each, 5 of 1 ml each, and 5 of 0.1 ml each. It is possible, however, to use larger or smaller portions. Thus, samples of 100, 10, or 1 ml could be used when the MPN is 0.1 times the values. Similarly, 1.0, 0.1, and 001 ml may be used when the MPN numbers should be ten times those on the table.

Table 7.5 Sample MPN table (From Anonymous 2006)

MPN index and 95% confidence limits for various combinations of positive results when five tubes are used per (10 ml, 1.0 ml, 0.1 ml)[a]

Combination of positives	MPN index/ 100 ml	Confidence limits		Combination of positives	MPN index/ 100 ml	Confidence limits	
		Low	High			Low	High
0-0-0	<1.8	-	6.8	4-0-3	25	9.8	70
0-0-1	1.8	0.090	6.8	4-1-0	17	6.0	40
0-1-0	1.8	0.090	6.9	4-1-1	21	6.8	42
0-1-1	3~6	0.70	10	4-1-2	26	9.8	70
0-2-0	3.7	0.70	10	.4-1-3	31	10	70
0-2-1	5.5	1.8	15	4-2-0	22	6.8	50
0-3-0	5.6	1.8	15	4-2-1	26	9.8	70
1-0-0	2.0	0.10	10	4-2-2	32	10	70
1-0-1	4.0	0.70	10	4-2-3	38	14	100
1-0-2	6.0	1.8	15	4-3-0	27	9.9	70
1-1-0	4.0	0.71	12	4-3-1	33	10	70
1-1-1	6.1	1.8	15	4-3-2	39	14	100
1-1-2	8.1	3.4	22	4-4-0	34	14	100
1-2-0	6.1	1.8	15	4-4-1	40	14	100
1-2-1	8.2	3.4	22	4-4-2	47	15	120
1-3-0	8.3	3.4	22	4-5-0	41	14	100
1-3-1	10	3.5	22	4-5-1	48	15	120
1-4-0	10	3.5	22	5-0-0	23	6.8	70
2-0-0	4.5	0.79	15	5-0-1	31	10	70
2-0-1	6.8	1.8	15	5-0-2	43	14	100
2-0-2	9.1	3.4	22	5-0-3	58	22	150
2-1-0	6.8	1.8	17	5-1-0	33	10	100
2-1-1	9.2	3.4	22	5-1-1	46	14	120
2-1-2	12	4.1	26	5-1-2	63	22	150
2-2-0	9.3	3.4	22	5-1-3	84	34	220
2-2-1	12	4.1	26	5-2-0	49	15	150
2-2-2	14	5.9	36	5-2-1	70	22	170
2-3-0	12	4.1	26	5-2-2	94	34	230
2-3-1	14	5.9	36	5-2-3	120	36	250
2-4-0	15	5.9	36	5-2-4	150	58	400
3-0-0	7.8	2.1	22	5-3-0	79	22	220
3-0-1	11	3.5	23	5-3-1	110	34	250
3-0-2	13	5.6	35	5-3-2	140	52	400
3-1-0	11	3.5	26	5-3-3	170	70	400
3-1-1	14	5.6	36	5-3-4	210	70	400
3-1-2	17	6.0	36	5-4-0	130	36	400
3-2-0	14	5.7	36	5-4-1	170	58	400
3-2-1	17	6.8	40	5-4-2	220	70	440
3-2-2	20	6.8	40	5-4-3	280	100	710
3-3-0	17	6.8	40	5-4-4	350	100	710
3-3-1	21	6.8	40	5-4-5	430	150	1,100
3-3-2	24	9.8	70	5-5-0	240	70	710
3-4-0	21	6.8	40	5-5-1	350	100	1,100
3-4-1	24	9.8	70	5-5-2	540	150	1,700
3-5-0	25	9.8	70	5-5-3	920	220	2,600
4-0-0	13	4.1	35	5-5-4	1,600	400	4,600
4-0-1	17	5.9	36	5-5-5	>1,600	700	
4-0-2	21	6.8	40				

[a]Results to two significant figures

Table 7.6 Procedure for the standard bacteriological analysis of water (Modified from *Standard Methods for the Examination of Water and Wastewater*; Anonymous 2006)

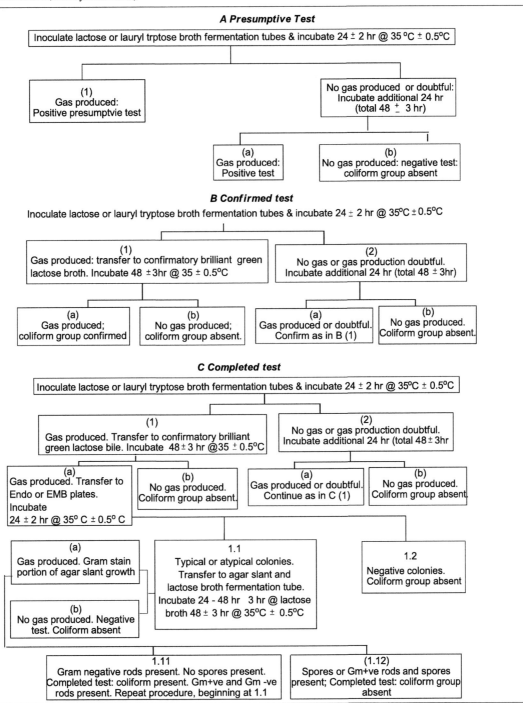

At least 100 ml of water are required for the MPN test. The volumes to be used in tests in liquid media will depend on the quality of the water to be examined, and the experience of the worker. When the water is expected to be of good quality one 50-ml volume and five 10-ml volumes are suitable. If it is of doubtful quality, one 50 ml, five 10 ml, and five 1-ml quantities could be used. When the water is heavily polluted the original water may have to be diluted by a factor of 100 or more in order to obtain some negative results in the series put up, and thus obtain a finite figure for the MPN. Whatever the series used, the volumes of water in the individual tubes and the number of tubes containing each volume of water should be such that an estimate of the MPN number of organisms present in 100 ml of the original water can be obtained from the table.

In the procedure outlined in *Standard Methods* for carrying out the presumptive test, water is inoculated into multiple tubes (chosen as indicated above) of lactose or lauryl tryptose broth and incubated at 35°C. The production of acid and gas within 48 h indicates a positive presumptive test.

In the confirmatory test, one to three loopfuls of broth are transferred from positive tubes to brilliant green lactose broth (BGLB) or streaked on endo or EMB agar (although the latter is not recommended). Gas formation after 48 h at 35°C indicates a positive Confirmed Test using the liquid medium.

On EMB agar, a typical positive colony of *E. coli* is nucleated with or without a metallic sheen. An atypical colony is opaque unnucleated, mucoid, and pink after 24 h in the incubator.

In the Completed Test, a positive growth from BGLB broth is streaked on endo or EMB agar. If plates were used in the confirmatory test typical and atypical colonies are transferred to lactose or lauryl tryptose broth, as well as to nutrient agar. Formation of gas in secondary fermentation and the occurrence of Gram-negative rods on staining is a positive Completed Test and indicates that *E. coli* is present. It ought to be pointed out that the media and some aspects of the techniques set out in Table 7.4 and which are available in *Standard Methods* are not universally adopted. On the European continent, the media used are different.

(b) Membrane filtration method

After filtration the pad is picked up with sterile forceps and placed on an absorbent pad soaked in lauryl tryptose broth (or agar) if enrichment is desired or directly on to M-Endo soaked pad or M-Endo agar. The result is expressed as coliform density (total coliform colonies per 100 ml). This is calculated as follows:

$$\frac{\text{Coliform colonies counted} \times 100}{\text{ml sample filtered}}$$

3. *Fecal coliforms*

The above tests are for general coliforms; they do not distinguish coli forms of animal origin from others. They are used for the examination of potable water since no coliform of any kind should be tolerated in treated water. For investigations of stream pollution, raw water sources, sewage treatment systems, bathing water, and general water-quality monitoring, it is important to know whether or not the coliform is of fecal origin, that is, whether or not from the intestine of warm-blooded animals.

The procedure for detecting fecal coliform is similar to the above in both the multiple tube or the membrane technique. In each the confirmatory test is carried out at an elevated temperature of 44.5°C for 24 h (i.e., the Eijkmann test). Gas production indicates coliform of warm-blooded animals. Prior to the confirmatory test, however, the coliform must be enriched in the lactose broth used in the presumptive test. A loopful is then transferred from all positive tubes to the confirmatory broth (Brilliant Green Lactose Bile Broth at 44.5°C for 24 h or Boric Acid Lactose Broth at 45°C for 48 h). When the membrane filter method is used, the incubation of the filters at 44.5°C may be done directly after the filtration of water.

Fecal Streptococci

(a) Multiple-tube method

Multiple portions of water are inoculated into tubes of glucose azide broth and incubated at 37°C for 72 h. When acidity is observed a heavy inoculation is introduced into further tubes of glucose azide broth and incubated at 44.5°C for 48 h. Tubes showing acidity at this temperature contain fecal streptococci.

Multiple portions of water may also be inoculated into tubes of buffered azide-glucose-glycerol broth (BAGG) and incubated at 44.5°C for 48 h. Acid production is indicative of fecal streptococcal presence, but the broth should be checked with Gram stain to confirm the presence of Gram-negative rods.

(b) Membrane filter technique

After filtration, the filter is placed on a plate of glucose-azide agar. It is then incubated at 37°C for 4 h and then at 44°C for 44 h. Red or maroon colonies are fecal streptococci.

4. *C. perfringens*

Multiple portions of water previously heated at 75°C for 10 min to kill nonspore formers are put into well-filled screw-cap bottles containing Differential Reinforced Clostridiral Medium (DRCM) and incubated at 37°C for 48°h. A positive reaction is shown by a blackening of the medium due to a reduction of sulfite and precipitation of ferrous sulfate. Any clostridium will show this reaction, but when inoculations are made from positive tubes into litmus milk a "stormy clot" is produced, that is, gas bubbles break up the clot of milk.

7.2.1.4 Standard Water Analysis

The procedure for standard water analysis is summarized in Table 7.6 and discussed below. It is divided into the presumptive test, the confirmed test, and the completed test (Anonymous 2006).

The Presumptive Test

In the presumptive test, a series of lactose broth tubes are inoculated with measured amounts of the water sample to be tested. The series of tubes may consist of three or four groups of three, five, or more tubes. The more tubes utilized, the more sensitive the test. Gas production in any one of the tubes is presumptive evidence of the presence of coliforms. The most probable number (MPN) of coliforms in 100 ml of the water sample can be estimated by the number of positive tubes (see Table 7.5).

The Confirmed Test

If any of the tubes inoculated with the water sample produce gas, the water is presumed to be unsafe. However, it is possible that the formation of gas may not be due to the presence of coliforms. In order to confirm the presence of coliforms, it is necessary to inoculate EMB (eosin methylene blue) agar plates from a positive presumptive tube. The methylene blue in EMB agar inhibits Gram-positive organisms and allows the Gram-negative coliforms to grow. Coliforms produce colonies with dark centers. *E. coli* and *E. aerogenes* can be distinguished from one another by the size and color of the colonies. *E. coli* colonies are small and have a green metallic sheen, whereas *E. aerogenes* forms large pinkish colonies.

If only *E. coli* or both *E. coli* and *E. aerogenes* appear on the EMB plate, the test is considered positive. If only *E. aerogenes* appears on the EMB plate, the test is considered negative. The reason for this interpretation is that, as previously stated, *E. coli* is an indicator of fecal contamination, since it is not normally found in water or soil, whereas *E. aerogenes* is widely distributed in nature and occurs outside the intestinal tract.

The Completed Test

The completed test is made using the organisms which grew on the confirmed test media. These organisms are used to inoculate a nutrient agar slant and a tube of lactose broth. After 24 h at 37°C, the lactose broth is checked for the production of gas, and a Gram stain is made from organisms on the nutrient agar slant. If the organism is a Gram-negative, nonspore-forming rod and produces gas in the lactose tube, then it is positive that coliforms are present in the water sample (Table 7.6).

7.2.1.5 Total Maximum Daily Loads and Microbial Source Tracking in Water Pollution

The US Clean Water Act (the actual official title is The Federal Water Pollution Control Act, 1972), in section 303(d)(1)(C) established the water quality standards and the Total Maximum Daily Loads (TMDL) programs thus (Anonymous 2007): (see Fig. 7.1).

(c) Each State shall establish for the waters identified in paragraph (1)(A) of this subsection, and in accordance with the priority ranking, the total maximum daily load, for those pollutants which the Administrator identifies under section 304(a)(2) as suitable for such calculation. Such load shall be established at a level necessary to implement the applicable water quality standards with seasonal variations and a margin of safety which takes into account any lack of knowledge concerning the relationship between effluent limitations and water quality.

A Total Maximum Daily Load (TMDL) is thus a calculation of the maximum amount of a pollutant that a water body can receive, from point and nonpoint sources, and still meet water quality standards, and an allocation of that amount to the pollutant's sources. The calculation must include a margin of safety to ensure that the water body can be used for the purposes the State has designated, account for seasonal variation in water quality, and natural background conditions.

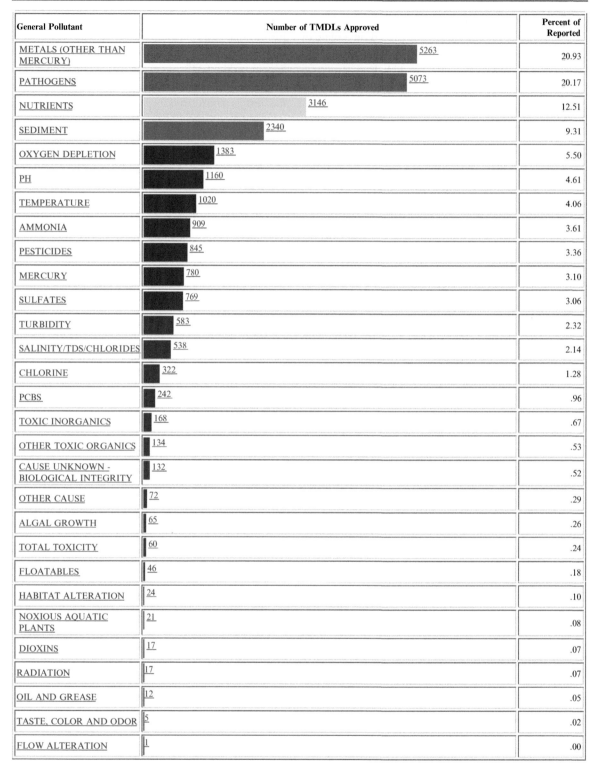

General Pollutant	Number of TMDLs Approved	Percent of Reported
METALS (OTHER THAN MERCURY)	5263	20.93
PATHOGENS	5073	20.17
NUTRIENTS	3146	12.51
SEDIMENT	2340	9.31
OXYGEN DEPLETION	1383	5.50
PH	1160	4.61
TEMPERATURE	1020	4.06
AMMONIA	909	3.61
PESTICIDES	845	3.36
MERCURY	780	3.10
SULFATES	769	3.06
TURBIDITY	583	2.32
SALINITY/TDS/CHLORIDES	538	2.14
CHLORINE	322	1.28
PCBS	242	.96
TOXIC INORGANICS	168	.67
OTHER TOXIC ORGANICS	134	.53
CAUSE UNKNOWN - BIOLOGICAL INTEGRITY	132	.52
OTHER CAUSE	72	.29
ALGAL GROWTH	65	.26
TOTAL TOXICITY	60	.24
FLOATABLES	46	.18
HABITAT ALTERATION	24	.10
NOXIOUS AQUATIC PLANTS	21	.08
DIOXINS	17	.07
RADIATION	17	.07
OIL AND GREASE	12	.05
TASTE, COLOR AND ODOR	5	.02
FLOW ALTERATION	1	.00

Fig. 7.1   Total Nationwide Number of TMDLs Approved since October 1, 1995 reported to EPA: 25,147 (Source: Anonymous 2007)

The Clean Water Act, requires States, Territories, and authorized Tribes every 2 years to prepare a list of impaired waters (i.e., water bodies for which water quality standards set for them by the States will not be met after application of technology-based controls), and establish priorities for action among the listed water bodies. These impaired waters do not meet water quality standards that states, territories, and authorized tribes have set for them for activities such as drinking water supply, contact recreation (swimming), and aquatic life support (fishing), and the scientific criteria to support that use. The States must then establish total maximum daily loads (TMDLs) for each listed water body, which are the sum of waste load allocations for point sources, load allocations for nonpoint sources, natural background contributions, and a margin of safety.

Until recently many states, territories, authorized tribes, and EPA had not developed many TDMLs' although they had been required by the Clean Water Act since 1972. Citizen organizations began bringing legal actions against EPA seeking the listing of waters and development of TMDLs. Legal actions have been instituted by these organizations in nearly 80% of the states and EPA is under court order or consent decrees in many states to ensure that TMDLs are established, either by the state or by EPA.

The processes for establishing a TDML are as follows:

1. Identify waters that do not meet water quality standards. In this process, the state identifies the particular pollutant(s) causing the water not to meet standards.
2. Prioritize waters that do not meet standards for TMDL development (e.g., waters with high naturally occurring "pollution" will fall to the bottom of the list).
3. Establish TMDLs (set the amount of pollutant that needs to be reduced and assign responsibilities) for priority waters to meet state water quality standards. A separate TMDL is set to address each pollutant with concentrations over the standards.
4. Develop strategies for reducing water pollution and assess progress made during implementation of the strategy. This is when a watershed partnership, including citizen groups, most likely will want to get involved. If the partnership has already developed a plan of action, it should be shared with the state. In fact, several states have incorporated watershed partnership plans in the state's strategy for specific TMDLs. The advantage is that local citizen groups usually know the topography and aspects of a watershed (i.e., the catchment area), which feeds water into a stream, river, lake or other body of water.

Microbial Source Tracking

Approximately 13% of surface waters in the USA do not meet designated use criteria as determined by high densities of fecal indicator bacteria. Although some of the contamination is attributed to point sources such as confined animal feeding operation and wastewater treatment plant effluents, nonpoint sources are believed to contribute substantially to water pollution (Anonymous 2005; Stockel and Harwood 2007).

The Clean Water Act in Section 101 (sub-sections 5 and 7) states that policy be developed in each state "to assure adequate control of *sources of pollutants* in each State"; and programs "for the control of nonpoint *sources of pollution* be developed and implemented in an expeditious manner so as to enable the goals of this Act to be met through the control of both point and nonpoint sources of pollution."

One of the requirements of the Clean Water Act is the identification of sources of pollution so as to adequately and rationally control their introduction into water. Accordingly, much effort has gone over the past years to develop methods for tracking the sources of pollutants. With regards to pathogens, the indicators of fecal pollution are used for reasons discussed above.

When a body of water has been listed as impaired by fecal bacteria, a TMDL study must be conducted to determine how the impairment can be remedied so the water body will meet appropriate water quality standards. It is required that the source of the pollutant bacteria be determined. Microbial source tracking (MST) methods have recently been used to help identify nonpoint sources responsible for the fecal pollution of water systems. MST tools are now being applied in the development of TMDLs as part of Clean Water Act requirements and in the evaluation of the effectiveness of best management practices.

Microbial source tracking (MST) includes a group of methodologies that are aimed at identifying, and in some cases quantifying, the dominant source(s) of fecal contamination in resource waters, including drinking, ground, recreational, and wildlife habitat waters. MST is transitioning from the realm of research to that of application. It is being discussed so that the student is aware of its existence and understands it as

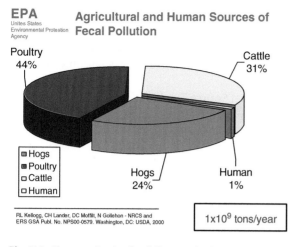

Fig. 7.2 Feces production by different animals and man: total, 1×10^{12} kg/annum (From Shanks 2005)

some of current procedures may become routine, while others are dropped (Long and Plumme 2004).

Methodologies Employed in MST

Present-day bacterial indicators of fecal pollution in water are total coliforms, fecal coliforms, *E. coli*, and enterococci. These are normally prevalent in the intestines and feces of warm-blooded mammals, including wildlife, livestock, and humans. It is not possible to tell which is from humans or from nonhuman sources. On the other hand, nonhuman feces spread pathogens such as *Salmonella* spp., *E. coli* O157, *Campylobacter jejuni*, *Leptospira interrogans*, *Giardia* spp., *Cryptosporidium parvum*, and hepatitis E virus. Figure 7.2 shows the annual feces production of humans and various animals in the USA. Humans produce only 0.7% of the total indicating the possibility that most of the fecal microorganisms in water could be from nonhuman sources, and indicating for adequate methodologies to accurately identify human fecal pollution (Stockel et al. 2004; Stewart-Pullaro et al. 2006).

Previously, fecal coliform/fecal streptococci (FC/FS) ratios were used to assess the general source of non-point fecal pollution: with FC/FS higher than 4 indicated human fecal pollution; FC/FS between 0.1 and 0.6 indicated domestic animals, and FC/FS less than 0.1 indicated fecal pollution from wild animals. This approach has, however, been abandoned because the FC/FS ratio was difficult to use in agricultural settings.

Current methods for MST fall into three groups: molecular, biochemical, and chemical. The molecular

or genotypic methods are based on the genetic make-up of different strains of fecal bacteria. The biochemical methods can be referred to as the phenotypic in contrast to the molecular methods, because these are based on the activities of the genes in secreting biochemical compounds. The chemical methods do not assess the microorganisms; rather they are based on finding in wastewaters chemical compounds associated with humans and are used to determine if the source of the pollution is human. A fourth method, the *immunological source tracking* method is being studied by some workers. It is based on identifying the unique proteins (antigens) peculiar to each animal and which are shed into feces. When developed this technique should enable the proper identification of the animal from which a fecal bacterium has come. Just as with chemical methods, immunological methods would not require a library, but it is yet under development.

MST methods can also be divided into two: library-dependent and library-independent methods. Library-dependent methods are culture based and rely on isolate-by-isolate typing of bacteria cultured from various fecal sources and from water samples. The isolates are then matched to their corresponding source categories by direct subtype matching or by statistical means. On the other hand, library-independent methods are based on sample-level detection of a specific, host-associated genetic marker in a DNA extract by PCR. All cultivation-independent methods are library independent as are chemical and immunological methods (when perfected). Some cultivation-dependent methods such as the use of phages are also library independent.

1. *Molecular methods*

   The molecular methods of MST are based on DNA patterns that are unique to each source due to variables such as the food consumption and health of the individual. For molecular methods, fecal bacteria are isolated from water samples and a DNA pattern or "fingerprint" is obtained. The pattern is then compared to the library of patterns from specific sources to identify the source(s) of isolates in a sample. Many molecular methods are being used and most require known-source libraries, although a few do not. It is thought that the distinctions between fecal bacteria from different animals (including humans) occur because the intestinal environments (selective pressures) are not the same, and fecal bacteria develop with detectable differences that can be related to sources. The key to molecular MST methods is finding these differences against a large background of similarity. Molecular methods are believed to be more accurate for individual sources (e.g., deer, muskrat, chickens, and horses). However, the molecular methods are also considerably more expensive and slower to perform in the laboratory. Only a selected number of molecular methods will be discussed and only the outline will be given.

(a) Repetitive PCR (Rep PCR)

Rep-PCR genomic fingerprinting makes use of DNA primers complementary to naturally occurring, highly conserved, repetitive DNA sequences, present in multiple copies in the genomes of most Gram-negative and several Gram-positive bacteria. DNA between adjacent repetitive extragenic elements is amplified using PCR to produce various size DNA fragments. The PCR products are then run on agarose-gel electrophoresis to produce specific DNA fingerprint patterns, which can then be analyzed using a pattern recognition computer software. The rep-PCR genomic fingerprints generated from bacterial isolates permit differentiation to the species, subspecies, and strain level.

(b) Random amplification of polymorphic DNA (RAPD)

This is a type of PCR, but the segments of DNA that are amplified are random. To perform RAPD creates several arbitrary, short primers (8–12 nucleotides) are produced, then PCR is performed using a large template of genomic DNA, hoping that fragments will amplify. By resolving the resulting patterns, a semi-unique profile can be gleaned from an RAPD reaction.

No knowledge of the DNA sequence for the targeted gene is required, as the primers will bind somewhere in the sequence, but it is not certain exactly where. This makes the method popular for comparing the DNA of biological systems that have not had previous attention, or in a system in which relatively few DNA sequences are compared (it is not suitable for forming a DNA databank). Due to the fact that it relies on a large, intact DNA template sequence, it has some limitations in the use of degraded DNA samples. This method requires screening primers (there are over 1,200 commercially available) to find sets of polymorphisms (i.e., differences in

nucleotide sequence among individuals) that are either unique to fecal bacteria from a given source or occur in a given source to a large and predictable degree. Once such sets of polymorphisms have been found, fecal bacteria can be "sourced" by comparison. It has been used in differentiating human and nonhuman *E. coli* sources.

(c) Ribotyping

Ribotyping, also referred as molecular finger printing, involves the bacterial genes that code for 16S ribosomal RNA. Because such genes are highly conserved (i.e., change very slowly evolutionarily) in microorganisms, ribotyping has been widely accepted for microbial identification. Ribotyping involves cutting the total genomic bacterial DNA with different DNAases, or restriction enzymes, followed by gel electrophoresis. Following electrophoresis, Southern blotting is performed to blot the DNA bands onto nylon membranes from the gels. DNA probes must be prepared for bacterial 16S and 23S rRNA and labeled with some type of detection system, such as fluorescent dye or radioactively). Membrane hybridization is then performed to hybridize the probes with the appropriate DNA bands on the nylon filter. Difference in the size and location of the ribosomal RNA bands on the filters can then be used to differentiate between the sources of the fecal bacteria.

Strains of *E. coli* are adapted to their own specific environment (intestines of host species), and as a result differ from other strains found in other host species. To use this MST method, collections of potential source material (fecal samples of all potential sources in the environment) must be collected and subtyped. The genetic fingerprints of the bacterial isolates from the water samples can then be compared to those of the suspected source.

Ribotyping has been widely used in microbial source tracking studies. It has been found that database size, geographic distribution of the isolated bacteria, and the presence of replicate isolates in the bacterial source library impact the ability of ribotyping to differentiate among bacteria at the host species level

(d) Pulse-field gel electrophoresis (PFGE)

PFGE is a DNA "fingerprinting" technique that uses restriction enzymes on the entire DNA genome. The large genomic fragments are placed in a unique gel apparatus where electric current is passed through a gel in different directions at low voltage for 10–12 h to achieve the best level of band separation possible. PFGE is similar to ribotyping, but instead of analyzing rRNA, it uses the whole DNA genome.

In another system of PFGE, the bacterial DNA is embedded in an agarose plug and digested while in the plug; the plugs are placed in hollow gel combs and become part of the gel as the gel is cast around the combs. The gels are stained and photographed after electrophoresis. It is better for discriminating host sources compared to the other fecal coliforms.

(e) Amplified fragment length polymorphism (AFLP-PCR)

Amplified Fragment Length Polymorphism (AFLP) is a polymerase chain reaction (PCR) based genetic fingerprinting technique. AFLP uses restriction enzymes to cut genomic DNA, followed by ligation of complementary double stranded adaptors to the ends of the restriction fragments. A subset of the restriction fragments are then amplified using two primers complementary to the adaptor and restriction site fragments. The fragments are visualized on denaturing polyacrylamide gels either through autoradiographic or fluorescence methodologies. AFLP-PCR is a highly sensitive method for detecting polymorphisms in DNA. The procedure of this technique is divided into three steps:

- Digestion of total cellular DNA with one or more restriction enzymes and ligation of restriction half-site specific adaptors to all restriction fragments.
- Selective amplification of some of these fragments with two PCR primers that have corresponding adaptor and restriction site-specific sequences.
- Electrophoretic separation of amplicons on a gel matrix, followed by visualization of the band pattern.

(f) Gene-specific PCR

Gene-specific PCR methods are based on the discovery that certain enterotoxin genes are carried almost exclusively by *E. coli* that infect individual species of warm-blooded mammals including humans, cattle, and swine. *E. coli* samples

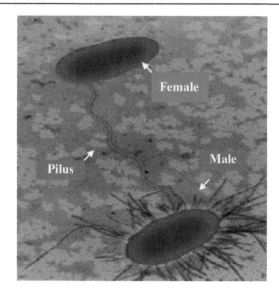

Fig. 7.3 Pilus linking a male to a female *Escherichia coli* cell in conjugation (Image reproduced with permission)

are isolated from water and the DNA extracted. The samples are amplified and probed with the genes specific for man and other animals. The procedure is relatively simple and can be performed within two working days. The advantages of the gene-specific methods are that they are highly specific and library independent.

(g) The use of bacteriophages

Bacteriophages (viruses) that infect *E. coli* and possibly other closely related coliform bacteria are called coliphages (Stewart-Pullaro et al. 2006). Coliphages were first proposed as indicators of the presence of *E. coli* bacteria and are taxonomically very diverse, covering the following six virus families: three families of double-stranded DNA viruses (*Myoviridae*, *Styloviriae*, *Podoviridae*), two families of single-stranded DNA phages (*Microviridae* and *Inoviridae*), and one family of single-stranded RNA viruses (*Leviviridae*). Coliphages that infect via the host cell wall of *E. coli* are called somatic coliphages (including families *Myoviridae*, *Styloviriae*, *Podoviridae*, and *Microviridae*). Male-specific (also called F+) coliphages (*Inoviridae* and *Leviviridae*) infect by attaching to hair-like appendages called F-pili protruding from the host bacterium surface (Fig. 7.3).

Male-specific coliphages have been studied extensively as fecal indicators and for water/

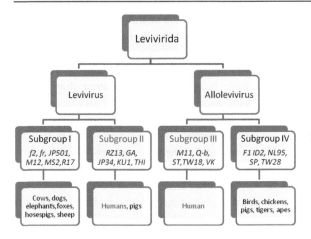

Fig. 7.4 Spectrum of F+ coliphages (Bacteriophage Family *Leviviridae*) (Modified from Smith 2006)

wastewater treatment/disinfection efficacy (Stewart-Pullaro et al. 2006). F⁺ RNA coliphages can be distinguished genetically (via nucleic acid detection methods) or antigenically (via immunological methods) into four distinct subgroups: I, II, III, and IV. Groups II and III are associated primarily with human fecal waste and Groups I and IV are associated primarily with animal fecal waste (see Fig. 7.4).

Strengths of the use of coliphages as indicators

The advantages of F⁺ coliphages as fecal indicators include their

- Presence in relatively high concentrations in sewage
- Relatively high persistence through wastewater treatment plants, compared to typical bacterial indicators like *E. coli* and fecal coliforms (coliphages may behave similarly to human viruses during wastewater treatment)
- Ability to be detected in relatively small (100 ml) to medium (1,000 ml) volumes of fecally contaminated water

The use of coliphage typing for microbial source tracking is library independent, but can only be used to broadly distinguish human and animal fecal contamination. However, there is a problem with separation between human serotypes and serotypes associated with pigs, which also contain group II. Furthermore, not all animals have FRNA coliphage associated with their respective *E. coli*. The coliphage is persistent in

the environment for less than a week and survival is a function of sunlight and water temperature. Ultraviolet light denatures the virus and below 25°C, F-pilus synthesis ceases. The coliphage does not replicate in the environment, only in the presence of F-pilus *E. coli*, and is not found in sediments, just in the water column.

2 *The biochemical methods*

(a) Antibiotic resistance analysis (ARA)

ARA is a method that is based on patterns of antibiotic resistance of bacteria from human and animal sources. The premise behind this method is that human fecal bacteria will have greater resistance to specific antibiotics followed by livestock and wildlife, and that livestock will have greater resistance to some antibiotics not used in humans. These differences occur because humans are exposed to different antibiotics as against cattle or pigs, poultry, or wildlife. Human fecal bacteria will be expected to have the greatest resistance to antibiotics normally used for humans and that domestic and wildlife animal fecal bacteria will have significantly less resistance (but still different) to the battery of antibiotics and concentrations used. Isolates of fecal streptococci and/or *E. coli* are taken from various sources (human, livestock, and wildlife), and these isolates are grown on a variety of antibiotics. Following incubation, isolates are scored as "growth/no growth" for each concentration of an antibiotic. The resistance pattern of an organism is used to identify its source. A database of antibiotic resistant patterns from known sources within a watershed is prepared as a basis for comparing and identifying isolates in that watershed.

Three approaches have been followed in the use of antibiotic resistance in MST: antibiotic resistance analysis (ARA), multiple antibiotic resistance (MAR), and the filter paper disc (or Kirby-Bauer). In ARA studies, different concentrations of each antibiotic being tested are used; in MAR studies, bacteria are tested for resistance to one concentration of different antibiotics. In the filter paper disc approach, small filter disks impregnated with antibiotics are placed on the test organism on agar; the zone of growth inhibition around the disks is used to quantify resistance. Some workers believe that ARA provides the most information of the three antibiotic-based approaches. The extent of the accuracy of this method is measured by the average rate of correct classification (ARCC). With wild animals the ARCC is between 98% and 100%, whereas with livestock it is between 34% and 89%. The methods are low cost and easy to perform.

(b) Carbon utilization profiles (CUP)

The CUP is based on differences among bacteria in their use of a wide range of carbon and nitrogen sources. The method compares differences in the utilization of several carbon and nitrogen substrates by different bacterial isolates. The working hypothesis behind CUP is that various animal populations have different diets; therefore, fecal bacteria have evolved in the various guts to utilize different food sources. Substrate utilization can be rapidly scored by the formation of a purple color due to the reduction of a tetrazolium dye included with the substrates and automatically detected using a microplate reader. The method is rapid, simple, and requires little technical expertise, and has been simplified by the availability of commercial microwell plates containing substrates, one of the most commonly used being Biolog microplates. The Biolog system allows the user to rapidly perform, score, and tabulate 96 carbon source utilization tests per isolate and is widely used in the medical field for identification of clinical isolates. The bacterial isolates are first grown in liquid culture and suspension of cells at a standardized turbidity is used to inoculate the microplates. After incubation at 37°C for 24 h, presence or absence of growth is indicated by purple dye formation and is assessed manually or automatically using a plate reader.

3. *Chemical methods*

The use of chemical targets has been suggested as an alternate approach to biological markers based on the premise that certain chemicals are only found in fecal samples. Chemical methods do not detect fecal bacteria; they detect chemical compounds that are associated with humans. If the compound(s) are found in water body, then there is a likely human source. Different chemical compounds have been recently used as tools to predict sources of human fecal pollution. Most chemical markers have been used primarily to trace human contamination. For example, caffeine, fragrance materials, and fluorescent whitening agents (laundry detergent brighteners) have been under investigation due to their exclusive use by humans. Fecal sterols and fecal stanols found in humans are also promising sewage pollution markers. The main problem is that the long-term fate of these organic chemicals in environmental waters is yet unknown.

(a) Optical brighteners

This method detects the optical brighteners that are in all laundry detergents. They are persistent in the environment and are detected using mass spectroscopy. Sample collection is accomplished by placing optical brightener-free cotton in a wire mesh trap and placing the trap in the stream for a few days. After the trap is recovered the cotton is examined with a black light to see if it glows. The fluorescent cotton can then be examined with mass spectroscopy to verify the presence of the compounds. If they can be detected, then there must be a human source. Laundry detergents such as fluorescent whitening agents, sodium tripolyphosphate, and linear alkyl benzenes have been used to predict human impact; however, these chemicals cannot reliably be traced to sewage or fecal pollution and can only be attributed to general human or industrial sources.

(b) Caffeine detection

Caffeine is present in several beverages, including coffee, tea, soft drinks, and in many pharmaceutical products. It is excreted in the urine of individuals who have ingested the substance, and subsequently, it has been suggested that the presence of caffeine in the environment would indicate the presence of human sewage. Levels of caffeine in domestic wastewater have been measured to be between 20 and 300 g/l (68). Caffeine could thus be used as an indicator chemical. A major problem is that it is expensive trying to detect caffeine in the environment. Furthermore, some other plants such as watermelon have significant levels of caffeine and could obscure the results. Finally, caffeine is easily degraded by soil microbes; hence, the detectable quantities could be greatly reduced.

(c) Coprostanol

Coprostanol is a fecal stanol that is formed during catabolism of cholesterol by indigenous bacteria present in the gut of humans and higher animals and is the primary stanol detected in domestic wastewater. Coprostanol has been found to make up about 60% of the total stanols in human, whereas feces from pigs and cats contained coprostanol at a tenfold lower quantity. Other fecal stanols, such as 24-ethyl-coprostanol, were found to be predominant in herbivores, such as cows, horses, and sheep, suggesting the potential use of these chemicals for MST.

Choice of the MST Method to Use

Although the impetus for MST comes from the TDML requirements of the Clean Water Act (1972) in the USA, tracing the source of water contamination is now a worldwide event, and is practiced in countries with developed economies such as those of the European Union and Japan.

At present, the various methods for MST are being examined in these countries and have not yet been selected. The purpose of this chapter is to expose the students to the various methods, most especially their scientific underpinning, so as to understand any methods, which are eventually selected in any one country, state within a country, or a water authority.

Table 7.7 gives the abbreviated procedure and the advantages and disadvantages of each method. Some of the factors, which will weigh in the minds of those deciding on which method to use are the following:

1. MST methods are still very new, and no one method will do for all situations; the method(s) to be used will depend of the fecal material be dealt with.
2. As many methods suitable for a given watershed as possible should be used. If similar results are obtained from different methods, then this helps reinforce their credibility.
3. MST methods have been used primarily with *E. coli* and the fecal streptococci, and to a lesser degree with bifidobacteria, *Bacteroides*, and coliphages. Although *E. coli* is widely used, other indicator organisms have shown good results, and are sometimes even more reliable than *E. coli*, in some circumstances. Wherever possible, therefore, the range of indicator organisms should be used (Fig. 7.5).

7.3 Pollution by Petroleum in Oceans and Seas: Role of Microorganisms in Oil Degradation and Remediation

7.3.1 Composition of Crude Oil

Crude petroleum is a complex combination of hydrocarbons. It consists predominantly of aliphatic, alicyclic, and aromatic hydrocarbons. It may also contain small amounts of nitrogen, oxygen, and sulfur compounds. This category encompasses light, medium, and heavy petroleums, as well as the oils extracted from tar sands (Anonymous 2003). Hydrocarbonaceous materials requiring major chemical changes for their recovery or conversion to petroleum refinery feedstocks such as crude shale oils, upgrade shale oils, and liquid coal fuels are not included in this definition of crude petroleum.

Crude oil contains hydrocarbons in the carbon number range from C1 to C60+. It also contains organometallic complexes, notably of sulfur and vanadium, and dissolved gases such as hydrogen sulfide. Crude oils range from thin, light-colored oils consisting mainly of gasoline quality stock to heavy, thick tar-like materials. An "average" crude oil has the following general composition: carbon 84%, hydrogen 14%, sulfur 1–3%, nitrogen 1%, and oxygen 1%. Minerals and salts make up 0.1%. The chemical composition of crude oils can vary tremendously from different producing regions (see Table 7.8) and even from within a particular formation (Anonymous 2003).

Crude oils are made up of a wide spectrum of hydrocarbons. They vary greatly in appearance depending on their composition. They are usually black or dark brown but may be yellowish greenish. In the ground crude oil is sandwiched between natural gas on top and salt water at the bottom. It may sometimes be in a semisolid form mixed with sand and is known as bitumen.

7.3.1.1 Categorization of Crude Petroleum

Crude petroleum may be further categorized thus as follows:

Paraffinic versus Naphthenic

Crude oils contain both paraffinic and naphthenic hydrocarbons but if there is a preponderance of paraffinic hydrocarbons present, the crude oil is referred to as paraffinic crude. These crudes would be rich in straight and branched chain paraffins. Conversely, a crude in which naphthenic hydrocarbons are predominant is

Table 7.7 Advantages and disadvantages of the various microbial source tracking (MST) methods (From *Microbial Source Tracking Guide Document*, Center for Disease Control [CDC]; Anonymous 2005)

Method	Advantages	Disadvantages
Antibiotic resistance	• Rapid; easy to perform	• Require reference library
		• Requires cultivation of target organism
	• Requires limited training	• Libraries geographically specific
	• May be useful to differentiate host source	• Libraries temporally specific
		• Variations in methods in different studies
Carbon utilization profiles (CUP)	• Rapid; easy to perform	• Require reference library
	• Requires limited training	• Requires cultivation of target organism
		• Libraries geographically specific
		• Libraries temporally specific
		• Variations in methods in different studies
		• Results often inconsistent
Repetitive PCR (rep-PCR)	• Highly reproducible	• Requires reference library
	• Rapid; easy to perform	• Requires cultivation of target organism
	• Requires limited training	• Libraries may be geographically specific
	• May be useful to differentiate host source	• Libraries may be temporally specific
Random amplification of polymorphic DNA (RAPD)	• Rapid; easy to perform	• Requires reference library
	• May be useful to differentiate host source	• Requires cultivation of target organism
		• Libraries may be geographically specific
		• Libraries may be temporally specific
		• Has not been used extensively for source tracking
AFLP	• Highly reproducible	• Labor-intensive
	• May be useful to differentiate host source	• Requires cultivation of target organism
		• Requires reference library
		• Requires specialized training of personnel
		• Libraries may be geographically specific
		• Libraries may be temporally specific
		• Variations in methods used in different studies
PFGE	• Highly reproducible	• Labor-intensive
	• May be useful to differentiate host source	• Requires cultivation of target organism
		• Requires specialized training of personnel
		• Requires reference library
		• Libraries may be geographically specific
		• Libraries may be temporally specific
Ribotyping	• Highly reproducible	• Labor-intensive (unless automated system used)
	• Can be automated	
	• May be useful to differentiate host source	• Requires cultivation of target organism
		• Requires reference library
		• Requires specialized training of personnel
		• Libraries may be geographically specific
		• Libraries may be temporally specific
F+ RNA coliphage	• Distinguishes human from animals	• Requires cultivation of coliphages
	• Subtypes are stable characteristics	• Subtypes do not exhibit absolute host specificity
	• Easy to perform	• Low in numbers in some environments
	• Does not require a reference library	

(continued)

Table 7.7 (continued)

Method	Advantages	Disadvantages
Gene-specific PCR	• Can be adapted to quantify gene copy number	• Require enrichment of target organism
	• Virulence genes may be targeted; providing direct evidence that potentially harmful organisms present	• Sufficient quantity of target genes may not be available requiring enrichment or large quantity of sample
	• Does not require reference library	• Requires training of personnel
		• Primers currently not available for all relevant hosts
Host-specific PCR	• Does not require cultivation of target organism	• Little is known about survival and distribution in water systems
	• Rapid; easy to perform	• Primers currently not available for all relevant hosts
	• Does not require a reference library	
Virus specific PCR	• Host specific	• Low in numbers, requires large sample size
	• Easy to perform	• Not always present even when humans present
	• Does not require reference library	

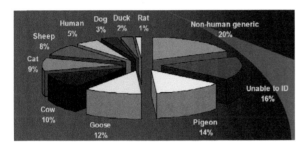

Fig. 7.5 Typical MST results (From Callaghan 2005)

referred to as a naphthenic crude. These crudes contain mainly naphthenic and aromatic hydrocarbons.

Sweet versus Sour

Crude oils may be referred to as either sweet or sour depending upon the level of hydrogen sulfide present. A sweet crude has very little H_2S, whereas a sour crude has larger quantities of H_2S present.

Light versus Heavy

Crude oil may be divided into light and heavy on the basis of their gravity. The API (American Petroleum Institute) gravity is determined as:

$$API = \frac{141.5}{\text{Specific gravity}} - 131.5$$

or

$$API = (141.5 / \text{Specific gravity}) - 131.5$$

Crude oils with gravity > 33°API are considered as light crudes. Such crudes with a high percentage composition of hydrogen are usually more suitable for processing for gasoline production. Heavy crudes, that is, those with gravity < 28°API tend to contain more asphaltenes and are usually rich in aromatics. These heavy crudes require more steps in their processing (Anonymous 2003).

The major components of crude oil are hydrocarbons, ranging from very volatile, light materials such as propane and benzene to more complex heavy compounds such as bitumens, asphaltenes, resins, and waxes. They are separable into four fractions: the saturates, the aromatics, the resins (pyridines, quinolines, carbozoles, sulfoxudes, and amides), and asphaltenes (phenols, fatty acids, ketones, esters, and porphyrins).

Saturates are hydrocarbons containing no double bonds. They are further classified according to their chemical structures into alkanes (paraffins) and cycloalkanes (naphthenes). Alkanes have either a branched or unbranched (normal) carbon chain(s), and have the general formula C_nH_{2n+2}. Cycloalkanes have one or more rings of carbon atoms (mainly cyclopentanes and cyclohexanes), and have the general formula C_nH_{2n}. Most of the cycloalkanes in crude oil have an alkyl substituent(s) (Fig. 7.6).

Aromatics have one or more aromatic rings with or without an alkyl substituent(s). Benzene is the simplest one (Fig. 7.6), but alkyl-substituted aromatics generally exceed the nonsubstituted types in crude oil. In contrast to the saturated and aromatic fractions, both the resin and asphaltene fractions contain non-hydrocarbon

Table 7.8 Gross compositions of crude oils from different parts of the world (From http://www.petroleumhpv.org/docs/crude_oil/111503_crude_robsumm_final.pdf. Reproduced courtesy of the American Petroleum Institute from *Robust Summary of Information on Crude Oil*, 2003; Anonymous 2003)

Country or origin of crude	Paraffin vol%	Naphthalenes vol%	Aromatics vol%	Sulfur wt%	API gravity (°API)
Light crudes					
Saudi Light	63	18	19	20	34
South Louisiana	79	45	19	0	35
Nigerian Light	37	54	9	0.1	36
North Sea Brent	50	34	16	0.4	37
Beryl	47	34	19	0.4	32
Lost Hill	Nonaromatics	50	50	0.9	
Mid-range crudes					
Kuwait	63	26	24	24	31
Venezuela Light	52	34	14	1.5	30
USA West Texas Sour	46	32	22	1.9	32
Heavy crudes					
Prudhoe Bay	27	36	15	2.1	28
Saudi Heavy	60	20	15	2.1	28
Venezuela Heavy	35	53	12	23	24
Belridge Heavy	Nonaromatics	37	63	1.1	–

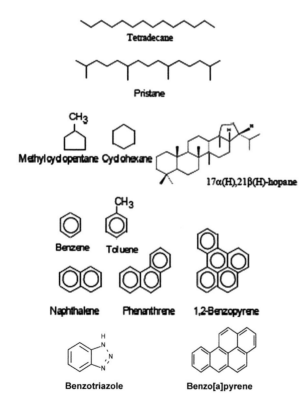

Fig. 7.6 Various hydrocarbons (From Harayama et al. 1999)

polar compounds. Their elements contain, in addition to carbon and hydrogen, trace amounts of nitrogen, sulfur, and/or oxygen. These compounds often form complexes with heavy metals.

Asphaltenes consist of high-molecular-weight compounds, which are not soluble in a solvent such as *n*-heptane, while resins are *n*-heptane-soluble polar molecules.

Resins contain heterocyclic compounds, acids, and sulfoxides. The components of petroleum in crude oil have been analyzed mainly by using gas chromatography in combination with mass spectrometry (GC/MS). Consequently, the chemical structures of the higher-molecular-weight components (the heavy fractions) that cannot be identified by GC are mostly unknown. Furthermore, the compositions of many branched alkanes and alkyl cycloalkanes have not been determined because their isomers are numerous and cannot be resolved by gas chromatography. Many compounds in crude oil are yet to be identified.

During refining, many products are obtained from crude oil by fractional distillation, yielding different fractions or cuts. Alkenes, unsaturated hydrocarbons including ethylene, are not found in crude oil, but are produced during the refining or cracking of crude oil.

Various hydrocarbons and fractions of crude oil are given in Fig. 7.6.

7.3.2 Oil Spills

Petroleum has a wide range of applications in the form of vehicle fuel, heating source for homes and industry, for electricity generation, and in basic industrial operations; there is, therefore, an enormous demand for the production and movement of oil. Due to its high-energy density, easy transportability, and relative abundance, crude petroleum has become the world's most important source of energy since the mid-1950s (American Petroleum Institute 2010).

About 16% of the petroleum produced is used as raw material for many chemical products, including pharmaceuticals, solvents, fertilizers, pesticides, and plastics. Oil is thus being continuously pumped from the ground, refined, transported, and stored, sometimes resulting in spilling during "operations" or "accidents."

World oil production peaked in 2008 at 81.73 million barrels/day (mbd) including crude oil, lease condensate, oil sands, and natural gas plant liquids. If natural gas plant liquids are excluded, then the production peak remains in 2008 but at 73.79 mbd. About half of this is transported by sea. The international transport of petroleum by tankers is frequent and the potential for spills is therefore great. Tanker accidents such as the well-known one of the T/V Exxon Valdez in Prince William Sound, Alaska, severely affect the local marine environment. Tankers take on ballast water, which contaminates the marine environment when it is subsequently discharged. Off-shore drilling is another source of petroleum marine pollution. The largest source of marine contamination by petroleum appears to be the runoff from land. Annually, more than two million tons of petroleum are estimated to end up in the sea. This section will discuss oil spills and their occurrence, the fate of oil in spills (see Table 7.9), and some aspects of remediation in spills.

Oil spills have been a major cause of concern as they pose a danger to public health, devastate natural resources, and disrupt the economy. Marine life can be affected by both the physical and chemical properties of the oil spilled, the main threat posed by the residues on the contaminated sea surface. A short-term exposure can render unpleasant tastes and smells to aquatic life, but a prolonged exposure can impair the ability of marine organisms to reproduce, grow, feed, or perform other functions. Beaches and shorelines may be adversely affected from the aesthetic and recreational points of view (Anonymous 2003).

In freshwaters, oil contamination can result in severe impacts on the habitat because the movement associated with water is minimal, as compared to marine environment. Stagnant water bodies cause the oil to remain in the environment for long, resulting in prolonged exposure of the plants and animals. In the case of flowing streams and rivers, the oil not only tends to collect on plants and grasses growing on the banks, but also interacts with sediments, thereby affecting the organisms.

Oil spills are usually thought of as occurring in takers in oceans only. However, they can be due to accidents within the industry, or to natural causes. Accidents occur within the industry during the processes of generating and moving oil in tankers, barges, pipelines, refineries, and storage facilities. Natural causes of oil spillage include hurricanes, which affect tankers. Furthermore, oil lying in the ground under the sea may seep out and pollute sea and ocean water. Some of the causes of oil spills from 1970 to 2009 are given in Table 7.10.

The spills in oceans and seas are more dramatic and are usually carried by the news media. There have been numerous oil spills and three of the largest will be mentioned briefly. The world's largest so far (January 2011) is generally agreed to be the *Gulf War Oil Spill* (January 23, 1991), which occurred during the Iraq–Kuwait war. Most oil spills are accidents, but this was no accident at all. Rather, the Iraqi military had released oil from several tankers to prevent US forces from attacking. The Lakeview Gusher Oil Spill (March 14, 1910) is regarded as the largest recorded US oil well gusher, in the Sunset oil field in Kern County, California. It took 18 months to control and it is believed to be due to the infancy of oil technology at the time. The Ixtoc I oil Spill (June 3, 1979) was an exploratory oil well in the bay of Campeche in the Gulf of Mexico. It is the third largest oil spill and second largest accidental spill in the history. In recent times, the Gulf of Mexico oil spill (also known as Deepwater Horizon oil spill and the BP oil spill) resulted from a seafloor oil gusher and flowed for 3 months before being capped on July 15, 2010, but not before 4.9 million barrels or 205.8 million gallons of crude oil had gushed from it. With 627,000 t of crude oil, this would appear to be the largest spill in history, second only to the Lakeview Gusher, California, USA, March 1910–September 1911, which spilled 1,230,000 t. Table 7.11 gives some major oil spills in oceans and seas (see also Table 11.1).

Table 7.9 Degradation of benzene-extractable material from Colgate Creek sediment by microorganisms indigenous to the sediment (From Walker and Colwell 1977. With permission)

Class of hydrocarbon	Weight (mg) in:		Amount of hydrocarbons wt% degraded expressed as:	
	Un-degraded control	Degraded sample	%total hydrocarbon	% individual hydrocarbon
Alkanes	312.0	96.3	94.4	69.1
1-Ring cycloalkanes	417.7	134.2	92.3	67.9
2-Ring cycloalkanes	145.6	41.9	97.6	71.2
3-Ring cycloalkanes	112.7	31.8	98.2	71.8
Alkylbenzenes	350.1	77.2	95.5	77.9
Benzcycloparraffins	152.5	45.4	97.4	70.2
Benzdicycloparraffins	110.0	35.8	97.9	67.7
Naphthalenes	74.5	23.2	98.7	68.9
Acenaphthalenes	26.0	9.6	99.5	63.1
Fluorenes	13.9	4.0	99.8	71.2
Phenanthrenes	10.4	2.0	99.9	71.0
Cyclopentanaphenalenes	6.9	2.0	99.9	71.0

Table 7.10 Causes of oil spills, 1970–2009 (From http://www.itopf.com/information-services/data-and-statistics/statistics/; Anonymous 2010b. With permission)

	<7 t	7–700 t	>700 t	Total
Operations				
Loading/discharging	3,155	383	36	3,574
Bunkering	560	32	0	593
Other operations	1,221	62	5	1,305
Accidents				
Collisions	176	334	129	640
Groundings	236	265	161	662
Hull failures	205	57	55	316
Equipment failures	206	39	4	249
Fire and explosions	87	33	32	152
Other/unknown	1,983	44	22	2,049
Total	7,829	1,249	444	9,522

7.3.2.1 Behavior of Oil in an Oil Spill

Complex physical, chemical, and biological factors collectively known as weathering, determine the fate and behavior of oil during a spill leading to its dispersion and the degradation of its components and its disappearance. Below are the weathering phenomena associated with spilled oil (Anonymous 2009).

1. *Spreading of the oil*

   Oil floats on the water in a marine spill and quickly spreads. In as short as 10 min, a spill of 1 t of oil can disperse over a radius of 50 m, forming a slick 10-mm thick. The slick gets thinner (<1 mm) as oil continues to spread, covering an area of up to 12 km². When oil is spilled into the sea evaporation plays an important part in its disappearance and dur-

ing the first several days after the spill, the volatile components of the oil evaporate. The rate of this evaporation depends on the composition of the crude oil. Brent (North Sea) and Nigerian oil, which are known for their high content of the lighter components, would have, within a few days, lost about two-thirds of its content on being spilled into the sea. Venezuelan oil, on the other hand, would have lost only two-fifths because it has a larger quantity of the heavier components of crude oil. Furthermore, apart from the water-soluble components a considerable part of oil transforms into the gaseous phase. Besides volatile components, the slick rapidly loses water-soluble hydrocarbons. The rest – the more viscous fractions – slow down the slick spreading. The spread of the slick follows the direction of the wind, waves, and current. As the slick thins down to a thickness of about 0.1 mm, it disintegrates into separate fragments that spread over larger and more distant areas. It soon forms into droplets, which are transported to distances far removed the site of the spill.

2. *Solubility of some components*

   Many components are water soluble to a certain degree, especially low-molecular-weight aliphatic and aromatic hydrocarbons. Some compounds formed as a result of oxidation of some oil fractions in the marine environment also dissolve in seawater. In comparison with evaporation, dissolution is slower.

3. *Emulsification*

   The emulsification of oil in the marine environment depends on oil composition and the degree of turbulence in the water. The most stable water-in-oil

Table 7.11 Major oil spills, 1910–2010 (in chronological order) (Modified from The International Tankers Owners Pollution Federation (ITOPF), http://www.itopf.com/information-services/data-and-statistics/statistics/ and http://en.wikipedia.org/wiki/List_of_oil_spills; Anonymous 2010b, c. With permission)

S. No.	Dates	Vessel name or name of spill or cause of spill	Location	Spill size (t)
1.	January 21, 2011	India, Mumbai, Arabian Sea	Mumbai–Uran pipeline spill	55
2.	January 11, 2011	Italy, Sardinia, Porto Torres	Fiume Santo power station	15
3.	August, 2010	Mumbai oil spill	Mumbai, India, Arabian Sea	400
4.	July 26, 2010	Talmadge Creek oil spill	Kalamazoo River, Michigan, USA	3,250
5.	July 16, 2010	Xingang Port oil spill	Yellow Sea, China	90,000
6.	June 16, 2010	Jebel al-Zayt oil spill	Egypt, Red Sea	Unknown
7.	June 11, 2010	Red Butte Creek oil spill	Salt Lake City, Utah, USA	107
8.	May 25, 2010	Trans-Alaska Pipeline spill	Anchorage, Alaska, USA	1,200
9.	May 25, 2010	MT *Bunga Kelana 3*	Singapore Strait, Singapore	2,500
10.	May 1, 2010	ExxonMobil oil spill	Niger Delta, Nigeria	95,500
11.	April 20, 2010	*Deepwater Horizon*	Gulf of Mexico, USA	627,000
12.	January 23, 2010	Port Arthur oil spill	Port Arthur, Texas, USA	1,500
13.	January 5, 2010	Yellow River oil spill	China, Chishui River (Shaanxi)	130
14.	August 21, 2009	Montara oil spill	Timor Sea, Australia	30,000
15.	July 31, 2009	*Full City* oil spill	Rognsfjorden near Langesund, Norway	200
16.	April 8, 2009	2009 Lüderitz oil spill	Southern coast, Namibia	Unknown
17.	March 10, 2009	Queensland oil spill	Queensland, Australia	260
18.	February 2009	West Cork oil spill	Southern coast, Ireland	300
19.	July 28, 2008	2008 New Orleans oil spill	New Orleans, Louisiana, USA	8,800
20.	December 12, 2007	2007 *Statfjord* oil spill	Norwegian Sea, Norway	4,000
21.	December 7, 2007	2007 Korea oil spill	Yellow Sea, South Korea	10,800
22.	November 11, 2007	Kerch Strait oil spill	Strait of Kerch, Ukraine	1,000
23.	November 7, 2007	*COSCO Busan* oil spill	San Francisco, California, USA	188
24.	October 23, 2007	*Kab 101*	Bay of Campeche, Mexico	1,869
25.	August 11, 2006	Guimaras oil spill	Philippines	1,540
26.	July 14, 2006	Jiyeh power station oil spill	Lebanon	30,000
27.	June 19, 2006	Citgo refinery oil spill	Lake Charles, Louisiana, USA	6,500
28.	March 2, 2006	Prudhoe Bay oil spill	Alaska North Slope, Alaska, USA	689
29.	August 30, 2005	Bass Enterprises (Hurricane Katrina)	Cox Bay, Louisiana, USA	12,000
30.	August 30, 2005	Shell (Hurricane Katrina)	Pilottown, Louisiana, USA	3,400
31.	August 30, 2005	Chevron (Hurricane Katrina)	Empire, Louisiana, USA	3,200
32.	August 30, 2005	Murphy Oil USA refinery spill (Hurricane Katrina)	Meraux and Chalmette, Louisiana, USA	3,410
33.	August 30, 2005	Bass Enterprises (Hurricane Katrina)	Pointe à la Hache, Louisiana, USA	1,500
34.	August 30, 2005	Chevron (Hurricane Katrina)	Port Fourchon, Louisiana, USA	170
35.	December 8, 2004	MV *Selendang Ayu*	Unalaska Island, Alaska, USA	1,560
36.	November 26, 2004	*Athos 1*	Delaware River, New Jersey, USA	860
37.	September 16, 2004	MP-80 Delta 20" pipeline (Hurricane Ivan)	Louisiana, USA	963

(continued)

Table 7.11 (continued)

S. No.	Dates	Vessel name or name of spill or cause of spill	Location	Spill size (t)
38.	September 16, 2004	Nakika 18" pipeline (Hurricane Ivan)	Louisian, USA	618
39.	September 16, 2004	Chevron-Texaco tank collapse (Hurricane Ivan)	Louisiana, USA	423
40.	July 28, 2003	*Tasman Spirit*	Karachi, Pakistan	30,000
41.	April 27, 2003	Bouchard No. 120	Buzzards Bay, Massachusetts, USA	320
42.	November 13, 2002	*Prestige* oil spill	Galicia, Spain	63,000
43.	October 6, 2002	*Limburg* (bombing)	Gulf of Aden, Yemen	12,200
44.	November 23, 2001	Manguinhos refinery	Guanabara Bay, Rio de Janeiro, Brazil	97
45.	October 4, 2001	Trans-Alaska Pipeline gunshot spill	Alaska, USA	932
46	June 25, 2001	2001 Shell Ogbodo oil spill	Niger Delta, Nigeria	Unknown
47.	May 2001	2001 Shell Ogoniland oil spill	Niger Delta, Nigeria	Unknown
48.	March 15, 2001	*Petrobras 36*	, Campos Basin, Brazil,	274
49.	January 14, 2001	Amorgos oil spill	Southern coast, Taiwan	1,150
50.	January 2001	*Jessica*	Galapagos Islands, Ecuador,	568
51.	August 1, 2000	Pine River	Chetwynd, British Columbia, Canada,	850
52.	June 2000	Project Deep Spill	Helland Hansen ridge, Norway,	100
53.	June 2000	*Treasure*	Cape Town, South Africa,	1,400
54.	January 2000	Petrobras pipeline	Guanabara Bay, Rio de Janeiro, Brazil,	1,100
55.	December 12, 1999	*Erika*	Bay of Biscay, France,	25,000
56.	January 12, 1998	Mobil Nigeria oil spill	Niger Delta, Nigeria	5,500
57.	December 1997	*Nakhodka*	Sea of Japan, Japan	6,240
58.	September 27, 1996	*Julie N.*	Portland, Maine, USA	586
59.	February 15, 1996	*Sea Empress*	Pembrokeshire, UK	72,000
60.	January 19, 1996	North Cape	Rhode Island, USA	2,500
61.	March 31, 1994	*Seki* oil spill	United Arab Emirates	15,900
62.	January 7, 1994	*Morris J. Berman* oil spill	Puerto Rico	2,600
63.	January 5, 1993	MV *Braer*	,Shetland Islands, UK	85,000
64.	December 3, 1992	Aegean Sea	A Coruña, Spain	74,000
65.	April 26, 1992	*Katina P*	Maputo, Mozambique	72,000
66.	March 2, 1992	Fergana Valley	Uzbekistan	285,000
67.	July 21, 1991	*Kirki*	Coast of Western Australia, Australia,	17,280
68.	May 28, 1991	*ABT Summer*	Offshore, Angola	260,000
69.	April 11, 1991	MT *Haven*	Mediterranean Sea near Genoa, Italy	144,000
70.	January 23, 1991	Gulf War oil spill	Persian Gulf, Iraq	820,000
71.	June 8, 1990	*Mega Borg*	Gulf of Mexico, Texas, USA	16,501
72.	February 7, 1990	*American Trader*	Bolsa Chica State Beach, California, USA	981
73.	December 19, 1989	*Khark 5*	off Las Palmas de Gran Canaria, Spain,	80,000
74.	June 24, 1989	*Presidente Rivera*	Delaware River, Pennsylvania, USA	993

(continued)

Table 7.11 (continued)

S. No.	Dates	Vessel name or name of spill or cause of spill	Location	Spill size (t)
75.	March 24, 1989	*Exxon Valdez*	Prince William Sound, Alaska, USA	104,000
76.	November 10, 1988	*Odyssey*	off Nova Scotia, Canada,	132,000
77.	January 2, 1988	Ashland oil spill	Floreffe, Pennsylvania, USA	10,000
78.	December 6, 1985	*Nova*	Gulf of Iran, Kharg Island, Iran	70,000
79.	September 28, 1985	*Grand Eagle*	Delaware River, Pennsylvania, USA	1,400
80.	August 6, 1983	*Castillo de Bellver*	Saldanha Bay, South Africa	252,000
81.	February 4, 1983	Nowruz Field Platform	Persian Gulf, Iran	260,000
82.	March 7, 1980	*Tanio*	Brittany, France	13,500
83.	February 23, 1980	*Irenes Serenade*	Pylos, Greece	100,000
84.	November 15, 1979	MT *Independen a*	Bosphorus, Turkey	95,000
85.	November 1, 1979	*Burmah Agate*	Galveston Bay, Texas, USA	8,440
86.	July 19, 1979	*Atlantic Empress/Aegean Captain*	Trinidad and Tobago	287,000
87.	June 3, 1979	Ixtoc I oil spill	Bay of Campeche, Mexico,	480,000
88.	January 8, 1979	*Betelgeuse*	Bantry Bay, Ireland	64,000
89.	March 16, 1978	*Amoco Cadiz*	Brittany, France	227,000
90.	February 15, 1978	Trans-Alaska Pipeline	Alaska, USA	2,162
91.	April 22, 1977	Ekofisk oil field	North Sea, Norway	27,600
92.	February 26, 1977	*Hawaiian Patriot*	Honolulu, Hawaii	95,000
93.	February 7, 1977	*Borag*	Northern coast, Taiwan	34,000
94.	December 15, 1976	*Argo Merchant*	Nantucket Island, Massachusetts, USA	28,000
95.	May 12, 1976	*Urquiola*	A Coruña, Spain	100,000
96.	January 31, 1975	*Corinthos*	Delaware River, Pennsylvania, USA	35,700
97.	January 29, 1975	*Jakob Maersk*	Oporto, Portugal	88,000
98.	August 9, 1974	VLCC *Metula*	Strait of Magellan, Chile	51,000
99.	December 19, 1972	*Sea Star*	Gulf of Oman, Iran	115,000
100.	January 17, 1971	*Arizona Standard/Oregon Standard* collision	San Francisco Bay, USA	2,700
101.	March 20, 1970	*Othello*	Trälhavet Bay, Sweden	60,000
102.	January 28, 1969	1969 Santa Barbara oil spill	Santa Barbara, California, USA	14,000
103.	March 18, 1967	*Torrey Canyon*	Isles of Scilly, UK	119,000
104.	December 30, 1958	*African Queen* oil spill	Ocean City, Maryland, USA	21,000
105.	1950s–1994	Guadalupe Oil Field	Guadalupe, California, USA	29,000
106.	1940s–1950s	Greenpoint, Brooklyn oil spill	Newtown Creek, New York, USA	97,400
107.	March 6, 1937	SS *Frank H. Buck*/ SS *President Coolidge* collision	San Francisco Bay, California, USA	8,870
108.	March 14, 1910–September 10, 1911	Lakeview Gusher	Kern County, California, USA	1,230,000

emulsions contain from 30% to 80% water. The emulsions usually occur after strong storms in the zones of spills of heavy oils with an increased content of nonvolatile fractions, especially asphaltenes. They can exist in the marine environment for over 100 days in the form of peculiar "chocolate mousses." The stability of these emulsions usually increases with decreasing temperature. The reverse emulsions, such as oil-in-water, droplets of oil suspended in water, are much less stable because surface-tension forces quickly decrease the dispersion of oil. This process can be slowed with the help of emulsifiers – surface-active substances with strong hydrophilic properties used to eliminate oil spills. Emulsifiers help to stabilize oil emulsions and promote dispersing oil to form microscopic (invisible) droplets. This accelerates the decomposition of oil products in the water column.

4. *Oxidation of some oil components*

Chemical transformations of oil on the water surface and in the water column start to reveal themselves no earlier than a day after the oil enters the marine environment. They mainly have an oxidative nature and often occur as a result of the ultraviolet light of the sun. These processes are catalyzed by some trace elements (e.g., vanadium) and inhibited (slowed) by compounds of sulfur. The final products of oxidation (hydroperoxides, phenols, carboxylic acids, ketones, aldehydes, and others) usually have increased water solubility. The reactions of photooxidation initiate the polymerization and decomposition of the most complex molecules in oil composition. This increases the oil's viscosity and promotes the formation of solid oil aggregates.

5. *Sedimentation*

In the narrow coastal zone and shallow waters where particulates are abundant and water is subjected to intense mixing from 10% to 30% the oil in a spill is adsorbed on the suspended material and deposited at the bottom sediment. In deeper areas remote from the shore, sedimentation of oil, except for the heavy fractions, is an extremely slow process. Biological objects bring about "biosedimentation": plankton filtrators and other organisms absorb the emulsified oil and sediment it to the bottom with their metabolites and remainders. The suspended forms of oil and its components undergo intense chemical and microbial breakdown in the water column. When oil particles reach the bottom of seas and oceans, which are anaerobic their decomposition slows down considerably, if not totally stopped, the heavy oil fractions accumulated inside the sediments can be preserved for many months and even years.

6. *Aggregation*

Oil aggregates in the form of petroleum lumps, tar balls, or pelagic tar can be found both in the open and coastal waters as well as on the beaches. They derive from crude oil after the evaporation and dissolution of its relatively light fractions, emulsification of oil residuals, and chemical and microbial transformation. The chemical composition of oil aggregates is changeable, but most often, its base includes asphaltenes (up to 50%) and high-molecular-weight compounds of the heavy fractions of the oil.

Oil aggregates look like light gray, brown, dark brown, or black sticky lumps. They have an uneven shape and vary from 1 mm to 10 cm in size (sometimes reaching up to 50 cm). Their surface serves as a substrate for developing bacteria, unicellular algae, and other microorganisms. Many invertebrates (e.g., gastropods, polychaetes, and crustaceans), which are not affected by oil components often use them as shelter. Oil aggregates can exist from a month to a year in the enclosed seas and up to several years in the open ocean. They slowly degrade in the water column, on the shore (if they are washed there by currents), or on the sea bottom (if they lose their floating ability).

7. *Microbial degradation*

The fate of most petroleum substances in the marine environment is ultimately defined by their transformation and degradation due to microbial activity. About a hundred known species of bacteria and fungi are able to use oil components to sustain their growth and metabolism. In pristine areas, their proportions usually do not exceed 0.1–1.0% of the total abundance of heterotrophic bacterial communities. In areas polluted by oil, however, this portion increases to 1–10%.

Biochemical processes of oil degradation with microorganism participation include several types of enzyme reactions based on oxygenases, dehydrogenases, and hydrolases. These cause aromatic and aliphatic hydrooxidation, oxidative deamination, hydrolysis, and other biochemical transformations of the original oil substances and the intermediate products of their degradation.

The degree and rates of hydrocarbon biodegradation depend on the following (Walker and Cowell 1974, 1977):

(a) *The structure of the molecules*

The paraffin compounds (alkanes) biodegrade faster than aromatic and naphthenic substances. With increasing complexity of molecular structure, that is, increasing the number of carbon atoms and degree of chain branching, as well as with increasing molecular weight, the rate of microbial decomposition decreases.

(b) *The physical state of the oil*

The degree of the dispersion of the oil is important in deciding the rate of microbial breakdown; the more dispersed the oil the faster the rate of microbial breakdown because of the greater contact with the microorganisms.

(c) *Environmental factors*

The environmental factors affecting microbial breakdown include the temperature, the availability of oxygen, and the nutrients available in the water.

(d) *The nature and numbers of the oil-degrading microorganisms in the water*

The species, composition and abundance of oil-degrading microorganisms in the water is important in deciding the rate of oil breakdown. Complex and interconnected factors influence biodegradation in the marine environment. These factors will be examined in a little more detail below.

(e) *Biodegradation of oil*

Of all the various factors, which affect the weathering or disappearance of oil, microbial degradation is the most important.

Microorganisms, especially bacteria, definitely play a part in the breakdown of oil. Many isolations have been made from oil and water of microorganisms capable of using crude oil as the only source of carbon. Table 7.12 shows the bacteria (including cyanobacteria), algae, and fungi, which can degrade oil and oil components (Raghukumar et al. 2001). Microbes degrading oil usually account for less than 1% of natural populations of microbes, but can account for more than 10% of the population in waters polluted with oil. Obligate hydrocarbon-degrading bacteria, designated as "hydrocarbonoclastic" and belonging to genera in the gamma-proteobacteria, are isolated regularly in marine water when an oil spill has occurred. Crude oil is a complex mixture of different hydrocarbons, and different bacteria have been shown to attack specific components of oil (Leahy and Colwell 1990). In one study, for instance, the degradation of PAHs (polycyclic aromatic hydrocarbons) was shown to be carried out mainly by *Cycloclasticus*. PAHs are chemical compounds that consist of fused aromatic rings and do not carry substituents. PAHs occur in oil, coal, and tar deposits, and are believed to be carcinogenic.

Table 7.12 Bacteria and fungi degrading oil (Modified from Gordon 1994; Raghukumar et al. 2001)

Bacteria	Fungi
Achromobbacter	*Allescheria*
Acinetobacter	*Aspergillus*
Actinomyces	*Aureobasidium*
Aeromonas	*Botrytis*
Alcaligenes	*Candida*
Arthrobacter	*Cephaiosporium*
Bacillus	*Cladosporium*
Beneckea	*Cunninghamella*
Brevebacterium	*Debaromyces*
Coryneforms	*Fusarium*
Erwinia	*Gonytrichum*
Flavobacterium	*Hansenula*
Klebsiella	*Helminthosporium*
Lactobacillus	*Mucor*
Leucothrix	*Oidiodendrum*
Moraxella	*Paecylomyces*
Nocardia	*Phialophora*
Peptococcus	*Penicillium*
Pseudomonas	*Rhodosporidium*
Sarcina	*Rhodotorula*
Spherotilus	*Saccharomyces*
Spirillum	*Saccharomycopisis*
Streptomyces	*Scopulariopsis*
Vibrio	*Sporobolomyces*
Xanthomyces	*Torulopsis*
Oscillatoria *Plectonema*	*Trichoderma*
Aphanocapsa	*Trichosporon*
Thalassolituus	
Alcanivorax	
Cycloclasticus	

Note: The last six in the column for bacteria are Cyanobacteria

On the other hand, the degradation of the branched alkane, pristane, was carried out almost exclusively by *Alcanivorax*. Bacteria related to *Thalassolituus oleivorans* were the dominant degraders in *n*-alkane (C12–C32). The marine bacterium, *Roseobacter*, was present among the various groups but was not dominant. Figure 7.7 shows the pathways for the breakdown of some oil components. Table 7.12 shows a list of microorganisms degrading oil.

7.3.2.2 Remediation of Oil Spills

Remediation and cleanup techniques for oil spills depend largely on the type of oil and the conditions

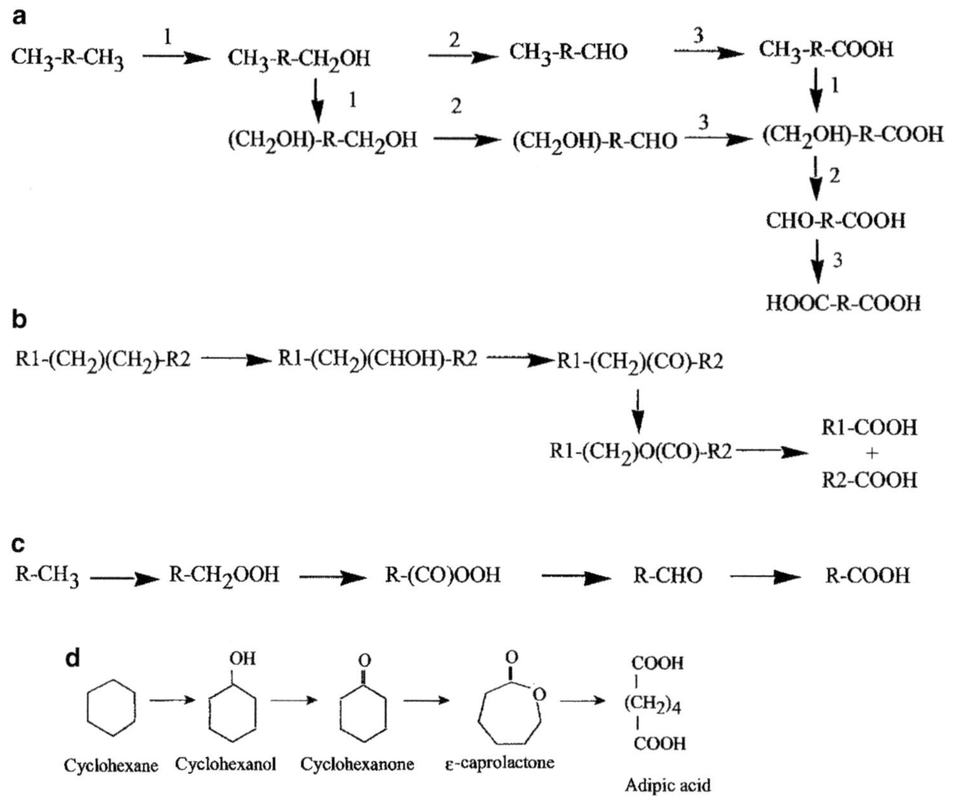

a

$$CH_3\text{-}R\text{-}CH_3 \xrightarrow{\ \ 1\ \ } CH_3\text{-}R\text{-}CH_2OH \xrightarrow{\ \ 2\ \ } CH_3\text{-}R\text{-}CHO \xrightarrow{\ \ 3\ \ } CH_3\text{-}R\text{-}COOH$$

$$\downarrow 1 \qquad\qquad\qquad\qquad\qquad\qquad\qquad\qquad \downarrow 1$$

$$(CH_2OH)\text{-}R\text{-}CH_2OH \xrightarrow{\ \ 2\ \ } (CH_2OH)\text{-}R\text{-}CHO \xrightarrow{\ \ 3\ \ } (CH_2OH)\text{-}R\text{-}COOH$$

$$\downarrow 2$$

$$CHO\text{-}R\text{-}COOH$$

$$\downarrow 3$$

$$HOOC\text{-}R\text{-}COOH$$

b

$$R1\text{-}(CH_2)(CH_2)\text{-}R2 \longrightarrow R1\text{-}(CH_2)(CHOH)\text{-}R2 \longrightarrow R1\text{-}(CH_2)(CO)\text{-}R2$$

$$\downarrow$$

$$R1\text{-}(CH_2)O(CO)\text{-}R2 \longrightarrow \begin{array}{c} R1\text{-}COOH \\ + \\ R2\text{-}COOH \end{array}$$

c

$$R\text{-}CH_3 \longrightarrow R\text{-}CH_2OOH \longrightarrow R\text{-}(CO)OOH \longrightarrow R\text{-}CHO \longrightarrow R\text{-}COOH$$

d

Cyclohexane   Cyclohexanol   Cyclohexanone   ε-caprolactone

$$\begin{array}{c} COOH \\ | \\ (CH_2)_4 \\ | \\ COOH \end{array}$$

Adipic acid

Fig. 7.7 Pathways for the breakdown of some oil components (From Harayama et al. 1999)

present at the location during the time of the spill. The method for the cleanup or remediation of oil spills may be physical or biological.

1. *Physical methods of remediation*

   (a) Containment and recovery using booms and skimmers

   Containment and recovery are the earliest methods used to remediate and clean up an oil spill. Long, floating plastic or rubber barriers or booms are placed around the floating oil slick. These act like fences, containing the oil and preventing it from further spreading. In addition, booms may be used to divert and channel oil slicks along desired paths, making them easier to remove from the surface of the water. Booms can be divided into several basic types. They are not very effective in rough waters, which flow over the barriers. Nonrigid or inflatable booms may be used; they are easy to clean and store, and they perform well in rough seas. They are, however, expensive, more complicated to use, and may puncture and deflate easily.

   After the oil is contained using booms, "skimmers" or boats that skim spilled oil from the water surface are used. In calm waters, vacuum skimmers are used to suck the oil and put it into storage tanks. In rough waters, floating disk and rope skimmers can be passed through the oil. The amount of oil recovered by booms and skimmers is small, and if used for long periods, may hamper the spread of oil, which helps its biodegradation.

   (b) Absorption

   Absorption is the technique employed in choppy or fast-moving waters, when methods like containment and removal fail. In this method, sorbent materials such as talc, straw, sawdust, and synthetic absorbents are added to the oil slick and they are removed when they have soaked up some of the oil. These sorbent materials act like a big sponge, removing oil but contaminated absorbent materials must be treated as toxic waste and present disposal problems. Furthermore, straw and sawdust can become waterlogged and difficult to remove.

(c) Dispersants

The adverse economic and environmental effects of offshore oil spills are greatest when the oil slick reaches the shoreline. On account of this, much effort is put into preventing offshore oil spills from reaching the shoreline. In calm seas, use of skimmers and booms to collect the oil at sea is the conventional method of cleanup and recovery. In situ burning (see below) is also used in such a situation. In rough seas, skimming or burning the oil is not effective, and the use of chemical dispersants appears to be the most suitable for cleaning up the oil spill.

Dispersants are chemicals that promote the formation of tiny oil droplets, and delay the reformation of slicks. They contain surfactants and/or solvent compounds that cause the oil sleek to break into small droplets in a process known in oil parlance as *dispersion*. After oil spills, oil droplets break down because of waves and currents during the dispersion process. Water and oil droplets then combined to form water-in-oil emulsions, which have high viscosity. Dispersants inhibit emulsion formation and promote oil dispersion. Thus, they can remove the spilled oil from the water surface and reduce the impact to environment, especially to the shoreline and sensitive habitats. The generated small oil droplets get transported or transferred into the water column due to wave action and sea turbulence. They subsequently move away from the contaminated area in accordance with the prevailing currents. They can then more easily adsorb onto suspended particulate matter and/or more easily biodegrade. After an oil spill, oil droplets break down because of waves and currents during the dispersion process.

Very light oils are, however, not easily dispersible because the formed droplets have to be very small to overcome buoyancy. For this reason, a high dosage of dispersant is required to cause the formation of such small droplets in very light oils. Very heavy oils on the other hand are much more resistant to dispersion because their high viscosity prevents the dispersant from penetrating them. Such penetration is necessary to produce dispersed oil droplets.

In very calm seas, the applied dispersant is not very effective as it tends to run off the oil and gathers in small pools within the slick. In very rough seas the use of large amounts of dispersants might not be needed because a high degree of dispersion occurs naturally due to the mixing effect of the waves.

The evaluation of dispersant effectiveness used for oil spills is commonly done using tests conducted in laboratory flasks, which attempt to mimic the conditions at sea using tests such as the swirling flask devised by the US EPA.

The advantages of dispersants over other types of remediation methods are that they can be used over a wide area through the use of aircraft and also in remote areas. Furthermore, in instances where other methods do not work well such as the use of booms in high seas, dispersants work excellently. By dispersing the oil it is brought in closer contact with microorganisms, which biodegrade it.

On the other hand, many dispersants are toxic and may have lethal or long-term effects on plankton, animals, or fish. They are ineffective on oils which are viscous because of their content of the heavier fractions of oil. Finally, they are not very effective in calm waters.

(d) In situ burning

In in situ burning of oil, the controlled burning of an oil spill takes place on the water's surface. It requires minimal equipment although some specialized equipment and training is required. In burning oil, no need exists for collecting and transporting recovered oil. The burning leads to the production of green house gases and may leave toxic residues in the water. Because the oil is gasified during combustion, the need for physical collection, storage, and transport of recovered product is reduced but can cause air pollution. At times, burning also leaves a toxic residue on the surface of water, thereby causing more pollution, rather than removing it from the natural environment.

2. *Biological methods of remediation (bioremediation)*

Bioremediation is the process of using biological means in remediating or ameliorating the effects of oil pollution. In general terms, it is the remediation of any pollution by the use of microorganisms in transforming harmful organic compounds, such as oil, into nontoxic and less dangerous compounds. Seawater contains a range of microorganisms that can partially or completely degrade oil to water-soluble compounds and eventually to carbon dioxide and water. Two approaches are adopted in using microorganisms to remediate oil pollution: (i) biostimulation, which seeks to increase the activity of oil-degrading indigenous organisms and (ii) bioaugmentation, which is the introduction of microorganisms, which degrade oil into water (Smith and Osborn 2009).

(i) *Biostimulation*

Biostimulation of the indigenous microorganisms involves the provision of materials, which increase the activity of the oil-degrading microorganisms. This includes providing those nutrients known to be deficient in the sea environment, or providing conditions, which increase the efficiency of the microorganisms to utilize what is available.

- Addition of nutrients:

  Usually, the rate-limiting factor for oil degradation in the marine environment is inorganic nutrient concentration, and in particular, nitrogen (N) and phosphorous (P). Biostimulation through the addition of nutrients containing these two elements has been the most widely used bioremediation strategy. Oil degradation is significantly enhanced by this method.

- Addition of surfactancts:

  The hydrocarbons present in crude oil are mostly insoluble in water; furthermore, being hydrophobic the microorganisms, which degrade them need to be in contact with the oil for the membrane-bound enzymes, which breakdown the oil to act. Many of the bacteria produce surfactants, which reduce the surface tension of, and emulsify, the oil and enable contact between the microorganisms and oil. For this reason, the addition of surfactants to supplement those produced naturally is a widely used method to enhance degradation of oil. Surfactants reduce surface tension and increase the surface area of hydrophobic compounds such as oil, therefore, increasing its bioavailability. Biosurfactants produced by bacteria have been used for bioremediation. They have the advantage of nontoxicity when compared with chemical surfactants. In cases where the surfactants is membrane bound the microorganism is added directly to the oil spill (Maneerat 2005).

(ii) *Bioaugmentation*

The addition of oil degraders has been done in many cases, especially in the early stages of the spill (about 5 days). Several studies have shown that bioaugmentation with *Alcanivorax* leads to enhanced degradation of the branched alkanes (pristane and phytane). Similar observations have also been made that bioaugmentation with *Thalassolituus*, which causes increased degradation of *n*-alkanes. There have also been a few reports of failures.

References

American Petroleum Institute (2010). http://www.petroleumhpv.org/docs/crude_oil/111503_crude_robsumm_final.pdf. Accessed 26 July 2010.

Anonymous. (2003). *Robust Summary Information on Crude Petroleum* (CAS No 8002-05-9).

Anonymous. (2004). *Indicators for waterborne pathogens.* Washington, DC: National Research Council.

Anonymous. (2005). *Microbial source tracking guide document.* National Risk Management Research Laboratory Office of Research and Development. Cincinnati: U.S. Environmental Protection Agency. US EPA/600/R-05/064 June 2005.

Anonymous. (2006). *Standard methods for the examination of water and wastewater.* Washington, DC: American Public Health Association, American Water Works Association, Water Environment Federation.

Anonymous. (2007). Total daily maximum loads. US EPA. http://iaspub.epa.gov/waters/national_rept.control%23TOP_IMP. Accessed 19 Sept 2007.

Anonymous. (2009). Overview of spills. http://www.temple.edu/environment/review_oil_spills.htm. Accessed 10 April 2009.

Anonymous. (2010a). Harmful algal blooms. http://www.cdc.gov/hab/cyanobacteria/facts.htm#cyanoform. Accessed 29 June 2010.

Anonymous. (2010b). The International Tanker Owners Pollution Federation Ltd. http://www.itopf.com/information-services/data-and-statistics/statistics/. Accessed 29 June 2010.

Anonymous. (2010c). List of oil spills. http://en.wikipedia.org/wiki/List_of_oil_spills. Accessed 3 Jan 2011.

Bonde, G. (1966). Bacteriological methods for the estimation of water pollution. *Health Laboratory Science, 3*, 124–128.

Callaghan, T. (2005). Microbial source tracking *science or scam?* 2005 Beach closure workshops. New England Interstate Water Pollution Control Commission. www.neiwpcc.org/beachworkshop/presentations.asp. Accessed 22 Sept 2007.

Gordon, R. (1994). Bioremediation and its application to Exxon Valdez oil spill in Alaska. http://www.geocities.com/Cape Canaveral/Lab/2094/bioremed.html. Accessed 1 Feb 2009.

Harayama, S., Kishira, H., Kasai, Y., & Shutsubo, K. (1999) Petroleum Biodegradation in Marine Environments. *Journal of Molecular and Microbiological Biotechnology, 1*, 63–70.

Leahy, J. G., & Colwell, R. R. (1990). Microbial degradation of hydrocarbons in the environment. *Microbiological Reviews, 54*, 305–315.

Long, S. C., & Plumme, J. D. (2004). Assessing land use impacts on water quality using microbial source tracking. *Journal of the American Water Resources Association, 40*, 1433–1448.

Maneerat, S. (2005). Biosurfactants from marine microorganisms. *Songklanakarin Journal of Science and Technology, 27*, 1263–1272.

Okafor, N. (1985). *Aquatic and waste microbiology*. Enugu: Fourth Dimension Publishers.

Raghukumar, C., Vipparty, V., David, J. J., & Chandramohan, D. (2001). Degradation of crude oil by marine cyanobacteria. *Applied Microbiology and Biotechnology, 57*, 433–436.

Shanks, O.C. (2005). *Microbial source tracking*. Geneticist US EPA Region 5 genomics training workshop, Chicago, IL, 28, April 2005.

Smith, D.C. (2006). Microbial source tracking using F-specific coliphages and quantitative PCR. http://ciceet.unh.edu/news/releases/springReports07/pdf/smith.pdf. Accessed 22 Sept 2007.

Smith, C.J., & Osborn, A.M. (2009). Advantages and limitations of quantitative PCR(Q-PCR)-based approaches in microbial ecology. *FEMS Microbiology Ecology, 67*, 6–20.

Stewart-Pullaro, J., Daugomah, J. W., Chestnut, D. E., Graves, D. A., Sobsey, M. D., & Scott, G. I. (2006). dF+RNA coliphage typing for microbial source tracking in surface waters. *Journal of Applied Microbiology, 101*, 1015–1026.

Stockel, D. M., & Harwood, V. J. (2007). Performance, design, and analysis in microbial source tracking studies. *Applied and Environmental Microbiology, 73*, 2405–2415.

Stockel, D. M., Mathes, M. V., Hyer, K. V., Hagerdon, C., Kator, H., Lukasik, J. O., Brien, T. L., Fenger, T. W., Samadpour, M., Strickler, K. M., & Wiggins, B. A. (2004). Comparison of seven protocols to identify fecal contamination sources using *Escherichia coli*. *Environmental Science & Technology, 38*, 6109–6117.

Walker, J. D., & Cowell, R. R. (1974). Microbial degradation of model petroleum at low temperatures. *Microbial Ecology, 1*, 63–95.

Walker, J. D., & Colwell, R. R. (1977). Role of autochthonous bacteria in the removal of spilled oil from sediment. *Environmental Pollution, 12*, 51–56.

Disease Transmission in Water

8

Abstract

Disease transmission through water can occur through drinking water contaminated with microorganisms causing disease, through inhalation of aerosols of water at swimming pools, or through air-conditioners, or finally through contact in swimming pools. Diseases transmitted through drinking water include bacterial diseases (cholera, dysentery, typhoid, gastrointestinal disorders by *E. coli*, *Yersinia*, and *Campylobacter*), viruses (hepatitis A and E), and protozoan and helminth parasites (*Cryptosporidium*, *Giardia*, *Entamoeba*). Inhalation of aerosols can lead to Legionnaires' disease and tuberculosis, while persons with open wounds can contact infections by *Aeromonas* and non-tubercular mycobacteria.

Keywords

Atypical tuberculosis • Cholera • Cryptosporidiosis • Cyanotoxins • Disease transmission drinking water • Disease transmission recreational waters • Enteroviruses • Legionella • Parasitic worms • Shigellosis • Tuberculosis

8.1 Disease Transmission Through Drinking Water

Long before the demonstration that water was a vehicle of disease, man sometimes suspected it. Following an outbreak of cholera in 1854 in London, England, a commission was set up under the chairmanship of John Snow, an anesthetist. This commission which reported a year later, in 1855, for the first time established a casual relationship between water and the transmission of bacterial disease. It was found by the commission that the epidemic was restricted to a particular area of London where the inhabitants drank from a well into which sewage entered from a nearby sewer (Cliff and Haggett 1988). A similar and more recent study (Pineo and Subrahmanyam 1975) carried out in Malawi in March 1974 showed that by simply directing water in a piped system, which though untreated came from high in the mountain, a dramatically different picture of cholera distribution occurred (see Fig. 8.1). The great majority of evident water-related health problems are the result of microbial (bacteriological, viral, protozoan, or other biological) contamination (see Table 8.1). Nevertheless, as will be seen below, an appreciable number of serious health concerns may also occur as a result of the chemical contamination of drinking-water.

Crude sewage is one of the most important pollutants of water. Besides supplying organic nutrients, it frequently contains the agents causing enteric infectious diseases in man. In general terms, the greatest microbial enteric disease risks are associated with ingestion of water that is contaminated with human or animal (including bird) feces. Feces can thus be a

N. Okafor, *Environmental Microbiology of Aquatic and Waste Systems*,
DOI 10.1007/978-94-007-1460-1_8,

Fig. 8.1 Map showing
cholera-free zone in Malawi,
March 1974 (Note: The
cholera-free zone had piped
water, whereas other areas
obtained their water from
the river or well) (From Pineo
and Subrahmanyam 1975.
With permission)

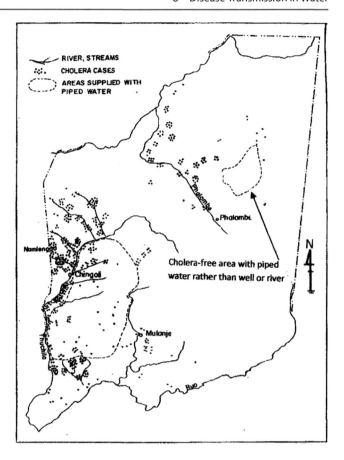

source of pathogenic bacteria, viruses, protozoa, and helminthes.

There are four ways by which water, or the lack of it, may be associated with disease:

Waterborne diseases are caused by the ingestion of water contaminated by human or animal feces or urine containing pathogenic bacteria or viruses. The diseases which can be transmitted include cholera, typhoid, amoebic and bacillary dysentery, and other diarrheal diseases. This list can be stretched to include diseases which occur by the ingestion of a toxin produced by an organism growing in drinking water such as toxins produced by cyanobacteria (Anonymous 2006a).

Water-washed diseases are caused by poor personal hygiene and skin or eye contact with contaminated water. They include scabies, trachoma, and flea, lice, and tick-borne diseases.

Water-based diseases are caused by parasites found in intermediate organisms living in water. They include dracunculiasis, schistosomiasis, and diseases caused by other helminths.

Water-related diseases are caused by insect vectors which breed in water. They include dengue, filariasis, malaria, onchocerciasis, trypanosomiasis, and yellow fever.

Diseases may thus be transferred through water in ways outside drinking (see Fig. 8.2).

In this chapter, we shall look at diseases transmitted through drinking water, through recreational waters such as swimming pools and through the eating of aquatic invertebrates, shellfish, which ingest aquatic microorganisms.

8.1.1 Communicable Diseases Transmitted Through Drinking Water

Communicable diseases are those which are brought about by microorganisms. Drinking-water-borne outbreaks of any kind are particularly to be avoided because of their capacity to result in the simultaneous infection of a large number of persons and potentially a high proportion of the community. Thus, in the

Table 8.1 Waterborne pathogens and their significance in water supplies (From Anonymous 2006a. With permission)

Pathogen	Health significance	Persistence in water supplies[a]	Resistance to chlorine[b]	Relative infectivity[c]	Important animal source
Bacteria					
Burkholderia pseudomallei	Low	May multiply	Low	Low	No
Campylobacter jejuni, C. coli	High	Moderate	Low	Moderate	Yes
Escherichia coli – Pathogenic[d]	High	Moderate	Low	Low	Yes
E. coli – Enterohaemorrhagic	High	Moderate	Low	High	Yes
Legionella spp.	High	Multiply	Low	Moderate	No
Non-tuberculous mycobacteria	Low	Multiply	High	Low	No
Pseudomonas aeruginosa[e]	Moderate	May multiply	Moderate	Low	No
Salmonella typhi	High	Moderate	Low	Low	No
Other salmonellae	High	May multiply	Low	Low	Yes
Shigella spp.	High	Short	Low	Moderate	No
Vibrio cholerae	High	Short	Low	Low	No
Yersinia enterocolitica	High	Long	Low	Low	Yes
Viruses					
Adenoviruses	High	Long	Moderate	High	No
Enteroviruses	High	Long	Moderate	High	No
Hepatitis A virus	High	Long	Moderate	High	No
Hepatitis E virus	High	Long	Moderate	High	Potentially
Noroviruses and sapoviruses	High	Long	Moderate	High	Potentially
Rotaviruses	High	Long	Moderate	High	No
Protozoa					No
Acanthamoeba spp.	High	Long	High	High	No
Cryptosporidium parvum	High	Long	High	High	Yes
Cyclospora cayetanensis	High	Long	High	High	No
Entamoeba histolytica	High	Moderate	High	High	No
Giardia intestinalis	High	Moderate	High	High	Yes
Naegleria fowleri	High	May multiply[f]	High	High	No
Toxoplasma gondii	High	Long	High	High	Yes
Helminths					
Dracunculus medinensis	High	Moderate	Moderate	High	No
Schistosoma spp.	High	Short	Moderate	High	Yes

Note: Waterborne transmission of the pathogens listed has been confirmed by epidemiological studies and case histories. Part of the demonstration of pathogenicity involves reproducing the disease in suitable hosts. Experimental studies in which volunteers are exposed to known numbers of pathogens provide relative information. As most studies are done with healthy adult volunteers, such data are applicable to only a part of the exposed population, and extrapolation to more sensitive groups is an issue that remains to be studied in more detail

[a]Detection period for infective stage in water at 20°C: short, up to 1 week; moderate, 1 week to 1 month; long, over 1 month

[b]When the infective stage is freely suspended in water treated at conventional doses and contact times. Resistance moderate, agent may not be completely destroyed

[c]From experiments with human volunteers or from epidemiological evidence

[d]Includes enteropathogenic, enterotoxigenic and enteroinvasive

[e]Main route of infection is by skin contact, but can infect immunosuppressed or cancer patients orally

[f]In warm water

10 year period between 1991 and 2000, nearly half a million people were affected by all categories of water-associated diseases system in the USA in a total of 155 outbreaks (Anonymous 2003). Table 8.1 gives the causes of waterborne outbreaks in the USA by type of water system between 1991 and 2000, while Fig. 8.3 summarizes all the outbreaks of diseases linked to drinking water in the same period.

Fig. 8.2 Routes of transmission of water-associated diseases (From Anonymous 2006a. With permission)

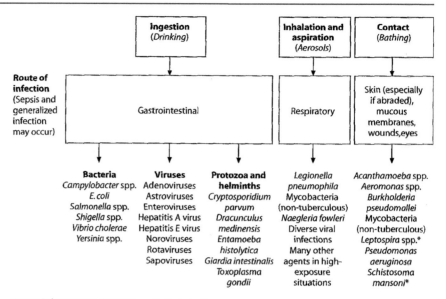

| Ingestion (Drinking) | Inhalation and aspiration (Aerosols) | Contact (Bathing) |

Route of infection (Sepsis and generalized infection may occur)

| Gastrointestinal | Respiratory | Skin (especially if abraded), mucous membranes, wounds, eyes |

Bacteria	**Viruses**	**Protozoa and helminths**	*Legionella pneumophila*	*Acanthamoeba* spp.
Campylobacter spp.	Adenoviruses	*Cryptosporidium*	Mycobacteria	*Aeromonas* spp.
E. coli	Astroviruses	*parvum*	(non-tuberculous)	*Burkholderia*
Salmonella spp.	Enteroviruses	*Dracunculus*	*Naegleria fowleri*	*pseudomallei*
Shigella spp.	Hepatitis A virus	*medinensis*	Diverse viral	Mycobacteria
Vibrio cholerae	Hepatitis E virus	*Entamoeba*	infections	(non-tuberculous)
Yersinia spp.	Noroviruses	*histolytica*	Many other	*Leptospira* spp.*
	Rotaviruses	*Giardia intestinalis*	agents in high-	*Pseudomonas*
	Sapoviruses	*Toxoplasma*	exposure	*aeruginosa*
		gondii	situations	*Schistosoma*
				*mansoni**

* Primarily from contact with highly contaminated surface waters.

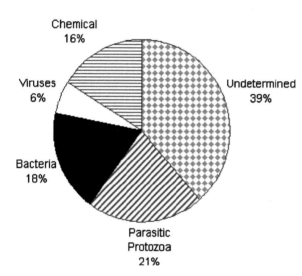

Fig. 8.3 Summary of the causes of outbreaks of diseases linked to drinking water in the US, 1991–2000 (From the American Chemistry Council; Anonymous 2003. With permission)

of certain diseases. That water-borne diseases are important is not in doubt. Even in a country of high economic development such as the USA, where treated water is easily available, data show (Fig. 8.3) a good number of cases annually (Anonymous 2003).

As seen in Fig. 8.3, the cause of water-associated diseases recorded in the US between 1991 and 2000 could not be determined in 39% of the cases. Where the cause was known, protozoa topped the list followed by bacteria, chemicals, and viruses (Fig. 8.3, Table 8.1). Indeed the number of cases appears to be rising, although this could be a reflection of greater reporting and more investigation than of absolute increase in cases.

In addition to fecally borne pathogens, other microbial hazards (e.g., guinea worm *Dracunculus medinensis*, toxic cyanobacteria, and *Legionella*) may be of public health importance under specific circumstances. The infective stages of many helminths, such as parasitic roundworms and flatworms, can be transmitted to humans through drinking water. As a single mature larva or fertilized egg can cause infection, these should be absent from drinking water. However, the water route is relatively unimportant for helminth infection, except in the case of the guinea worm.

Public health concern regarding cyanobacteria relates to their potential to produce a variety of toxins, known as "cyanotoxins." In contrast to pathogenic bacteria, cyanobacteria do not proliferate within the human body after uptake; they proliferate only in the

The type and incidence of water borne diseases varies according to the economic circumstance of the country. This is reflected in the ability of a society to provide clean potable water for its members. Thus, while cholera is endemic in Asian countries, it does not exist at all in the USA. Figures for many developing countries are not available, but if they were they would most probably be unreliable as many of these countries do not have laws which compel the reporting

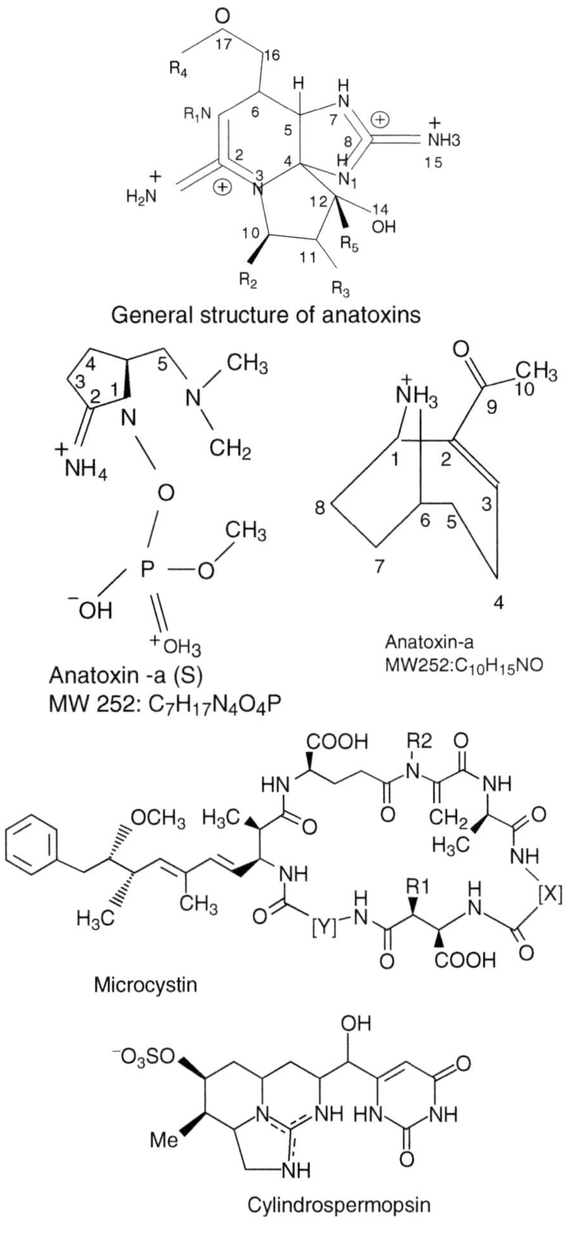

General structure of anatoxins

Anatoxin -a (S)
MW 252: $C_7H_{17}N_4O_4P$

Anatoxin-a
MW252: $C_{10}H_{15}NO$

Microcystin

Cylindrospermopsin

Fig. 8.4 Akaloid neurotoxins from Cyanobacteria (From Falconer and Humpage 2005. With permission)

8.1.2 Disease Outbreaks in Drinking Water Due to the Presence of Chemicals, and Biotoxins

It should be pointed out that water-borne diseases are not restricted to communicable diseases (Fig 6.2); some water-associated diseases are caused by chemicals or lack of them. For example, it has been shown that mortality due to cardio-vascular disease incidence is higher by at least 24% in towns with soft water than in those with hard water. Cities with high amounts of magnesium (Mg^{2+}) and calcium (Ca^{2+}), the ions responsible for hardness, show lower incidence of cardiovascular disease by 17% and 22% respectively. Goiter and tooth decay have been attributed to the absence of iodine and fluoride respectively, in drinking water. On the other hand, nitrate levels have been incriminated with methanogloblinemia in infants; while yet unknown chemicals in water have been implicated with the high incidence of bladder cancer in some cities. Although the short-term or long-term presence or absence of certain amounts of some chemicals in drinking water is important, the incidence of water borne diseases due to micro-organisms and biological agents appears far more important, especially in developing countries.

Biotoxins are sometimes produced in drinking water by cyanobacteria. Examples are the neurotoxin, anatoxins and the cytotoxin, cylindrospermopsin (Fig. 8.4).

8.1.2.1 Brief Notes on Some Water-Borne Diseases

This section will give some insight into the nature of waterborne diseases (Anonymous 2006b).

Cholera

Frequently called Asiatic cholera or epidemic cholera, the disease is caused by the Gram-negative, comma-shaped bacterium *Vibrio cholera* with a single polar flagellum. Cholera is characterized by a diarrhea with profuse watery stool, vomiting, rapid dehydration, fall of blood pressure, subnormal temperature and collapse may occur within 48 h unless medical care is given. The bacterium produces cholera toxin, an enterotoxin, whose action on the mucosal epithelium lining of the small intestine is responsible for the characteristic massive diarrhea of the disease. In its most severe forms, cholera is one of the most rapidly fatal illnesses known; a healthy person may become hypotensive within an hour of the onset of symptoms and may die

aquatic environment before intake. While the toxic peptides (e.g., microcystins) are usually contained within the cells (Falconer and Humpage 2005) and thus may be largely eliminated by filtration, toxic alkaloids such as cylindrospermopsin and neurotoxins are also released into the water and may break through filtration systems (see Fig. 8.4).

within 2–3 h if no treatment is provided. More commonly, the disease progresses from the first liquid stool to shock in 4–12 h, with death following in 18 h to several days, if rehydration treatment is not given.

Members of the species are typed according to their O antigens. There are a number of pathogenic species, including *V. cholerae*, *V. parahaemolyticus*, and *V. vulnificus*. However, *Vibrio cholerae* is the only pathogenic species of significance from freshwater environments. While a number of serotypes can cause diarrhea, only the serological varieties (serovars) O1 and O139 currently cause the classical cholera symptoms in which a proportion of cases suffer fulminating and severe watery diarrhea. The O1 serovar has been further divided into "classical" and "El Tor" biotypes. The latter is distinguished by features such as the ability to produce a dialyzable heat-labile haemolysin, active against sheep and goat red blood cells. The classical biotype is considered responsible for the first six cholera pandemics, while the El Tor biotype is responsible for the seventh pandemic that commenced in 1961. Strains of *V. cholerae* O1 and O139 that cause cholera produce an enterotoxin (cholera toxin) that alters the ionic fluxes across the intestinal mucosa, resulting in substantial loss of water and electrolytes in liquid stools. Other factors associated with infection are an adhesion factor and an attachment pilus. Not all strains of serotypes O1 or O139 possess the virulence factors, and they are rarely possessed by non-O1/O139 strains.

These types are further sub-divided into the *Inaba* (A,C) and *Ogawa* (A,B).and *Hikojima* (A,B,C) serotypes on somatic O1 biotypes antigens, and each biotype may display the "classical" or El Tor phenotype.

Healthy carriers (i.e., individuals who have suffered from cholera and carry the vibrio without manifesting illness) vary from 2% to 9% while for the El Tor biotype it could be as high as 25% for up to about 1 month. Chronic carriers may carry and shed the organisms in their stools for up to 1 year.

A large number of non-cholera vibrios are found in water. They are distinguished from the cholera vibrio because they lack the somatic O1 antigen, hence they are known as non-agglutinating vibrios (NAG) or non-cholera vibrios.

Salmonellosis

Salmonellae are motile, gram-negative, lactose-fermenting rod-shaped bacteria of the family *Enterobacteriaceae*. They are named after Daniel E.

Salmon, who first isolated the organism from porcine intestine. Salmonellae have been implicated in a wide spectrum of other diseases, including enteric or typhoid fever (primarily *Salmonella typhi* and *Salmonella paratyphi*), bacteremia, endovascular infections, focal infections (e.g., osteomyelitis), and enterocolitis (typically *Salmonella typhimurium*, *Salmonella enteritidis*, and *Salmonella heidelberg*).

All salmonellae are grouped into a single species, *Salmonella choleraesuis*, which is divided into seven subgroups based on DNA homology and host range. Most of the salmonellae that are pathogenic in humans belong to a single subgroup (subgroup I). Additionally, each of the salmonellae can be serotyped according to their particular complement of somatic O, surface Vi, and flagellar H antigens. Presently, more than 2,300 *Salmonella* serovars exist.

The transmission of salmonellae to a susceptible host usually occurs by consumption of contaminated foods or water. Although nontyphoid salmonellae generally precipitate a localized response, *S typhi* and other especially virulent strains invade deeper tissues via lymphatics and capillaries and elicit a major immune response. In the USA, in 1997, the estimated annual incidence of salmonellosis was 13.8 cases per 100,000 people, especially among persons who traveled to developing countries. In 1994, the most frequently isolated *Salmonella* strains causing human disease reported to the US Centers for Disease Control and Prevention was *S enteritidis*. Carriage is usually from 2% to 5%.

Cryptosporidiosis

This is a disease characterized by the stomach cramps, fever, diarrhea, dehydration, and low grade fever and lasts for 1–2 weeks in healthy persons. However, in immune-compromised persons, e.g., AIDS patients, it produces a debilitating, cholera-like diarrhea of up to 20 l/day, severe abdominal cramps, weight loss and tiredness.

Between March 23 through April 8, 1993, about 400,000 persons came down with cryptosporidiosis in the Wisconsin city of Milwaukee, out of which about 100 died. It was the greatest incidence of water borne infection in US history. It turned out that raw sewage had leaked into one of the two reservoirs serving the city.

It is caused by a Protozoon belonging to the Sporozoa (see Chap. 4). However, it is atypical in that

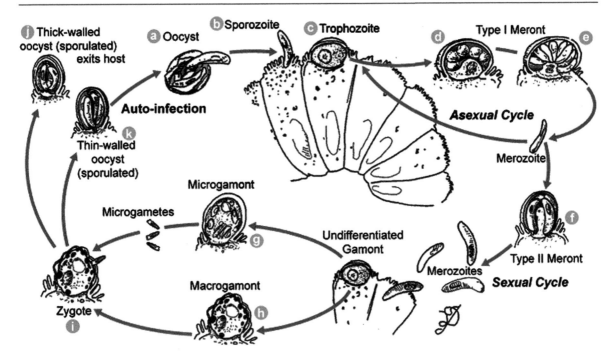

Fig. 8.5 Life history of *Cryptosporidium parvum* (Note the oocysts whose thick walls shield them from disinfectants) (From http://www.dpd.cdc.gov/dpdx/html/frames/A-/Cryptosporidiosis/ body_Cryptosporidiosis_life_cycle_lrg.htm; Credit: Center for Disease Control [CDC])

it completes its life history in a single host, unlike other Sporozoans which complete their life history in two hosts, such as the malaria parasite, *Plasmodium* spp. Members of this group produce different kinds of spores (hence their name, *Sporozoa*). *Cryptosporidium* isolated from humans is now referred to as *C. parvum*. *Cryptosporidium* infections have been reported from a variety of wild and domesticated animals, and in the last 6 or 7 years, literally hundreds of human infections have been reported, including epidemics in several major urban areas in the United States. Cryptosporidiosis is now recognized as an important opportunistic infection, especially in immunocompromised hosts. The disease is transmitted when feces of infected persons or animals enter drinking water. The thick walled oocytes of *Cryptosporidium parvum*, are known to be resistant to chlorine at concentrations typically applied for water treatment (Fig. 8.5).

Shigellosis

Shigellosis, also known as bacillary dysentery is a water borne illness caused by infection by bacteria of the genus *Shigella*. The causative organism is frequently found in water polluted with human feces, and is transmitted via the fecal-oral route. An estimated 18,000 cases of shigellosis occur annually in the United States, mainly affecting infants, the elderly, and especially those suffering from AIDS. It is characterized by diarrhea, often with blood or mucus in the watery stool, cramps, fever, and sometimes vomiting. It sets in a day or 2 after contact.

Discovered over a 100 years ago, the organism is named Shiga, the Japanese discover of the organism. There are four species of *Shigella: Boydii, dysenteriae, flexneri, and sonnei*. The genus *Shigella* is divided into four groups, A, B, C, and D, with the common nomenclature of *Shigella dysenteriae, S. flexneri, S. boydii,* and *S. sonnei*, respectively. There are 49 recognized serological types or serovars, representing subtypes from three of the four groups. Infection and outbreaks associated with this organism are prominent in developing countries and are strongly associated with overcrowding and poor hygienic conditions. Shigellosis is also travel associated, and several outbreaks have been reported in developed countries such as Canada and the United States which are travel-related. *Shigella sonnei* ("Group D" *Shigella*) accounts for over two-thirds of the shigellosis in the United States. *Shigella flexner* ("group B" *Shigell*) accounts for almost all of the rest. Other types of *Shigella* are rare in this country, although they are important causes of disease in the developing world. One type, *Shigella dysenteriae* type 1, causes deadly epidemics in many developing regions and nations. Following the invasion of the walls of the large intestine or colon, blood leaks into the colon hence the subsequent leakage of blood; due to the colon inflammation, mucus and pus may also be seen in the stool.

Tuberculosis

The tuberculous or "typical" species of *Mycobacterium*, such as *M. tuberculosis, M bovis, M. africanum,* and *M. leprae,* have only human or animal reservoirs and are not transmitted by water. In contrast, the non-tuberculous or "atypical" species (see below) of *Mycobacterium* are natural inhabitants of a variety of water environments. These aerobic, rod-shaped, and acid-fast bacteria grow slowly in suitable water environments and on culture media. Mycobacteria are called acid-fast because the high lipid content of their walls absorbs the red dye basic fuchsin and therefore the walls retain the dye when decolorized with dilute acid.

Tuberculosis, pulmonary tuberculosis (abbreviated as TB for *Tubercle Bacillus*), is a common and deadly infectious disease that is caused by *Mycobacterium tuberculosis.* Tuberculosis most commonly affects the lungs (as pulmonary TB) but can also affect the central nervous system, the lymphatic system, the circulatory system, the genitourinary system, bones, joints, and even the skin.

In 2004, 14.6 million people had active TB and there were 8.9 million new cases and 1.7 million deaths, mostly in developing countries. In addition, a rising number of people in the developed world are contracting tuberculosis because their immune systems are compromised by immunosuppressive drugs, substance abuse or HIV/AIDS.

Pulmonary (or lung) tuberculosis caused by *Mycobacterium tuberculosis* can be transmitted via water although this is uncommon. The difficulty of a categorical statement stems in part from the fact that the incubation period of tuberculosis is long, lasting from 4 to 6 weeks. Hence, it is often difficult to tell when contact was first made with the organisms. Children who had fallen into a river at a point where sewage from a chest clinic was discharged were shown

to have pulmonary tuberculosis after it was shown that the water in which the children had fallen contained mycobacteria. The organisms have been shown to persist in water for up to 36 days.

Atypical Tuberculosis

Atypical tuberculosis is also known as atypical mycobacterial disease, MOTT (mycobacteria other than tuberculosis), opportunist mycobacterial disease, environmental tuberculosis, and NTM TB. The organism that causes tuberculosis is *Mycobacterium tuberculosis* (*M. tb*). The atypical mycobacteria belong to the same family of mycobacterial organisms as *M.tb*, and it is also acid-fast but include other species such as *M. avium, M. intracellularae, M. kansasii, M. xenopi,* and *M. fortuitum.* Most infections with these organisms are believed to arise from environmental exposure to organisms in infected water, soil, dust, or aerosols. Person-to-person and animal-to-animal transmission of atypical mycobacteria is not an important factor in acquisition of infection with these organisms. Atypical tuberculosis is more common in immunosuppressed persons and in young children.

A simple TB skin test is the best way to address any concerns about infections. The test is only sensitive for mycobacterium tuberculosis; a positive test is usually followed by a chest x-ray to look for evidence of infections in the lungs.

Leptospirosis

Leptospirosis is an acute bacterial infection which affects the kidneys, liver, the central nervous system of man and other animals. It is caused by bacteria of the genus *Leptospira.* In humans, it causes a wide range of symptoms, and some infected persons may have no symptoms at all. Symptoms of leptospirosis include high fever, severe headache, chills, muscle aches, and vomiting, and may include jaundice (yellow skin and eyes), red eyes, abdominal pain, diarrhea, or a rash. When jaundice is involved, it is known as Weir's disease after the German doctor who first described it. If the disease is not treated, the patient could develop kidney damage, meningitis (inflammation of the membrane around the brain and spinal cord), liver failure, and respiratory distress. In rare cases, death occurs. Many of these symptoms can be mistaken for other diseases. Leptospirosis is confirmed by laboratory testing of a blood or urine sample.

The spirochete enters recreational waters and lakes directly through urination of infected cattle, swine, rodents, pigs, and frogs in whose kidneys the organisms lodge. Once inside the water, the spirochete enter the human body through abrasions, cuts, etc.

Enteropathogenic *Escherichia coli*

E. coli is normally a harmless commensal in the alimentary canals of man and other animals. However, some sero-types frequently cause a gastroenteritis characterized by severe diarrhea with little mucus or blood, and with dehydration but usually without fever. Children, especially the newborn, are usually affected but increasing cases of adult diarrhea caused by EEC are also being noted. The cases have usually been due to contaminated drinking water.

Currently, there are four recognized classes of enterovirulent *E. coli* (collectively referred to as the EEC group) that cause gastroenteritis in humans. Among these are the enteropathogenic (EPEC) strains. EPEC are defined as *E. coli* belonging to serogroups epidemiologically implicated as pathogens but whose virulence mechanism is unrelated to the excretion of typical *E. coli* enterotoxins. Humans, bovines, and swine can be infected, and the latter often serve as common experimental animal models. *E. coli* are present in the normal gut flora of these mammals. Acute infantile diarrhea is usually associated with EPEC.

Aeromonas

Species of *Aeromonas* are Gram-negative, non-spore-forming, rod-shaped, facultatively anaerobic bacteria that occur ubiquitously and autochthonously in aquatic environments. The aeromonads share many biochemical characteristics with members of the Enterobacteriaceae, from which they are primarily differentiated by being oxidase-positive. There are about 17 species. Among them is the psychrophilic *A. salmonicid*, which is a fish pathogen and has not been associated with humans. On the other hand, the mesophilic species, including mesophilic *A. hydrophila, A. caviae, A. sobria, A. veronii*, and *A. schubertii*, have been associated with a wide range of infections in humans. The species principally associated with gastroenteritis are *A. caviae, A. hydrophila*, and *A. veronii* biovar sobria ; *A. caviae* is particularly associated with young children (under 3 years of age). Numerous studies have resulted in the isolation of several species of *Aeromonas* from patients with gastroenteritis, but attempts to artificially infect volunteers have not been clear cut. They are also associated with sepsis and wounds, and with eye, respiratory tract, and other systemic infections. Many of the systemic infections arise following contamination of lacerations and fractures with *Aeromonas*-rich waters.

Amoebiasis

Amoebiasis is a parasitic infection of the large intestine caused by *Entamoeba histolytica*. Dysentery with mucus and blood, sometimes alternating with constipation, is a common feature. It is usually contracted by ingesting water or food contaminated with amoebic cysts. Amoebiasis is an intestinal infection that may or may not be symptomatic. When symptoms are present, it is generally known as invasive amoebiasis.

Contaminated drinking water or consumption of vegetables, fruits contaminated with feces are common sources and the organism survives adverse conditions by the production of cysts. Amoebiasis is usually transmitted by contamination of drinking water and foods with fecal matter, but it can also be transmitted indirectly through contact with dirty hands or objects as well as by sexual intercourse.

Giardiasis

Giardiasis is caused by *Giardia lamblia*, a flagellated protozoon. The clinical manifestations may vary from the passage of cysts without major malfunctions to severe malabsorption. Its typical symptoms are diarrhea, gas or flatulence, greasy stools that tend to float, stomach cramps, and upset stomach or nausea. It has become more important in the United States, where it has reportedly been spread from tourists returning to that country after holidays abroad. The cysts of the protozoon are not destroyed – as also is the case with ·*E. histolytica* – by chlorination at the level of contact normally employed in water treatment. Once an animal or person has been infected with *Giardia intestinalis*, the parasite lives in the intestine and is passed in the stool. Because the parasite is protected by an outer shell, it can survive outside the body and in the environment for long periods of time.

During the past two decades, *Giardia* infection has become recognized as one of the most common causes of waterborne disease (found in both drinking and recreational water) in humans in the United States. *Giardiasis* is found worldwide and within every region of the United States.

Table 8.2 Causes of waterborne outbreaks in the US by type of water system, 1991–2000 (From Anonymous 2003. With permission)

Etiological agent	Community water systems[a]		Noncommunity water systems[b]		Individual water systems[c]		All systems	
	Outbreaks	Cases	Outbreaks	Cases	Outbreaks	Cases	Outbreaks	Cases
Giardia	11	2,073	5	167	6	16	22	2,256
Cryptosporidium[d]	7	407,642	2	578	2	39	11	408,259
Campylobacter	1	172	3	66	1	102	5	340
Salmonellae, nontyphoid	2	749	0	0	1	84	3	833
E. coli	3	208	3	39	3	12	9	259
E. coli O157:H7/C. jeuni	0	0	1	781	0	0	1	781
Shigella	1	83	5	484	2	38	8	605
Plesiomonas shigelloides	0	0	1	60	0	0	1	60
Non-01 V. cholerae	1	11	0	0	0	0	1	11
Hepatitis A virus	0	0	1	46	1	10	2	56
Norwalk-like viruses	1	594	4	1,806	0	0	3	2,400
Small, round-structured virus	1	148	1	70	0	0	2	218
Chemical	18	522	0	0	7	9	25	531
Undetermined	11	10,162	38	4,837	11	238	60	15,237
Total	57	422,364	64	8,934	34	548	155	431,846

Data in Table 8.2 are compiled from CDC Morbidity and Mortality Weekly Report Surveillance Summaries for 1991–1992, 1993–1994, 1995–1996, 1997–1998 and 1999–2000. Figures include adjustments to numbers of outbreaks and illness cases originally reported, based on more recent CDC data

[a]Community water systems are those that serve communities of an average of at least 25 year-round residents and have at least 15 service connections

[b]Noncommunity water systems are those that serve an average of at least 25 residents and have at least 15 service connections and are used at least 60 days per year

[c]Individual water systems are those serving less than 25 residents and have less than 15 service connections

[d]There were 403,000 cases of illness reported in Milwaukee in 1993

Viruses

Viruses which are excreted in feces may be contacted through drinking water. Over 100 viruses which cause a variety of illnesses in man have been encountered in water contaminated by sewage (Bosch 1998). These include infectious hepatitis virus, adenoviruses, enteroviruses (i.e., polio-virus, coxsackie viruses and ECHO entero cytopathic human orphans) (see Table 8.2). However, the virus which has been definitely shown to be transmissible through drinking water is the infections hepatitis virus. The transmission of poliomyelitis virus appears to be mainly by contact and only secondarily by water. Some of the viruses in sewage-contaminated are described below.

1. *Enteroviruses*

   These are small RNA viruses and are divided into two: The polioviruses and the non-polio viruses.

The polio viruses have been eliminated in the western hemisphere and most parts of the world through extensive vaccination. A few pockets of polio endemicity remain in parts of Africa.

There are 62 non-polio enteroviruses that can cause disease in man: 23 Coxsackie A viruses, 6 Coxsackie B viruses, 28 echoviruses, and 5 other enteroviruses. They are second only to the "common cold" viruses, the rhinoviruses, as the most common viral infectious agents in humans. The enteroviruses cause an estimated 10–15 million or more symptomatic infections a year in the United States, especially in the summer and fall. Most people who are infected with an enterovirus have no disease at all. Those who become ill usually develop either mild upper respiratory symptoms (a "summer cold"), a flu-like illness with fever and muscle aches, or an illness with rash. In rare cases, some persons have "aseptic" or viral meningitis, illnesses that affect the heart (myocarditis) or the brain (encephalitis) or causes paralysis. Enterovirus infections are suspected to play a role in the development of juvenile-onset diabetes mellitus (sugar diabetes).

2. *Hepatoviruses*

A genus of Picornaviridae (non-enveloped, positive-stranded RNA viruses with an icosahedral capsid) causing infectious hepatitis naturally in humans and it is transmitted through fecal contamination of food or water. Hepatitis A virus (HAV) is the type species. Hepatitis A virus (HAV) is readily transmitted through water. HAV causes infectious hepatitis, an illness characterized by inflammation and necrosis of the liver. HAV can be removed from drinking water through coagulation, flocculation, and filtration.

3. *Reovirus*

A group of viruses that contain double-stranded RNA and are associated with various diseases in animals, including human respiratory and gastrointestinal infections. They derive their name from the acronym [r(espiratory)+e(nteric)+o(rphan)+viru]. They are suspected to cause respiratory and enteric illness.

4. *Rotavirus*

Rotaviruses are non-enveloped, icosahedral, double stranded (ds) RNA viruses with double capsid. Their electron microscopic appearance shows a 60–80 nm wheel with radiating spokes (Latin, rota, ≡ wheel).

Rotaviruses are the major cause of childhood gastroenteritis world-wide. In developing countries, deaths are common among children<5 years. Although the disease occurs in all age groups, it is mild and inapparent in adults. Infection is generally not recognized as food borne but outbreaks associated with food and water have been reported in a number of countries.

5. *Mastaldenovirus*

They are a genus of adenoviruses that infects mammals including humans and causes a wide range of diseases, including enteric and respiratory. The type species is Human adenovirus C. Of the many types of adenovirus, only two types, 40 and 41, are generally associated with fecal–oral spread and gastroenteritis (especially in children). Most infections are subclinical or mild.

6. *Astroviruses, Caliciviruses, Parvoviruses*

The above are all diarrhea viruses; they are of great economic importance, causing millions of lost working days each year, as well as much discomfort. Diarrhea continues to be a major cause of morbidity and mortality worldwide resulting in an estimated 1,000 deaths among children each day, the highest incidence being in developing countries of the world. One well documented source of infection is the consumption of shellfish (polluted by sewage) – and therefore, they also have economic consequences for fishermen and the food industry. Such viruses often cause mini-epidemics in families, hospital wards, etc. and are potentially very dangerous to seriously ill hospital patients. More importantly, these viruses contribute to the massive mortality caused by infantile diarrhea in developing countries and are responsible for uncounted millions of deaths each year.

The Norwalk-like viruses (NLV, now renamed as Norovirus) and Hepatitis E virus belong to the calicivirus group. Noroviruses are believed to be the most common causative agent for community gastroenteritis world-wide. On the other hand, the Hepatitis E virus appears to occur widely in Asia, Africa, and Latin America, where waterborne outbreaks are common. It has rarely been identified elsewhere. The virus infects the liver and symptoms of hepatitis.

Parasitic Worms

Various parasitic worms may be transmitted through water, including *Schistosoma* spp., *Taenia* spp., *Ascaris* spp., and *Enterobius* spp.

1. Guinea worm disease (GWD, *Dracunculiasis*) is an infection caused by the parasite *Dracunculus medinensis*. It affects poor communities around the world, including parts of Africa that do not have safe water to drink. In 2003, only 32,193 cases of GWD were reported. Most (63%) of those cases were from Sudan where the ongoing civil war makes it impossible to eradicate the disease. All affected countries except Sudan are aiming to eliminate Guinea worm disease as soon as possible.

2. Roundworm (*Ascaris lumbricoides*) is symptomless in many people. It is estimated that over one billion people in the world may be infected with roundworm. The source of infection is contamination of soil and vegetables with feces. Adult roundworms live in the small intestines and can exit through the mouth or nose of the infected person.

Occasionally, there is obstruction of the pancreatic or bile duct, appendix, or small intestines. Dry cough, fever, and sleep disturbance may occur. Diagnosis is by stool exam for eggs and blood test.

3. Hookworms (*Necator americanus*) is transmitted through unbroken skin by walking barefoot. Hookworms travel into blood and through the lung and intestines. Hookworm infection is usually symptomless. There may be itching at the area of skin penetration. There can be digestive symptoms. The worms attach to and suck the blood from the mucous of the small intestines, leading to iron deficiency anemia, low energy, and peptic ulcer-like symptoms in severe infections.

4. Pinworm (*Enterobius vermicularis*) infection is common in the United States. It is transmitted through contaminated food and water. The worms live in the intestines near the rectum and travel at night outside to the skin around the anus. From there, it can be transmitted through person to person contact. It can be symptomless. There is often itching at night around the anus. There can also be unusual symptoms such as hyperactivity, vision problems, vaginitis, and psychological disturbances. Tape is often applied to the anal area at night. When the tape is removed, adult worms may be seen with the unaided eye. At least five to seven tests are required to rule out infection.

5. Whipworm (*Trichuris trichiura*) is a large intestine parasite that rarely shows symptoms. It is transmitted by ingestion of the eggs in soil or on vegetables. Symptoms of heavy infection include diarrhea, stomach pain, rectal prolapse, and stunted growth.

8.2   Disease Transmission in Recreational Waters

Apart from regular swimming pools, there are a number of other types of recreational waters: *Hot tubs* is the term used to encompass a variety of facilities that are designed for sitting in (rather than swimming), contain water usually above 32°C, are generally aerated, contain treated water and are not drained, cleaned, or refilled for each user. They may be domestic, semi-public, or public and located indoors or outdoors. A wide range of names is used for them, including spa pools, whirlpools, whirlpool spas, heated spas, bubble baths, and Jacuzzi, a term that is used generically but is in fact a trade name.

Plunge pools are usually used in association with saunas, steam rooms, or hot tubs and are designed to cool users by immersion in unheated water. They are usually only large enough for a single person, but can be larger. They may be considered to be the same as swimming pools.

Natural spa is the term used to refer to facilities containing thermal and/or mineral water, some of which may be perceived to have therapeutic value and because of certain water characteristics may receive minimal water quality treatment.

Some of the microorganisms found in swimming pools and similar recreational water environments, have caused disease outbreaks. In the case of drinking water, which is taken into the human body, the concern is the consumption of water containing fecal material and hence the possibility of being infected by the pathogenic microorganisms or helminthic parasites present in such contaminated water. With recreational waters, however, while some of the disease causing agents could come from fecal material, some do not. Therefore, in looking at disease transmission through recreational waters, the disease agents will be looked at in two ways: The fecally transmitted and the non-fecally transmitted. Such a distinction will facilitate the control of diseases acquired through the use of recreational waters.

Fecal and non-fecal sources of disease causing agents in recreational waters (see Fig. 8.6):

(a) Fecal contamination of a recreational water can occur when
   1. The source water used in the recreational is pre-contaminated, and remains untreated.
   2. Fecal material is accidentally released by bathers as formed stool or diarrhea.
   3. There is residual fecal material on swimmers' bodies and this is washed into the pool.
   4. When feces from animals and birds get into recreational waters, e.g., in outdoor pools.

(b) Non-fecal contamination of recreational waters can occurs when
   1. There is nonfecal introduction of body fluids into the water, though vomiting, mucus, saliva, or skin peelings.
   2. Infected users directly contaminate pool waters and the surfaces of objects or materials at a facility with pathogens (notably viruses or fungi), which may lead to skin infections in other patrons who come in contact with the contaminated water or surfaces.

Fig. 8.6 Disease transmission hazards in recreational waters (Reproduced from http://www.who.int/water_sanitation_health/bathing/srwe2begin.pdf; Anonymous 2006c. With permission)

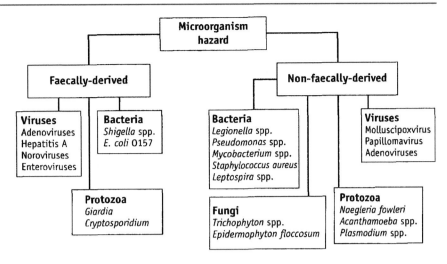

3. Some free-living aquatic bacteria and amoebae which can grow in pool, natural spa, or hot tub waters, in pool or hot tub components or facilities (including heating, ventilation, and air-conditioning (HVAC) or on other wet surfaces within the facility, develop to a point at which some of them may cause a variety of respiratory, dermal, or central nervous system infections or diseases.

8.2.1   Disease Transmission in Recreational Waters Through Fecal Material

8.2.1.1   Fecal Bacteria Which Have Caused Disease Outbreaks in Recreational Waters

Shigella species and *Escherichia coli* O157 are two related bacteria that have been linked to outbreaks of illness associated with swimming in pools or similar environments. *Shigella* has been responsible for outbreaks related to artificial ponds and other small bodies of water, where water movement has been very limited. The lack of water movement means that these water bodies behave very much as if they were swimming pools, except that chlorination or other forms of disinfection are not being used. Similar non-pool outbreaks have been described for *E. coli* O157, although there have also been two outbreaks reported where the source was a children's paddling pool. The contaminations which led to the outbreaks were due mostly to accidental fecal releases, and incidentally, the waters were not treated or had received inadequate chlorine treatment. The two organisms are controlled by chlorine and when

accidental fecal release occurs and is detected, the water should be emptied and disinfected with chlorine.

8.2.1.2   Fecal Protozoa Which Have Caused Disease Outbreaks in Recreational Waters

Giardia and particularly *Cryptosporidium* spp. are fecally derived protozoa that have been linked to outbreaks of illness in swimming pools and similar environments. These two organisms are similar in a number of respects in that they have a cyst or oocyst form that is highly resistant to both environmental stress and disinfectants; they both have a low infective dose and they are both shed in high densities by infected individuals. There have been a number of outbreaks of disease attributed to these pathogens, but about five times as many outbreaks have occurred with *Cyptosporidium* than with *Giardia*. Most outbreaks have resulted from accidental fecal discharges. Outbreaks of the former have been recorded in the US, Canada, Australia, and UK and generally affect children, perhaps because they swallow swimming pool water more than adults. Chlorine is not very effective eliminating the two protozoa; ozone and UV light are better. However, these later two do not, like chlorine, have residual doses in water.

The most practical approach to eliminating cysts and oocysts is through the use of filtration. *Cryptosporidium* oocysts are removed by filtration where the porosity of the filter is less than 4 μm. *Giardia* cysts are somewhat larger and are removed by filters with a porosity of 7 μm or less, Removal and inactivation of cysts and oocysts occur only in the fraction of water passing through treatment and the rate of

Table 8.3 Non-fecally derived bacteria found in swimming pools and similar environments and their associated diseases (Reproduced from http://www.who.int/water_sanitation_health/bathing/srwe2begin.pdf; Anonymous 2006c. With permission)

Organism	Infection/disease	Source
Legionella spp.	Legionellosis (Pontiac fever and Legionnaires' disease)	Aerosols from natural spas, hot tubs and HVAC systems
		Poorly maintained showers or heated water systems
Pseudomonas aeruginosa	Folliculitis (hot tubs)	Bather shedding in pool and hot tub waters and on wet surfaces around pools and hot tubs
	Swimmer's ear (pools)	
Mycobacterium spp.	Swimming pool granuloma	Bather shedding on wet surfaces around pools and hot tubs
		Aerosols from hot tubs and HVAC systems
Staphylococcus aureus	Skin, wound and ear infections	Bather shedding in pool water
Leptospira spp.	Haemorrhagic jaundice	Pool water contaminated with urine from infected animals
	Aseptic meningitis	

HVAC heating, ventilation and air conditioning

reduction in concentration in the pool volume is slow. Most outbreaks of giardiasis and cryptosporidiosis among pool swimmers have been linked to pools contaminated by sewage, accidental fecal releases, or suspected accidental fecal releases. Pool maintenance and appropriate disinfection levels are easily overwhelmed by accidental fecal releases or sewage intrusion; therefore, the only possible response to this condition, once it has occurred, is to prevent use of the pool and physically remove the oocysts by draining or by applying a long periods of filtration, as inactivation in the water volume (i.e., disinfection) is impossible. However, the best intervention is to prevent accidental fecal releases from occurring in the first place, through education of pool users about appropriate hygienic behavior. Immunocompromised individuals should be aware that they are at increased risk of illness from exposure to pathogenic protozoa.

8.2.1.3   Fecal Viruses Which Have Caused Disease Outbreaks in Recreational Waters

Viruses cannot multiply in water, and therefore their presence must be a consequence of pollution. Viruses which have been linked to outbreaks of virus diseases derived from fecal viruses in swimming pool water are linked to Adenoviruses (3, 4 7, 7a), Hepatitis A Norovirus, and Echovirus 30. Some adenoviruses may also be shed from eyes and the throat and are responsible for swimming pool conjunctivitis. Viruses of six types (rotavirus, norovirus, adenovirus, astrovirus, enterovirus, and hepatitis A virus) are all shed following infection. Ordinarily, rotaviruses are by far the most prevalent cause of viral gastroenteritis in children, and noroviruses cause the most cases of viral diarrhea in adults. However, few waterborne pool outbreaks have been associated with these agents. Even when outbreaks are detected, the evidence linking the outbreak to the pool is generally circumstantial.

Sampling water for viruses is not done routinely and is only carried out if the swimming pool is suspected to be the source of an infection. The control of viruses in swimming pool water and similar environments is usually by the application of disinfectants. Episodes of gross contamination of a swimming pool due to an accidental fecal release or vomit from an infected person cannot be effectively controlled by normal disinfectant levels. The only approach to maintaining public health protection under conditions of an accidental fecal release or vomit is to prevent the use of the pool until the contaminants are inactivated, through draining and the application of a disinfectant. The education of parents/caregivers of small children and other water users about good hygienic behavior at swimming pools is another approach that may contribute to the health of pool users. People with gastroenteritis should be advised not to use public or semipublic pools and hot tubs while ill and for at least a week after their illness, in order to avoid transmitting the illness to other pool or hot tub users.

8.2.2   Disease Transmission in Recreational Water Through Non-fecal Material

8.2.2.1   Disease Transmission by Bacteria in Recreational Water Through Non-fecal Material

Infections and diseases associated with non-enteric pathogenic bacteria found in swimming pools and similar recreational water environments are summarized in Table 8.3. A number of these bacteria may be shed by bathers or may be present in biofilms. Biofilms may form on the lining of pipes, e.g., in contact with water and may serve to protect the bacteria from disinfectants.

1. *Legionella* spp.

Legionella are Gram-negative, non-spore-forming, motile, aerobic bacilli, which may be free-living or living within amoebae and other protozoa or within biofilms. *Legionella* spp. are heterotrophic bacteria found in a wide range of water environments and can proliferate at temperatures above 25°C. They may be present in high numbers in natural spas using thermal spring water, and they can also grow in poorly maintained hot tubs, associated equipment, and HVAC systems. *Legionella* spp. can also multiply on filter materials, namely granular activated carbon. However, exposure to *Legionella* is preventable through the implementation of basic management measures, including filtration, maintaining a continuous disinfectant residual in hot tubs (where disinfectants are not used, there must be a high dilution rate with freshwater) and the maintenance and physical cleaning of all natural spa, hot tub, and pool equipment, including associated pipes and air-conditioning units.

Legionella spp. cause legionellosis (or Legionnaires' disease), a range of pneumonic and non-pneumonic diseases are caused by *L. pneumophila*. Males are roughly three times more likely than females to contract Legionnaires' disease among people 50 or older; chronic lung disease, cigarette smoking, and excess consumption of alcohol are predisposing factors. Specific risk factors, in relation to pools and hot tubs, include frequency of hot tub use and length of time spent in or around hot tubs. Although the attack rate is often less than 1%, mortality among hospitalized cases ranges widely up to 50%. Risk of legionellosis from pools and similar environments is associated with proliferation of *Legionella* in spas or hot tubs, associated equipment, and HVAC systems. Inhalation of bacteria or aspiration following ingestion, during natural spa or hot tub use, may lead to disease; showers may sometimes present a greater risk of legionellosis than pool water. Most of the reported legionellosis associated with recreational water use has been associated with hot tubs and natural spas. Natural spa waters (especially thermal water) and associated equipment create an ideal habitat (warm,

nutrient-containing aerobic water) for the selection and proliferation of *Legionella*. Hot tubs used for display in retail/wholesale outlets are also potential sources of infection. In order to control the growth of *Legionella* in hot tubs and natural spas, physical cleaning of surfaces is critical, and high residual disinfectant concentrations should be required. Features such as water sprays, etc., in pool facilities should be periodically cleaned and flushed with a level of disinfectant adequate to eliminate *Legionella* spp. (e.g., by use of a solution of at least 5 mg of hypochlorite per liter). Bathers should be encouraged to shower before entering the water. This will remove pollutants such as perspiration, cosmetics, and organic debris that can act as a source of nutrients for bacterial growth and neutralize oxidizing biocides. Bather density and duration spent in hot tubs should also be controlled. Public and semipublic spa facilities should have programmed rest periods during the day. High-risk individuals (such as those with chronic lung disease) should be cautioned about the risks of exposure to *Legionella* in or around pools and hot tubs.

Legionella spp. are ubiquitous in the environment and can proliferate at the higher temperatures experienced at times in piped drinking-water distribution systems and more commonly in hot and warm water distribution systems. Exposure to *Legionella* from drinking-water is through inhalation and can be controlled through the implementation of basic water quality management measures in buildings and through the maintenance of disinfection residuals throughout the piped distribution system.

2. *Pseudomonas aeruginosa*

Pseudomonas aeruginosa is an aerobic, non-spore-forming, motile, Gram-negative, straight or slightly curved rod with dimensions 0.5–1 μm × 1.5–4 μm. It can metabolize a variety of organic compounds and is resistant to a wide range of antibiotics and disinfectants. *P. aeruginosa* is ubiquitous in water, vegetation, and soil. Although shedding from infected humans is the predominant source of *P. aeruginosa* in pools and hot tubs, the surrounding environment can be a source of contamination. The warm, moist environment on decks, drains, benches, and floors provided by pools and similar environments is ideal for the growth of *Pseudomonas*, and it can grow well up to temperatures of 41°C. *Pseudomonas* tends to accumulate in biofilms in filters that are poorly maintained and in areas where pool hydraulics are poor (under moveable floors, for example). It is also likely that bathers pick up the organisms on their feet and hands and transfer them to the water. It has been proposed that the high water temperatures and turbulence in aerated hot tubs promote perspiration and desquamation (removal of skin cells).

These materials protect organisms from exposure to disinfectants and contribute to the organic load, which, in turn, reduces the residual disinfectant level; they also act as a source of nutrients for the growth of *P. aeruginosa*. In hot tubs, the primary health effect associated with the presence of *P. aeruginosa* is folliculitis. Otitis externa and infections of the urinary tract, respiratory tract, wounds, and cornea caused by *P. aeruginosa* have also been linked to hot tub use. Infection of hair follicles in the skin with *P. aeruginosa* produces a pustular rash, which may appear under surfaces covered with swimwear or may be more intense in these areas. The rash appears 48 h (range 8 h–5 days) after exposure and usually resolves spontaneously within 5 days. It has been suggested that warm water supersaturates the epidermis, dilates dermal pores, and facilitates their invasion by *P. aeruginosa*. There are some indications that extracellular enzymes produced by *P. aeruginosa* may damage skin and contribute to the bacteria's colonization. Other symptoms, such as headache, muscular aches, burning eyes, and fever, have been reported. Some of these secondary symptoms resemble humidifier fever and therefore could be caused by the inhalation of *P. aeruginosa* endotoxins.

In swimming pools, the primary health effect associated with *P. aeruginosa* is otitis externa or swimmer's ear, although folliculitis has also been reported. Otitis externa is characterized by inflammation, swelling, redness, and pain in the external auditory canal. Risk factors reported to increase the occurrence of otitis externa related to water exposure include amount of time spent in the water prior to the infection, people less than 19 years of age, and a history of previous ear infections. Repeated exposure to water is thought to remove the protective wax coating of the external ear canal, predisposing it to infection. An indoor swimming pool with a system of water sprays has been implicated as the source of two sequential outbreaks of granulomatous pneumonitis among lifeguards. Inadequate chlorination led to the colo-

nization of the spray circuits and pumps with Gram-negative bacteria, predominantly *P. aeruginosa*. The bacteria and associated endotoxins were aerosolized and respired by the lifeguards when the sprays were activated. When the water spray circuits were replaced and supplied with an ozonation and chlorination system, there were no further occurrences of disease among personnel. An outbreak of pseudomonas hot-foot syndrome, erythematous plantar nodules, has been reported as a result of exposure to a community wading pool. The floor of the pool was coated in abrasive grit, and the water contained high concentrations of *P. aeruginosa*. Another outbreak occurred in Germany due to high concentrations of *P. aeruginosa* on the stairs to a water slide and resulted in some of the children being admitted to hospital.

Maintaining adequate residual disinfectant levels and routine cleaning are the key elements to controlling *P. aeruginosa* in swimming pools and similar recreational environments. Under normal operating conditions, disinfectants can quickly dissipate. Most hot tubs use either chlorine- or bromine-based disinfectants; chlorination was superior to bromine in controlling *P. aeruginosa*.

3. *Mycobacterium* spp.

Mycobacterium spp. are rod-shaped bacteria that are 0.2–0.6 μm × 1.0–10 μm in size and have cell walls with a high lipid content. This feature means that they retain dyes in staining procedures that employ an acid wash; hence, they are often referred to as acid-fast bacteria. Atypical mycobacteria (i.e., other than strictly pathogenic species, such as *M. tuberculosis*) are ubiquitous in the aqueous environment and proliferate in and around swimming pools and similar environments. In pool environments, *M. marinum* is responsible for skin and soft tissue infections in normally healthy people. Infections frequently occur on abraded elbows and knees and result in localized lesions, often referred to as swimming pool granuloma. The organism is probably picked up from the pool edge by bathers as they climb in and out of the pool. Respiratory illnesses associated with hot tub use in normally healthy individuals have been linked to other atypical mycobacteria. For example, *M. avium* in hot tub water has been linked to hypersensitivity pneumonitis and possibly pneumonia. Symptoms were flu-like and included cough, fever, chills, malaise, and headaches. The illness followed the inhalation of heavily contaminated aerosols generated by the hot tub. The reported cases relate to domestic hot tubs, many of which were located outdoors. In most instances, the frequency of hot tub use was high, as was the duration of exposure (an extreme example being use for 1–2 h each day), and maintenance of disinfection and cleaning were not ideal. It is likely that detected cases are only a small fraction of the total number of cases. Amoebae may also play a role in the transmission of *Mycobacterium* spp. Mycobacteria are more resistant to disinfection than most bacteria due to the high lipid content of their cell wall. Therefore, thorough cleaning of surfaces and materials around pools and hot tubs where the organism may persist is necessary, supplemented by the maintenance of disinfection at appropriate levels. In addition, occasional shock dosing of chlorine may be required to eradicate mycobacteria accumulated in biofilms within pool or hot tub components. In natural spas where the use of disinfectants is undesirable or where it is difficult to maintain adequate disinfectant levels, superheating the water to 70°C on a daily basis during periods of nonuse may help to control *M. marinum*.

4. *Staphylococcus aureus*

The genus *Staphylococcus* comprises nonmotile, non-spore-forming and nonencapsulated Gram-positive cocci (0.5–1.5 μm in diameter) that ferment glucose and grow aerobically and anaerobically. They are usually catalase positive and occur singly and in pairs, tetrads, short chains and irregular grape-like clusters. In humans, there are three clinically important species – *Staphylococcus aureus*, *S. epidermidis* and *S. saprophyticus*. *S. aureus* is the only coagulase-positive species and is clinically the most important. Humans are the only known reservoir of *S. aureus*, and it is found on the anterior nasal mucosa and skin as well as in the feces of a substantial portion of healthy individuals. *S. aureus* is shed by bathers under all conditions of swimming, and the bacteria can be found in surface films in pool water. Coagulase-positive *Staphylococcus* strains of normal human flora have been found in and around swimming pools and similar environments.

The presence of *S. aureus* in swimming pools is believed to have resulted in skin rashes, wound infections, urinary tract infections, eye infections, otitis externa, impetigo, and other infections. Infections of *S. aureus* acquired from recreational waters may not become apparent until 48 h after contact. Recreational waters with a high density of bathers present a risk of staphylococcal infection that is comparable to the risk of gastrointestinal illness involved in bathing in water considered unsafe because of fecal pollution. Fifty percent or more of the total staphylococci isolated from swimming pool water samples are *S. aureus*. Adequate inactivation of potentially pathogenic *S. aureus* in swimming pools can be attained by maintaining free chlorine levels. There is evidence that showering before pool entry can reduce the shedding of staphylococci from the skin into the pool. Continuous circulation of surface water through the treatment process helps to control the build-up of *S. aureus*. Pool contamination can also be reduced if the floors surrounding the pool and in the changing areas are kept at a high standard of cleanliness. Although it is not recommended that water samples be routinely monitored for *S. aureus*, where samples are taken, levels should be less than 100/100 ml.

5. *Leptospira interrogans sensu lato*

Leptospires are motile spirochaete (helically coiled) bacteria. Traditionally, the genus *Leptospira* consists of two species, the pathogenic *L. interrogans* sensu lato and the saprophytic *L. bifl exa* sensu lato. Serological tests within each species revealed many antigenic variations, and, on this basis, leptospires are classified as serovars. In addition, a classification system based on DNA relatedness is used. The current species determination is based on this principle. The serological and genetic taxonomies are two different systems with only little correlation.

Free-living strains (*L. bifl exa* sensu lato) are ubiquitous in the environment, but the pathogenic strains (*L. interrogans* sensu lato), however, live in the kidneys of animal hosts.

Pathogenic leptospires live in the proximal renal tubules of the kidneys of carrier animals (including rats, cows, and pigs) and are excreted in the urine, which can then contaminate surface waters. Humans and animals (humans are always incidental hosts) become infected either directly through contact with infected urine or indirectly via contact with contaminated water. Leptospires gain entry to the body through cuts and abrasions of the skin and through the mucosal surfaces of the mouth, nose, and conjunctiva. Diseases caused by *Leptospira interrogans* sensu lato have been given a variety of names, including swineherd's disease, Stuttgart disease, and Weil's syndrome, but collectively all of these infections are termed leptospirosis.

Table 8.4 Non-fecally transmitted viruses found in water and their associated diseases (Reproduced from http://www.who.int/water_sanitation_health/bathing/srwe2begin.pdf; Anonymous 2006c. With permission)

Organism	Infection	Source
Adenoviruses	Pharyngo-conjunctivitis (swimming pool conjunctivitis)	Other infected bathers
Molluscipoxvirus	Molluscum contagiosum	Bather shedding on benches, pool or hot tub decks, swimming aids
Papillomavirus	Plantar wart	Bather shedding on pool and hot tub decks and floors in showers and changing rooms

The clinical manifestations of leptospirosis vary considerably in form and intensity, ranging from a mild flu-like illness to a severe and potentially fatal form of the disease, characterized by liver and kidney failure and hemorrhages (Weil's syndrome). Severity is related to the infecting serovar as well as host characteristics, such as age and underlying health and nutritional status. Specific serovars are often associated with certain hosts. Compared with many other pathogens, leptospires have a comparatively low resistance to adverse chemical and physical conditions, including disinfectants. They are seldom found in water of pH below 6.8, and they cannot tolerate drying or exposure to direct sunlight. The majority of reported outbreaks of waterborne leptospirosis have involved fresh recreational waters, and only two outbreaks have been associated with non-chlorinated swimming pools. Domestic or wild animals with access to the implicated waters were the probable sources of *Leptospira*.

The risk of leptospirosis can be reduced by preventing direct animal access to swimming pools and maintaining adequate disinfectant concentrations. Informing users about the hazards of swimming in water that is accessible to domestic and wild animals may also help to prevent infections. Outbreaks are not common; thus, it appears that the risk of leptospirosis associated with swimming pools and hot tubs is low. Normal disinfection of pools is sufficient to inactivate *Leptospira* spp.

8.2.2.2 Disease Transmission by Viruses in Recreational Water Through Non-fecal Material

Infections associated with non-fecally derived viruses found in swimming pools and similar environments: Adenoviruses, Molluscipoxvirus and Papillomavirus (see Table 8.4):

1. *Adenoviruses*
The adenoviruses are same as had been discussed.

2. *Molluscipoxvirus*
Molluscipoxvirus is a double-stranded DNA virus in the Poxviridae family. Virions are brick-shaped, about 320 nm × 250 nm × 200 nm. The virus causes molluscum contagiosum, an innocuous cutaneous disease limited to humans. It is spread by direct person-to-person contact or indirectly through physical contact with contaminated surfaces. The infection appears as small, round, firm papules or lesions, which grow to about 3–5 mm in diameter. The incubation period is 2–6 weeks or longer. Individual lesions persist for 2–4 months, and cases resolve spontaneously in 0.5–2 years. Swimming pool-related cases occur more frequently in children than in adults. The total number of annual cases is unknown. Since the infection is relatively innocuous, the reported number of cases is likely to be much less than the total number. Lesions are most often found on the arms, back of the legs and back, suggesting transmission through physical contact with the edge of the pool, benches around the pool, swimming aids carried into the pool or shared towels. Indirect transmission via water in swimming pools is not likely. Although cases associated with hot tubs have not been reported, they should not be ruled out as a route of exposure.

The only source of molluscipoxvirus in swimming pool and similar facilities is infected bathers. Hence, the most important means of controlling the spread of the infection is to educate the public about the disease, the importance of limiting contact between infected and noninfected people and medical treatment. Thorough frequent cleaning of surfaces in facilities that are prone to contamination can reduce the spread of the disease.

3. *Papillomavirus*
Papillomavirus is a double-stranded DNA virus in the family Papovaviridae. The virions are spherical and approximately 55 nm in diameter. The virus

causes benign cutaneous tumors in humans. An infection that occurs on the sole (or plantar surface) of the foot is referred to as a verruca plantaris or plantar wart. Papillomaviruses are extremely resistant to desiccation and thus can remain infectious for many years. The incubation period of the virus remains unknown, but it is estimated to be 1–20 weeks. The infection is extremely common among children and young adults between the ages of 12 and 16 who frequent public pools and hot tubs. It is less common among adults, suggesting that they acquire immunity to the infection. At facilities such as public swimming pools, plantar warts are usually acquired via direct physical contact with shower and changing room floors contaminated with infected skin fragments. Papillomavirus is not transmitted via pool or hot tub waters.

The primary source of papillomavirus in swimming pool facilities is infected bathers. Hence, the most important means of controlling the spread of the virus is to educate the public about the disease, the importance of limiting contact between infected and noninfected people, and medical treatment. The use of pre-swim showering, wearing of sandals in showers and changing rooms, and regular cleaning of surfaces in swimming pool facilities that are prone to contamination can reduce the spread of the virus.

8.2.2.3 Disease Transmission by Protozoa in Recreational Water Through Non-fecal Material

The non-fecally derived protozoa found in or associated with swimming pools and similar environments and their associated infections are *Naegleria fowleri*, *Acanthamoeba* spp., and *Plasmodium* spp.

1. *Naegleria fowleri*
   Naegleria fowleri is a free-living amoeba (i.e., it does not require the infection of a host organism to complete its life cycle) present in freshwater and soil. The life cycle includes an environmentally resistant encysted form. Cysts are spherical, 8–12 μm in diameter, with smooth, single-layered walls containing one or two mucus-plugged pores through which the trophozoites (infectious stages) emerge. *N. fowleri* is thermophilic, preferring warm water and reproducing successfully at temperatures up to 46°C.

   N. fowleri causes primary amoebic meningoencephalitis (PAM). Infection is usually acquired by exposure to water in ponds, natural spas, and artificial lakes. Victims are usually healthy children and young adults who have had contact with water about 7–10 days before the onset of symptoms. Infection occurs when water containing the organisms is forcefully inhaled or splashed onto the olfactory epithelium, usually from diving, jumping, or underwater swimming. The amoebae in the water then make their way into the brain and central nervous system. Symptoms of the infection include severe headache, high fever, stiff neck, nausea, vomiting, seizures, and hallucinations. The infection is not contagious. For those infected, death occurs usually 3–10 days after onset of symptoms. Respiratory symptoms occur in some patients and may be the result of hypersensitivity or allergic reactions or may represent a subclinical infection. Although PAM is an extremely rare disease, cases have been associated with pools and natural spas.

2. *Acanthamoeba* spp.
   Several species of free-living *Acanthamoeba* are human pathogens (*A. castellanii*, *A.culbertsoni*, *A. polyphaga*). They can be found in all aquatic environments, including disinfected swimming pools. Under adverse conditions, they form a dormant encysted stage, measuring 15–28 μm, depending on the species. *Acanthamoeba* cysts are highly resistant to extremes of temperature, disinfection, and desiccation. The cysts will retain viability from −20°C to 56°C. When favorable conditions occur, such as a ready supply of bacteria and a suitable temperature, the cysts hatch (excyst) and the trophozoites emerge to feed and replicate. All pathogenic species will grow at 36–37°C, with an optimum of about 30°C. Although *Acanthamoeba* is common in most environments, human contact with the organism rarely leads to infection. Human pathogenic species of *Acanthamoeba* cause two clinically distinct diseases: Granulomatous Amoebic Encephalitis (GAE) and inflammation of the cornea (keratitis). GAE is a chronic disease of the immunosuppressed; GAE is either sub-acute or chronic but is invariably fatal. Symptoms include fever, headaches, seizures, meningitis, and visual abnormalities. GAE is extremely rare, with only 60 cases reported worldwide. The route of infection in GAE is unclear; although invasion of the brain may result from the blood following a primary infection elsewhere in the body, possibly the skin or lungs (Martinez 1991). The precise source of such infections is unknown because of the almost ubiquitous presence of *Acanthamoeba* in the environment.

Acanthamoeba keratitis affects previously healthy people and is a severe and potentially blinding infection of the cornea. In the untreated state, *Acanthamoeba* keratitis can lead to permanent blindness. Although only one eye is usually affected, cases of bilateral infection have been reported. The disease is characterized by intense pain and ring-shaped infiltrates in the corneal stroma. Contact lens wearers are most at risk from the infection and account for approximately 90% of reported cases. Poor contact lens hygiene practices (notably ignoring recommended cleaning and disinfection procedures and rinsing or storing of lenses in tap water or non-sterile saline solutions) are recognized risk factors, although the wearing of contact lenses while swimming or participating in other water sports may also be a risk factor. In noncontact lens related keratitis, infection arises from trauma to the eye and contamination with environmental matter such as soil and water

Although *Acanthamoeba* cysts are resistant to chlorine- and bromine-based disinfectants, they can be removed by filtration. Thus, it is unlikely that properly operated swimming pools and similar environments would contain sufficient numbers of cysts to cause infection in normally healthy individuals. Immunosuppressed individuals who use swimming pools, natural spas or hot tubs should be aware of the increased risk of GAE. A number of precautionary measures are available to contact lens wearers, including removal before entering the water, wearing goggles, post-swim contact lens wash using appropriate lens fluid and use of daily disposable lenses.

3. *Plasmodium* spp.

Swimming pools are associated not with *Plasmodium* spp., but with anopheline mosquito larvae, the insect vectors of *Plasmodium*. Many swimming pools in malaria-prong region harbor the larvae. The problem relates to the seasonal use of the pools. Before people leave their summer houses, it is common to drain the pool; however, rainwater accumulated during the rainy season provides a suitable habitat for mosquito breeding, with the attendant risks of malaria as a result. This especially occurs in out of use pools. Such pools should be grained or sprayed worth kerosene or other oils which stop the larvae getting oxygen.

During the rains, when the pools fill with water, they should be drained every 5–7 days to avoid mosquito larvae developing into adults. The swimming pools may also be treated with appropriate larvicides when not in use for long periods.

8.2.2.4 Disease Transmission by Fungi in Recreational Water Through Non-fecal Material: *Trichophyton* spp. *and Epidermophyton floccosum*

Epidermophyton floccosum and various species of fungi in the genus *Trichophyton* cause superficial fungal infections of the hair, fingernails, or skin. Infection of the skin of the foot (usually between the toes) is described as tinea pedis or, more commonly, as "athlete's foot" Symptoms include maceration, cracking and scaling of the skin, with intense itching. Tinea pedis may be transmitted by direct person-to-person contact; in swimming pools, however, it may be transmitted by physical contact with surfaces, such as floors in public showers, changing rooms, etc., contaminated with infected skin fragments. The fungus colonizes the stratum corneum when environmental conditions, particularly humidity, are optimal. From in vitro experiments, it has been calculated that it takes approximately 3–4 h for the fungus to initiate infection. The infection is common among lifeguards and competitive swimmers, but relatively benign; thus, the true number of cases is unknown.

The sole source of these fungi in swimming pool and similar facilities is infected bathers. Hence, the most important means of controlling the spread of the fungus is to educate the public about the disease, the importance of limiting contact between infected and noninfected bathers, and medical treatment. The use of pre-swim showers, wearing of sandals in showers and changing rooms, and frequent cleaning of surfaces in swimming pool facilities that are prone to contamination can reduce the spread of the fungi. People with severe athlete's foot or similar dermal infections should not frequent public swimming pools, natural spas or hot tubs. Routine disinfection appears to control the spread of these fungi in swimming pools and similar environments.

8.3   Disease Transmission Through Shellfish Growing in Fecally Contaminated Water

8.3.1   Description of Shellfish

Shellfish is a culinary term for aquatic invertebrates used as food: Molluscs, crustaceans, and echinoderms. Shellfish come from both saltwater and freshwater.

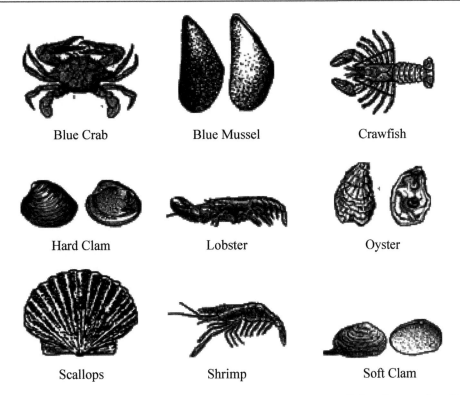

Fig. 8.7 Illustrations of different shellfish (Illustrations courtesy of Delaware Sea Grant College Program; http://deseagrant.org/outreach/seafood; www.decoastday.org; Anonymous 2009. With their permission)

The term finfish is sometimes used to distinguish them from ordinary fish which are vertebrates. Molluscs commonly used as food include the clam, mussel, oyster, eye winkles, and scallop. Some crustaceans commonly eaten are shrimps, prawn, lobster, crayfish, and crab. Echinoderms are not eaten as commonly as mollusks and crustaceans. In Asia, sea cucumber and sea urchins are eaten. Edible cephalopods, such as squid, octopus, and cuttlefish and terrestrial snails, though all molluscs, are sometimes classified as shellfish (Fig. 8.7).

8.3.2 Monitoring the Aquatic Environment of Shellfish Growth

Shellfish thrive best in coastal areas of the sea including estuaries. These locations are close to human habitations and are thus influenced by urbanization. There is great trend toward coastal habitation in the US and around the world and this trend impacts on the quality of water in which shellfish grow. Thus, in the United States, coastal counties cover only 17% of the total land area yet are home to more than 53% of the nation's population. Estimated at 139 million people in 1994,

the nation's coastal population is projected to reach 165 million people and an average density of 327 people per square mile by 2015. Globally, approximately 37% of the world's population lives within 100 km of the coastline and 50% within 200 km. This means the favorite sites for the growth of shellfish is subject to pollution, including microbial contamination, by human activities.

Microorganisms are discharged to shellfish growing areas from a variety of pollution sources along three main pathways: (1) Direct discharges from sewage outfalls, boaters, marine mammals, and other sources; (2) Subsurface flows from such sources as shoreline on-site sewage systems; and (3) Overland flows in the form of storm water runoff, stream flows, and other surface runoff. These sources and pathways are determined by a variety of human activities and land uses that tend to exert a progressively greater influence on the landscape and environmental conditions as development intensifies over time.

Oysters, clams, mussels, and scallops are filter feeders that pump large quantities of water through their bodies when actively feeding. During this process, molluscan shellfish can concentrate microorganisms, toxigenic micro-algae, and poisonous or deleterious substances from the water column when they are present in the growing waters. Concentrations in the shellfish may be as much as 100 times that found in the water column. If human pathogens are concentrated to an infective dose, and if the shellfish are consumed raw or partially cooked, human disease can result. If toxigenic micro-algae are present and producing toxin, human illness or death can occur, and cooking is not reliable as an effective barrier against intoxication.

Therefore, bodies of water where commercial shellfish are grown are monitored by the States bordering the sea where the shellfish industry contributes in an important way to the states' economy. Such States include Florida, Washington, and North Carolina to name a few.

Following the oyster-borne typhoid outbreaks during the winter of 1924–1925 in the United States, which threatened the collapse of the oyster industry, the National Shellfish Certification Program, now the National Shellfish Sanitation Program (NSSP), was initiated by the states, the Public Health Service, and the shellfish industry. The NSSP is today administered by the U.S. Food and Drug Administration. It has established guidelines for water quality in shellfish growing areas to decrease the risk of illness associated with these the consumption of shellfish; it has also put in place, sanitary controls over all phases of the growing, harvesting, shucking, packing, and distribution of fresh and fresh-frozen shellfish.

Other countries also monitor shellfish growing areas. Thus Canada, the European Union, and Japan all have agencies which regulate and control provide guidelines for the microbial content of water where shellfish is harvested.

8.3.2.1 Procedure for Monitoring Shellfish Growing Areas in the US

Monitoring the procedure by which the NSSP controls sites for the growing of shellfish is the *sanitary survey*. The principal components of a sanitary survey are:

1. *Identification and evaluation of the pollution sources that may affect the areas*. A pollution source survey (also known as a shoreline survey) must be conducted of the growing area shoreline and watershed to locate direct sewage discharges (e.g., municipal and private sewage and industrial waste discharges, sewage package treatment units, malfunctioning septic tanks, and animal manure treatment lagoons) etc.

2. *An evaluation of the meteorological factors*, since climate and weather can affect the distribution of pollutants or can be the cause of pollutant delivery to a growing area. Rainfall patterns and intensity can affect water quality through pollutant delivery in runoff or cause flooding which can affect the volume and duration of pollutant delivery.

3. *An evaluation of hydrographic factors that may affect distribution of pollutants throughout the area*. Examples of hydrographic factors are tidal amplitude and type (most shellfish are grown in estuaries), water circulation patterns, and the amount of freshwater. These factors, along with water depths and stratification caused by density (salinity and temperature) differences, and wastewater and other waste flow rates are used to determine dilution, and time of transport.

4. *An assessment of water quality; the assessment of water quality is based on the total coliform status of the water*. The National Shellfish Sanitation Program (NSSP) allows growing areas to be classified using either a total or fecal coliform standard. The guidelines set up by the NSSP require that the total coliform geometric mean MPN of the water sample results for each sampling station shall not exceed 70 MPN per 100 ml; and not more than 10% of the samples shall exceed an MPN of:
   (a) 230 MPN per 100 ml for a 5-tube, decimal dilution test
   (b) 330 MPN per 100 ml for a 3-tube, decimal dilution test
   (c) 140 MPN per 100 ml for the 12-tube, single dilution test

The goal of the NSSP is to control the safety of shellfish for human consumption by preventing harvest from contaminated growing waters. In implementing this concept, the NSSP uses five classifications for growing areas: Approved, conditionally approved, restricted, conditionally restricted, and prohibited.

All shellfish growing areas are surveyed every 3 years to document all existing or potential pollution sources, to assess the bacteriological quality of the water, and to determine the hydrographic and meteorological factors that could affect water quality. Water samples are collected at least six times a year from each growing area and tested for fecal coliform bacteria, which are an indicator that human or animal wastes are present in the water. In addition, reviews of bacteriological data and pollution sources are conducted annually.

Classification of Shellfish Growing Areas

The information gathered from the sanitary survey, including the probable presence or absence of pathogenic microorganisms, marine biotoxin or other poisonous or deleterious substances in growing area waters is used to classify each shellfish growing area as either approved, conditionally approved, restricted, or prohibited.

The major industrialized countries and regions of the world categorize the sites from where shellfish is grown to protect the health of consumers. This is based, in general, on the hygienic (i.e., the number of *E. coli* found in the monitoring of the areas. Thus, the codified Shellfish Waters Directive of the European Union (2006/113/EC), adopted on 12 December 2006 specifies standards based on traditional bacteria. The guidelines of the National Shellfish Sanitation Program (NSSP), a unit of the United States Food and Drug Administration, and the Fisheries Agency of the Japanese Ministry of Agriculture, Forestry and Fisheries, responsible for preserving and managing marine biological resources and fishery production, all use the traditional indicators i.e., *E. coli* and fecal coliforms.

Directive 91/492/EEC specifies standards based on traditional bacterial indicators i.e., *E. coli* and fecal coliforms; however, it is now widely recognized that these sanitary controls, whilst they play an important role in protecting public health, do not guarantee protection against viral infection. In Europe and elsewhere, viruses are responsible for the majority of outbreaks associated with shellfish, and many outbreaks have occurred when shellfish have been fully compliant with the Directive. The most common viruses that cause illness are Norwalk-like viruses, which cause diarrhea and vomiting. Hepatitis A virus occurs in a very small number of cases. Within the UK, these problems were highlighted in 1998 by the Advisory Committee on the Microbiological Safety of Foods report on Foodborne Viral Infection (Table 8.5).

In many cases, the risk of illness or infection has been linked to fecal contamination of the water. The fecal contamination may be due to feces released by bathers or a contaminated source water or, in outdoor pools, may be the result of direct animal contamination (e.g., from birds and rodents). Fecal matter is introduced into the water when a person has an accidental fecal release – AFR (through the release of formed stool or diarrhea into the water) or residual fecal material on swimmers' bodies is washed into the pool. Many of the outbreaks related to swimming pools would have been prevented or reduced if the pool had been well managed. Non-fecal human shedding (e.g., from vomit, mucus, saliva, or skin) in the swimming pool or similar recreational water environments is a potential source of pathogenic organisms. Infected users can directly contaminate pool or hot tub waters and the surfaces of objects or materials at a facility with pathogens (notably viruses or fungi), which may lead to skin infections in other patrons who come in contact with the contaminated water or surfaces. "Opportunistic pathogens" (notably bacteria) can also be shed from users and transmitted via surfaces and contaminated water (Anonymous 2007).

Some bacteria, most notably non-fecally derived bacteria may accumulate in biofilms and present an infection hazard. In addition, certain free living aquatic bacteria and amoebae can grow in pool, natural spa or hot tub waters, in pool or hot tub components or facilities (including heating, ventilation, and air-conditioning [HVAC] systems) or on other wet surfaces within the facility to a point at which some of them may cause a variety of respiratory, dermal, or central nervous system infections or diseases. Outdoor pools may also be subject to microorganisms derived directly from pets and wildlife.

8.4 Recent Developments Regarding Knowledge of Pathogens in Drinking Water

Today, in most industrialized countries, while outbreaks of waterborne infections from the classical pathogens like *Vibrio cholerae* and *Salmonella typhi* do not occur (except for sporadic, imported cases), *S. typhi*, *V. cholerae* O1, and *Shigella* spp. are rarely found in drinking water distribution systems, and their appearance points to a major failure of the systems (insufficient treatment or secondary fecal contamination of the distribution systems). "New" or "emerging" pathogens seem to have occurred, including bacteria, viruses, and parasites.

In recent years, several so-called "new or emerging pathogens" have arisen as problems in drinking water production and distribution from two sources. Firstly, newly recognized pathogens from fecal sources have occurred like *Campylobacter jejuni*, pathogenic

Table 8.5 Composite table showing classification criteria of shellfish harvest sites by the US Food and Drugs Administration and the EU Parliament and Council (Modified from Anonymous 2006d, 2008)

Classification by the National Shellfish Sanitation Program (NSSP), of the Food and Drugs Administration (FDA) (Anonymous 2008)			Classification by the Directive 2006/113/EC of the European Parliament and of the Council (Anonymous 2006d)		
Grade	Description	Method: Water sampled (per ml of water)	Grade	Description	Method: Shellfish sampled (cfu /100 g shellfish)
A	Sanitation survey finds growing areas safe found for the direct marketing of shellfish or is not subject to human or animal fecal pollution.	Fecal coliform MPN <14/100 ml	A	No restriction. Shellfish acceptable for immediate consumption	<230 *E. coli* or <300 Fecal *No Salmonella* in 25 g
CA	Sanitation survey finds growing areas are in the open status for a reasonable period of time and when pollution factors are predictable. There may be direct potential for distribution of pollutants based on unusual conditions or specific times of the year when bacterial numbers are increased by heavy water runoff that affects wastewater treatment plant function		B	Shellfish must be depurated or relayed until they meet category A standard	<4,600 *E. coli* or <6,000 Fecal coliforms in 90% of samples
R	Sanitation survey finds growing areas have suffered a limited degree of pollution and the levels of fecal pollution, human pathogens, or poisonous or deleterious pollutants are at such levels that shell stock can be made safe through either relaying or depuration.		C	Shellfish must be relayed over a long period (>2 months) until they meet category A standard	<60,000 Fecal coliforms
P	Sanitation survey finds no current sanitary survey exits for the growing area or when the survey determines that the growing area is adjacent to a sewage treatment plant outfall or other point source with public health significance or when the water is polluted because of previous or current sources of contamination.	Fecal coliform MPN <88/100 ml			

The approaches of the US FDA and the EU are different: the FDA samples water in the shellfish growing area and the EU samples actual shell fish. Both methods have their advantages and disadvantages
US FDA: *A* acceptable, *CA* conditionally acceptable, *R* restricted, *P* prohibited

Escherichia coli, Yersinia enterocolitica, new enteric viruses like rotavirus, calicivirus, small round-structured virus, astrovirus, and the parasites *Giardia lamblia, Cryptosporidium parvum,* and microsporidia. Secondly, some new pathogens comprise species of environmental bacteria that are able to grow in water distribution systems and only recently were recognized as relevant pathogens, such as *Legionella* spp., *Aeromonas* spp., *Mycobacterium* spp., and *Pseudomonas aeruginosa.*

There are a number of reasons for the emergence of these new pathogens. Many of these pathogens are not actually new and may have been causing disease for a long time, but they were not identified owing to a lack of detection methods. This is especially true for the viruses and parasites, but also for *Legionella* spp. Other new pathogens have been known to cause infections, but they were not associated with drinking water or were known only as animal pathogens. For example, *Campylobacter jejuni* had been known only as a rare opportunistic pathogen causing bloodstream infections, until it was recognized as a cause of diarrhea, in the 1970s, and even later as a possible infectious agent in drinkingwater. *Cryptosporidium* sp. was first described in 1907 and first recognized as an animal pathogen in 1955, but not until about 20 years later was it recognized as a human pathogen. The emergence of new pathogens has also been promoted by a change in habits of water usage. The increasing use of heated drinking water with warm water reservoirs in houses has the disadvantage that these systems are an ideal habitat for *Legionella* spp. leading to an increased number of infections. Furthermore, *P. aeruginosa* and environmental mycobacteria are able to use the in-house installations and reservoirs as new habitats, posing new risks to susceptible water consumers. Another important factor for the emergence of new pathogens is the increasing number of people who are susceptible to infections with specific potential pathogens, including on one hand immunocompromised persons, such as AIDS patients and patients receiving cancer chemotherapy or undergoing organ transplantation, and on the other hand, elderly persons whose immune systems are not as active as in healthy young adults. These persons are subject to infections that do not occur in healthy adults or, if they do occur, are much less severe. Like young children, the elderly have, for instance, a higher risk of death from diarrhea. Several of the new pathogens were recognized because they caused severe infections in these subpopulations. Infections with environmental mycobacteria, e.g., are almost exclusively found in immunocompromised persons (e.g., *Mycobacterium avium* infections in AIDS patients).

Severe infections with *P. aeruginosa* are described for immunocompromised patients or persons with underlying diseases like diabetes or cystic fibrosis. Similarly, elderly and immunocompromised patients are at the highest risk of infection by *Legionella* spp., with smoking as an additional risk factor. Although infections with *C. parvum* do occur in healthy adults, the outcome in elderly or immunocompromised patients is much more severe and may be fatal. Very few of these new pathogens are really of recent origin. Besides emerging antibiotic-resistant strains, this is true for bacteria that have acquired new virulence factors. The most prominent example is pathogenic *E. coli* strains that are supposed to have taken up virulence genes by horizontal gene transfer, resulting in very potent new pathogens, the enterohemorrhagic *E. coli* (EHEC). In addition to new pathogens transmitted via drinking water, the production of toxins by cyanobacteria growing in water resources is an increasing problem in drinking-water supplies from surface water.

Among the new or emerging pathogens which are fecally related are: *Enteric viruses* including hepatitis A, rotaviruses, small round structured viruses, Norwalk virus, and caliciviruses, the protozoa, *Cryptosporidium parvum and Giardia lamblia*. Among the bacteria are *Campylobacter fetus*, *C. jejuni* and *C. col Enterohemorrhagic Escherichia coli* EHEC, *Yersinia enterocolitica, Helicobacter pylori. Microsporidia*, are very small (0.5–1.2 m), obligate intracellular parasites of vertebrates and invertebrates.

Of the new pathogens growing in the distribution system, several have a natural reservoir in the environment including water or soil, and are introduced from the surface water into the drinking-water system usually in low numbers. They are able to multiply in the water or in the adjoining biofilms, and hence their numbers increase in the distribution system. Such bacteria include *Legionella pneumophila, Pseudomonas aeruginosa*, environmental mycobacteria including *M. gordonae, M.avium, M. intracellulare, M. kansasii, M. chelonae*, and *M. fortuitum. Aeromonas* spp. (*A. hydrophila, A. caviae*, and *A. sobria*).

References

Anonymous (2003). Drinking water chlorination: A review of disinfection practices and issues, Chlorine Chemistry Division, American Chemistry Council. http://www.c3.org/chlorine_issues/disinfection/c3white2003.html. Accessed 6 Aug 2010.

Anonymous (2006a). *Guidelines for drinking-water quality* [electronic resource]:incorporating first addendum. Vol. 1, Recommendations. – 3 rd ed. 1. Potable water – standards. 2. Water – standards. 3. Water quality – standards. 4. Guidelines. I. Title. Geneva: World Health Organization.

Anonymous (2006b). *Guidelines for drinking-water quality*, 3rd edition (Incorporating first and second addenda). Vol. 1 – Recommendations. Chapter 11: Microbial fact sheet. Geneva: World Health Organization. http://www.who.int/water_sanitation_health/dwq/gdwq3rev/en/index.html. Accessed 21 July 2010.

Anonymous. (2006c). *Guidelines for safe recreational water environments* (Swimming pools and similar environments, Vol. 2). Geneva: World Health Organization.

Anonymous (2006d). Directive 2006/113/EC of the European Parliament and of the Council of 12 December 2006 on the quality required of shellfish waters. http://europa.eu/scadplus/leg/en/lvb/l28177.htm. Accessed 10 Sept 2010.

Anonymous (2007). Shellfish. http://en.wikipedia.org/wiki/Shellfish. Accessed 18 June 2007.

Anonymous (2008). How shellfish areas are classified. http://maine.gov/dmr/rm/public_health/howclassified.htm. Accessed 10 Sept 2010.

Anonymous (2009). Shellfish. www.ocean.udel.edu/mas/seafood/index.html. Accessed 3 Jan 2010; see also: http://deseagrant.org/outreach/seafood and www.decoastday.org.

Bosch, A. (1998). Human enteric viruses in the water environment: A mini review. *International Microbiology, 1*, 191–196.

Cliff, A., & Haggett, P. (1988). *Atlas of disease distributions, analytical approaches to epidemiological data.* Oxford: Blackwell.

Falconer, I. R., & Humpage, A. R. (2005). Health risk assessment of cyanobacterial (blue-green algal) toxins in drinking water. *International Journal of Environmental Research and Public Health, 2*, 43–50.

Martinez, A. J. (1991). Infections of the central nervous system due to *Acanthamoeba. Reviews of Infectious Diseases, 13*, S399–S402.

Pineo, C.S., & Subrahmanyam, D.V. (1975). *Community water supply and excreta disposal situation in the developing countries: A commentary.* Series: WHO offset publication, no. 15. Geneva: World Health Organization. ISBN: 9241700157. http://whqlibdoc.who.int/offset/WHO_Offset_15.pdf. Accessed 14 Mar 2010.

Municipal Purification of Water

9

Abstract

Drinking water is a basic necessity for the maintenance of good health in humans, but it is also a vehicle for the introduction of harmful biological agents such as bacterial and protozoan pathogens into the body. Therefore raw waters are purified to render them safe for drinking. The processes adopted in municipal water purification include the following: pretreatment (pre-coagulation, pre-disinfection), aeration, coagulation, filtration (slow, rapid, ultrafiltration, carbon filtration), disinfection (chloramines, ozonation, ultraviolet light, chlorination) miscellaneous treatments (Fe/Mn removal, deionization, reverse osmosis, algal/odor control, softening, ion-exchange, fluoridation, radioactivity removal, plumbosolvency removal). Which of the processes is actually employed depends on the quality of the raw water, the regulations of the appropriate authorities, and the budgetary considerations of the operator. The world over, regulatory authorities decide the maximum contaminants permissible in drinking water and recreational water; thus, the US Environmental Protection Agency (USEPA), the European Union Environmental Agency (EEA), the World Health Organization (WHO), and governmental agencies around the world all set standards, which differ from one another, and which reflect the level of economic, social, and technical expectations and accomplishments of the constituencies to which the standards are addressed.

Keywords

Water purification • Fecal indicators • Microbiological & chemical standards of drinking water • Standards for recreational waters • Shell fish growing waters • USA EPA, EU EEA, and WHO water standards • UV, reverse osmosis, plumbosolvency

9.1 The Need for Water Purification

Water is a basic necessity of life and, as has been seen in the previous chapter, many diseases can easily be transmitted through it. The purification of water is the removal from raw waters, undesirable materials, whether biological or chemical in nature, so as to produce water that is fit for human consumption and for other uses. Substances that are removed during water purification include bacteria, viruses, algae, fungi, parasites (such as *Giardia* or *Cryptosporidium*), minerals (including toxic metals such as lead, copper, etc.), and man-made chemical pollutants. Some of the materials removed are detrimental to health, whereas others do

N. Okafor, *Environmental Microbiology of Aquatic and Waste Systems*,
DOI 10.1007/978-94-007-1460-1_9, © Springer Science+Business Media B.V. 2011

not necessarily affect health but affect the smell, taste, or color of water (Bitton 2005).

Governments the world over enact laws which ensure that water consumed by humans is safe to drink. In July 1970, the White House and Congress worked together to establish the US Environmental Protection Agency (EPA) in response to the growing public demand for cleaner water, air, and land. In the USA, the EPA works to develop and enforce regulations that implement environmental laws enacted by Congress. EPA is responsible for researching and setting national standards for a variety of environmental programs, and delegates to states and tribes the responsibility for issuing permits and for monitoring and enforcing compliance. Where national standards are not met, EPA can issue sanctions and take other steps to assist the states and tribes in reaching the desired levels of environmental quality.

In the European Union, the body which coordinates environmental matters is the European Environmental Agency (EEA), which has its headquarters in Copenhagen, Denmark. The Agency currently has 32 member countries: 27 European Union Member States, and 5 other cooperating non-EU countries.

The Ministry of the Environment of Japan was formed in 2001 from the subcabinet level Environmental Agency established in 1971. The minister is a member of the Cabinet and is chosen by the Prime Minister, usually from the Diet (Parliament).

Worldwide, in Africa, South America, and Asia, governments have either ministries or agencies handling the affairs of the environment, including water.

Much of the world's interest in the environment can be traced to the book, *Silent Spring*, published in 1962 and written by Rachel Carson. The book claimed detrimental effects of pesticides on the environment, particularly on birds. Carson accused the chemical industry of spreading disinformation, and public officials of accepting industry claims uncritically. It inspired widespread public concerns with pesticides and pollution of the environment. *Silent Spring* facilitated the ban of the pesticide DDT in 1972 in the United States.

Public water supplies are not only a source of drinking water, but are also for recreation; the processes of purification must reflect these uses (Anonymous 2006a, b). Therefore water is purified for the following reasons:

(a) To protect the health of the consumers by eliminating waterborne infections
(b) For aesthetic reasons, i.e., the removal of qualities which, while not being harmful, are aesthetically unpleasant. For example, the removal of taste, color, odor, and turbidity, all of which are not, in themselves, necessarily harmful to man
(c) For economic reasons, e.g., in softening water and removing iron to reduce laundry costs and save the laundered materials
(d) For industrial purposes, such as in the preparation of water suitable for use in boilers, e.g., by removal of salts of calcium and magnesium, which would form scales in boilers and increase heating costs and time
(e) For other miscellaneous reasons, e.g., to reduce corrosiveness and to add desirable health-related elements, e.g., iodine and fluorides to combat goiter and teeth decay, respectively

As seen from the above, an important requirement of water is the protection of health. This chapter will discuss the processes for purifying water so as to make it safe for consumption, to meet aesthetic expectations, and eliminate diseases transmitted through drinking water or recreational waters. It will discuss standards set by various bodies for the protection of health in drinking water, in the water used for recreation, and water in which shellfish are grown, because some shellfish are eaten raw (Tebbutt 1992).

9.2 The Quality of the Raw Water to Be Purified

Water treatment involves physical, chemical, and biological changes which transform raw waters into potable waters. The treatment to be employed has to be worked out from knowledge of the quality of the raw water and also the purpose for which the waters is needed. Thus, if drinking water is obtained from deep wells with low loads of bacteria, no treatment, apart from aeration, and not even chlorination, may sometimes be necessary. On the other hand, where the source is river or stream water, extensive purification including chlorination may be required. It is therefore essential that natural water supplies from which potable water is eventually obtained are not so polluted that self-purification and water treatment processes cannot produce water of reliable potability in an economical manner. In other words, the raw water must not be so highly polluted that the cost of purifying it puts it beyond the reach of the consumer. In many countries, therefore, governments enforce regulations to protect catchment areas, i.e., areas from which raw waters emanate.

The US Public Health Association has, e.g., classified raw waters into the following four categories:

Group I: Waters requiring no treatment

Limited to underground waters not subject to any possibility of contamination and meeting the standards of drinking water in every way.

Group II: Waters requiring simple chlorination

Includes underground and low contamination surface waters, containing 50 coliform bacteria per 100 ml per month.

Group III: Waters requiring complete rapid sand filtration and continuous post-chlorination

This group includes waters requiring filtration treatment for turbidity and color removal; waters requiring high amounts of chlorination, waters polluted by sewage to such an extent that they contain an average coliform numbers of 50–5.000 per 100 ml per month and beyond this number in not more than 20% of the samples examined in any month.

Group IV: Waters requiring auxiliary treatment as well as complete filtration treatment and post-chlorination

Similar to Group III, but showing coliform numbers more than 5.000 per 100 ml in more than 20% of the samples during any month and not exceeding 20.000 per 100 ml in more than 5% of the samples examined during any month. Auxiliary treatment includes pre-sedimentation with coagulation, pre-chlorination, or mere storage for extensive periods (30 or more days) (Anonymous 2006a).

Some authors argue that these standards are too low and hence should be revised to make them more stringent.

9.3 Processes for the Municipal Purification of Water

All or (more usually) some of the following treatments are given to raw waters to make them suitable for drinking (Singley et al. 2006).

1. Pretreatment
   (a) Pre-filtration
   (b) Pre-chlorination
   (c) pH adjustment
2. Storage and sedimentation without coagulation
3. Aeration
4. Coagulation and Flocculation
5. Sedimentation
6. Filtration
   (a) Slow sand
   (b) Rapid sand (with pre-coagulation and sedimentation)
   (c) Carbon filtration
   (d) Ultrafiltration
7. Disinfection
   (a) Chlorination
   (b) Chloramines
   (c) Ozonation
   (d) Ultraviolet
8. Iron and Manganese removal
9. Softening of water (sand stabilization) or demineralization
10. Fluoridation
11. Algae control (taste and odor control)
12. Miscellaneous treatments
    (a) Plumbosolvency removal
    (b) Radium removal
    (c) Reverse osmosis
    (d) Ion exchange
    (e) Electrodeionization

A flow diagram of the various processes in water treatment is given in Fig. 9.1. As many of the known procedures as possible are shown in the figure. Which procedure is actually used in any given situation depends on the quality of the raw water and how much consumers are willing to pay for the finished water. Thus, while the raw water from a surface water such as a river must be filtered either by rapid or by slow filtration in order the reduce the bacterial load, there would be no effort to achieve electrodeionization, since deionized water is required only in very specialized situations in laboratories. Therefore, various combinations of the basic procedures given above exist from plant to plant and in different countries .

9.3.1 Pretreatments

The pretreatments given to a body of raw water depend on the nature of the water and the practice of the plant or the country. What are described as pretreatments below are used as such in some works, but are major activities in others.

1. Pre-filtration
   To ensure the efficient and reliable operation of the main units in a treatment plant, it is first necessary to remove the large floating and suspended solids which could obstruct flow. This is especially true where the

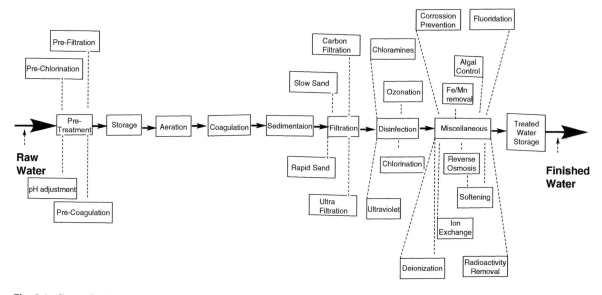

Fig. 9.1 Generalized methods for municipal purification of water

water supply is from surface waters such as rivers. The purpose is to remove large debris, including tree branches, rags, dead animals, fallen trees, etc., by passing the water through a 5–20 mm mesh.

2. Pre-chlorination

   It is becoming more and more the practice to pre-chlorinate raw waters, especially in the case of surface waters such as river water. Pre-chlorination is said to increase the efficiency of the downstream processes, including the elimination of bacteria and the removal of taste and odor. Sometimes this is combined with pre-ozonation.

3. pH adjustment

   If the water is acidic, the pH is raised using calcium oxide (lime, CaO). A slightly alkaline medium facilitates coagulation and flocculation. Furthermore, acid dissolves lead used in pipes, releasing it into the medium; lead is well known to have adverse health effects. The use of calcium could predispose the water to hardness; sodium carbonate is therefore often used in place of lime.

4. Pre-coagulation

   Pre-coagulation is used when the raw water is very turbid

9.3.2 Storage and Sedimentation Without Coagulation

When raw waters are impounded, they are stored in reservoirs. Usually, no treatment is initially made on such waters. They can be held for periods ranging from a few days to weeks or months. The mere fact of storage creates conditions favorable for the self-purification of the water through the activities of aerobic bacteria. Even the passage of water through a large lake will cause self-purification.

The factors of self-purification are interrelated and include physical, chemical, and biological factors and have been discussed in Chap. 7. They are, to a large extent, based on the activities of aerobic bacteria.

Sedimentation by gravity, if allowed to proceed for long enough, will remove all but the finest (colloidal) particles in the water (see Table 9.1).

Sunlight has a germicidal action in the upper 3 m (10 ft) in waters of low turbidity. Sunlight also induces photosynthesis in algae, thereby increasing the O_2 content of the water and hence the activity of the aerobes which break down organic matter. The oxidation of dissolved Mn^{++} and Fe^{++} compounds cause the oxides of these metals to precipitate. The breakdown of organic materials by aerobic bacteria is, however, the most important factor while the predatory activity of ciliates helps reduce the load of the bacteria themselves.

9.3.3 Aeration

Aeration is carried out in some waterworks but not in others. The purpose of aeration is as follows:

(a) To remove or reduce volatile taste, and odor, producing substances such as hydrogen sulfide,

Table 9.1 Sedimentation rate of objects of various diameters (Modified from Singley et al 2006)

Equivalent spherical radius	Approximate size	Sedimentation rate (time to settle 30 cm)
10 mm	Gravel	0.3 s
1 mm	Coarse sand	3 s
100 μm	Fine sand	38 s
10 μm	Silt	33 min
1 μm	Bacteria	55 h
100 nm	Colloid	230 h
10 nm	Colloid	6.3 years
1 nm	Colloid	63 years

methane, and other substances produced by bacteria or liberated by algal growth. Some taste- and odor-producing substances (such as geosmin produced by actinomycetes) are not volatile and are removed by other processes including coagulation and chlorination.

(b) To provide O_2 from the atmosphere for the oxidation of iron and manganese and thus to prevent corrosion and the staining of clothes.

(c) To restore "fresh" taste to water, as water devoid of dissolved air has a "flat" taste.

(d) In cases where water is obtained from very deep wells, where the water is hot, to cool the water.

During aeration, gases are dissolved by, or released from, the aerated water until an equilibrium is reached between the content of each individual gas in the atmosphere and its content in water. The diffusion of air into water is slow; hence, the need for agitation, which exposes various portions of the water to aeration. Since air normally contains little or no H_2S as well as other volatile gases, these are easily lost to the atmosphere during the aeration of water. On the other hand, when O_2 content is high because of algal activity, the excess O_2 is lost from water.

Temperature, as has been indicated earlier, also affects the amount of O_2 absorbed in water; the higher the temperature, the lower the O_2 dissolved.

Various designs of aerators are available; in some, water is allowed to fall by gravity over steps; in some, air is mechanically bubbled; in others, the water is forced into a fountain.

9.3.4 Coagulation and Flocculation

Coagulation and flocculation occur when colloidal particles clump together to such an extent that they settle out. Particles of the size of about 10-mm diameter will sediment unassisted in water, but smaller (or colloidal) particles will, for all practical purposes, not settle as shown in the Table 9.1. Because materials imparting color to water and a large proportion of the suspended materials in water are colloidal, they are best removed by coagulating them, hence causing them to sediment.

Several methods may ordinarily be used to settle out colloids:

(a) *Aging* allows colloids to collide in quiescent water by Brownian movement and hence to coagulate, but the method is too slow.

(b) *Heating* increases the movement of the particles, causing them to collide more often and to settle out.

(c) The use of *antagonistic colloids* – i.e., colloids with opposite charges. However, in practical water treatment these other methods are not used; rather coagulants are used.

Coagulants are electrolytes which, in water, form gelatinous flocs and collect or absorb colloidal particles. As the weight of the flocs increase, sedimentation takes place. Many of the suspended water particles have a negative electrical charge. The charge keeps particles suspended because they repel similar particles. Coagulation works by eliminating the natural electrical charge of the suspended particles so they attract and stick to each other. The joining of the particles to form larger settleable particles called flocs is a process known as flocculation. The coagulation chemicals are added in a tank (often called a rapid mix tank or flash mixer), which typically has rotating paddles. In most treatment plants, the mixture remains in the tank for 10–30 s to ensure full mixing. The amount of coagulant that is added to the water varies widely according to the load of colloids.

The coagulant is first rapidly mixed with water (in a *mixing basin*) to disperse it uniformly in water. It is next mixed at a slower rate by slow moving peddles to encourage flocculation or the massing of colloidal material and suspended particles. This takes place in a flocculation basin. The flocs sediment in special sedimentation tanks or clarifyers from which flocs are removed, intermittently or continuously.

The most popular coagulant used for water treatment is alum (aluminum sulfate). The reactions of alum in water are:

$$Al(SO_4)_3 + 6H_2O \rightarrow 2Al(OH)_3 + 3H_2SO_4$$

$$3H_2SO_4 + Ca(HCO_3) \rightarrow CaSO_4 + 3H_2SO_4$$

$$6H_2CO_3 \rightarrow 6CO_2 + 6H_2O$$

Over all:

$$Al_2(SO_4)_3 + 3Ca(HCO_3)_2 \rightarrow Al(OH)_3 + 6CO_2 + 3CaSO_4$$

For satisfactory coagulation, sufficient alkalinity must be available to react with the alum and also to leave a suitable residual alkalinity in the treated water.

The solubility of $Al(OH)_3$ is pH-dependent and is low at pH 5–7.5; in other words, for effective coagulation with alum, the pH of water should be in this range. Outside this range, coagulants sometimes used are ferrous sulfate $(FeSO_4)_3 \cdot 7H_2O$; also known as copperas, ferric sulfate $(Fe_2SO_4)_3$, ferric chloride $(FeCl_3)$, and sodium aluminate. Copperas may sometimes be treated with chlorine to give a mixture of ferric sulfate and ferric chloride known as chlorinated copperas. Ferric salts give satisfactory coagulation above pH 4.5 but ferrous salts are suitable only above pH 9.5. Iron salts are cheaper than alum but unless precipitation is complete, the former may stain clothes.

When colloidal matter is low, floc formation maybe encouraged by the addition of small amounts of coagulant aids, e.g., clay particles or, in some countries, heavy long-chain synthetic polymers.

These synthetic organic polymers ("polyelectrics") are long-chain carbon skeletons with recurring active sites, which absorb colloids. They have not come into general use, but are employed in some countries. Their advantages are that they are used in smaller volumes than the conventional coagulants, are simple to use because no pH maxima are involved, and they are cheaper and more efficient. Their disadvantages are that they require more vigorous mixing and also that they are yet to be as extensively studied as the better-known coagulants.

The amount of coagulants to be added has to be calculated from laboratory or jar tests. The pH of the raw water may be altered in the waterworks to suit the coagulant being used. The disposal of the sludge resulting from alum coagulation is sometimes a problem and in some countries, such as Japan, this is taken care of by various methods including lagooning (in which the sludge is allowed to settle in a lagoon and the supernatant recovered), drying beds, filter pressing, etc.

A method sometimes used is to draw about 10% of the clarified water, aerate it with air under pressure, and pump it to the bottom of the clarifying tank. The air bubbles float to the surface carrying with them the attached flocs. From time to time, this surface mat of flocs is scrapped and discarded. This method is called the dissolved air flotation method or DAF. Its advantage is that it reduces the load on the filters and is suitable for use with raw waters with a high amount of sediments.

9.3.5 Sedimentation

Water leaving the flocculation basin enters the sedimentation basin (settling basin or clarifier). Water flows through the tank slowly giving the flocs time to settle. It is usually a rectangle and so constructed that water flows through only at the top level. It typically takes 4 h for water to flow through a settling tank, but the longer the time in the sedimentation tank, the more the flocs that settle. The settling of flocs creates a bed of sludge, which could be between 2% and 5% of the total amount of water treated. The sludge must be removed from time to time.

9.3.6 Filtration

A minority of water personnel argue that if water is adequately stored and then chlorinated, it should be adequate for drinking. However, a great many waterworks still employ filtration (Anonymous 2007), of which there are two systems.

(a) Slow sand filtration

Filtration through sand was first developed in England, the earliest form having been used around 1829. The filters consist of 2–5 ft. of sand underlain by gravity. Particles in the raw water are filtered out near the top of the filter and provide a source of nutrients to microorganisms, which therefore grow to form a film. The slimy material formed by microorganisms, mud, and silt forms an efficient strainer.

After a period of use, depending on the nature of the raw water, the upper layers of the filter become clogged and have to be cleared by scraping. The sand may then be washed and reused. Slow sand filters can filter 3–6 million gal/acre/day. They have the advantage that they are relatively simple, and do not require the expense of pre-coagulation. Second, because no coagulation is employed, the water is less corrosive and more uniform in quality. Finally, they are very efficient and 99.99% bacteria in water may be removed by a properly organized slow sand filter.

The disadvantages of the slow sand filter are that:

1. A large area is required and hence a large sand volume and consequent high initial costs are involved
2. Unless the raw water is allowed to reduce its load of suspended material by plain sedimentation, they are not very efficient for purifying turbid waters containing more than 30–50 ppm for prolonged periods.
3. They are not very effective in removing color.
4. Unless pretreatment is given, they give poor results due to clogging effect in waters of high algal content.

Ordinarily, slow sand filters are used without pretreatment. However, in some waterworks, all or some of these pretreatments may be given:

1. Sedimentation
2. Chlorination
3. Addition of $CuSO_4$ for algae removal

The in-flowing and out-flowing waters should be examined bacteriologically regularly to ensure that the filter is not clogged.

(b) Rapid sand filtration

Rapid sand filters were introduced in the USA in about 1893. Other synonyms of rapid sand filters are *mechanical filters* (since originally sand was mechanically agitated during washing), *pressure filters* (since sometimes filter units were enclosed in steel tanks and *pressure* applied), and *gravity filters* (since water flows by gravity). They were introduced in the attempt to increase the rate of filtration to a calculated optimum of 125 million gal water /acre per day (2 gal/sq. ft/per min). The water to be filtered at this rate had to undergo prior coagulation and sedimentation to remove colloidal materials, which would otherwise block the filter. The present-day rapid sand filtration procedure consists of coagulation, flocculation, and sedimentation (discussed earlier), followed by filtration.

The rapid sand filter is a tank or box containing 20–30 in. of filter sand of 0.35–0.45 mm diameter overlying 16–24 in. of gravel ranging in diameter from 1/8 to 2½ in. The gravel is usually arranged in three to five layers, each layer containing material twice the size of that above it. The underdrain system collects the water for distribution and consists of perforated pipes with brass strainers. The coagulated water which at this stage carries a turbidity of about 10 ppm is distributed uniformly on the filter bed. Suspended material carried over from the sedimentation tanks as well as silt, clay, algae, and bacteria soon clog the filters and such materials are washed out in the flushing action of a back current of water, operated at about ten times the rate of filtration of the water. Sometimes the filter beds may be made of anthracite coal particles instead of sand.

To clean the filter, water is passed quickly upward through the filter, opposite the normal direction (called *backflushing* or *backwashing*) to remove embedded particles. Prior to this, compressed air may be blown up through the bottom of the filter to break up the compacted filter media to aid the backwashing process. Plate counts for bacteria should be done regularly before and after filtration to check the efficiency of the filter. Some plants use a slow sand filter after the rapid sand.

9.3.7 Chlorination (and Other Methods of) Disinfection

Even after the various treatments given to water – aeration, coagulation, sedimentation, and filtration – the possibility still exists that water may contain some bacteria, hence the need for disinfection. It should be noted, however, that complete sterilization is not aimed at in water treatment, even if it were possible to achieve this.

A number of chemicals have been used for the disinfection of water. The most common and most widely used are chlorine and its compounds, e.g., chloride of lime or calcium hypochlorite.

Historically, the introduction of the use of chlorine for disinfecting water has been credited to Sir Alexander H. Houston, who has also been called the "Father of Chlorination." He and a Dr. McGowan in 1904–1905 began the first continuous chlorination process designed to disinfect the municipal water supply of the city of Lincoln, England. They had used a 10–15% sodium hypochlorite solution for the purpose. Prior to

Table 9.2 Usage of chlorine as a water disinfectant in comparison with others (From The American Chemistry Council. http://www.americanchemistry.com/s_chlorine/sec_content.asp?CID=1133&DID=4530&CTYPEID=109. With permission) (Anonymous 2010a)

Disinfection	Large systems (>10,000 persons)	Small systems (<10,000 persons)
Chlorine gas	84%	61%
Sodium hypochlorite	20	34
Calcium hypochlorite	<1	5
Chloramines	29	–
Ozone	6	–
UV	–	–
Chlorine dioxide	8	–

that, between 1824 and 1826, chlorine had been introduced into hospitals, especially in obstetrical wards for the prevention of puerperal fever. A little later, in about 1831, it had been used in water (but not on a continuous scale) during the great cholera epidemic in Europe. Today, most water disinfection around the world is done with chlorine (see Table 9.2).

Chlorine has a number of advantages:

(a) It is easily available as a gas (or in compounds as liquid or powder).
(b) It is reasonably cheap.
(c) It is highly soluble (700 mg/l).
(d) It leaves a residue in solution which, while not harmful to man, provides protection in the distribution system.
(e) It is highly toxic to microorganisms.
(f) It has several important secondary uses, e.g., oxidation of iron, manganese, and H_2S, destruction of some taste- and odor-producing compounds, and acts as an aid to coagulation.

However, some of its shortcomings must be borne in mind:

(a) It is a poisonous gas and must be carefully handled.
(b) It can itself give rise to taste and odor problems, particularly in the presence of phenol (hence ammonia was once added to form chloramines, which have less odor problems).
(c) Suspended materials may shield bacteria from the action of chlorine.
(d) Chlorine is a powerful oxidizing agent and will attack a wide range of compounds, including unsaturated organic compounds as well as reduce substances, which are found in water, thereby making it less available to attack microorganisms.

(e) The effectiveness of chlorination is pH-dependent; chlorination is more effective at pH values of 7.2 and below than above pH 7.6.

There are alternative methods of disinfecting water and they are compared with chlorine in Table 9.3.

9.3.7.1 Reactions of Chlorine with Chemicals Found in Water

When water is chlorinated, a large number of chemicals may be present in it, especially if the water is wastewater or effluents from sewage treatments. These chemicals include nitrogenous compounds (especially ammonia), carbonaceous compounds, nitrites, iron, manganese, hydrogen sulfide, and cyanides. Chlorine combines with many of these compounds in the following manner.

(a) Nitrogen containing compounds

The compounds formed by the reaction of chlorine and a nitrogen-containing compound are chloramines, which could be either inorganic or organic.

1. *Ammonia*: This is the most important inorganic compound, which reacts with chlorine; others are nitrates and nitrites. Chlorine reacts in dilute aqueous solutions (1–50 ppm.) to form three choramines:

$$Cl_2 + NH_3 \rightarrow NH_2Cl + HCl$$
Monochloramine

$$NH_2 + Cl_2 \rightarrow NHCl_2 + HCl$$
Dichloramine

$$NH_2 + Cl_2 \rightarrow NCl_3 + HCl$$
Trichloramine (Nitrogen trichloride)

These chloramines, also known as combined residuals, have disinfectant properties, but they are far less effective than chlorine, requiring about ten times the contact time of chlorine. Furthermore, they confer odor and taste to water.

The pH of the water determines the relative amounts of the three kinds of chloramines. At pH 8.5, monochloramine is the major product. If it is low, e.g., below pH 4.4, virtually all the chloramine is in the form of nitrogen trichloride (NCl_3), which imparts a bad taste to drinking water and causes eye irritation in swimming pools. At the same time, it hardly disinfects.

Table 9.3 Comparison of chlorine and other water disinfectants (From The American Chemistry Council. http://www.americanchemistry. com/s_chlorine/sec_content.asp?CID=1133&DID=4530&CTYPEID=109. With permission) (Anonymous 2010c)

Disinfectant	Advantages	Limitations
Chlorine Gas	• Highly effective against most pathogens • Provides "residual" protection required for drinking water • Operationally, the most reliable • Generally the most cost-effective option	• By-product formation (THMs, HAAs[a]) • Special operator training needed • Additional regulatory requirements (EPA's Risk Management Program) • Not effective against Cryptosporidium
Sodium hypochlorite	• Same efficacy and residual protection as chlorine gas • Fewer training requirements than chlorine gas • Fewer regulations than chlorine gas	• Limited shelf life • Same by-products as chlorine gas, plus bromate and chlorate • Higher chemical costs than chlorine gas • Corrosive; requires special handling
Calcium hypochlorite	• Same efficacy and residual protection as gas • Much more stable than sodium hypochlorite, allowing long-term storage • Fewer Safety Regulations	• Same byproducts as chlorine gas • Higher chemical costs than chlorine gas • Fire or explosive hazard if handled improperly
Chloramines	• Reduced formation of THMs, HAAs • More stable residual than chlorine • Excellent secondary disinfectant	• Weaker disinfectant than chlorine • Requires shipment and use of ammonia gas or compounds • Toxic for kidney dialysis patients and tropical fish
Ozone	• Produces no chlorinated THMs, HAAs Fewer safety regulations • Effective against Cryptosporidium • Provides better taste and odor control than chlorination	• More complicated than chlorine or UV systems • No residual protection for drinking water • Hazardous gas requires special handling • By-product formation (bromate, brominated organics and ketones) • Generally higher cost than chlorine
UV	• No chemical generation, storage, or handling • Effective against Cryptosporidium • No known by-products at levels of concern	• No residual protection for drinking water • Less effective in turbid water • No taste and odor control • Generally higher cost than chlorine
Chlorine dioxide	• Effective against Cryptosporidium • No formation of THMs, HAAs • Provides better taste and odor control than chlorination	• By-product Formation (chlorite, chlorate) • Requires on-site generation equipment and handling of chemicals • Generally higher cost than chlorine

[a] *THMs* trihalomethanes, *Haas* haloacetic acids

2. *Organic nitrogen*: Organic nitrogen compounds include proteins and its various breakdown products – peptones, polypeptides, and amino acids. These compounds greatly retard chlorine with which they may react over several days. Known as chloro-organic compounds, they contribute to the odors in water. Furthermore, they produce a series of unstable residuals. Chloro-organic compounds, which titrate as combined chlorine, are also believed to have no germicidal action. Apart from all these, organic nitrogen is undesirable because it is a fairly good indication of recent pollution. It has therefore been suggested that its amount in raw waters be limited to 0.3 mg/1. The smaller the amount of protein, the more available chlorine is for disinfection.

(b) Hydrogen sulfide

H_2S is frequently dissolves in underground water and is common in waters where anaerobic decomposition has occurred. At a pH value of 6.4 and below, the H_2S is completely oxidized, giving rise to sulfuric acid and hydrochloric acid:

$$H_2S + 4Cl_2 + 4H_2O \rightarrow H_2SO_4 + 8HCl$$

At higher pH values, the reaction is thus

$$H_2S + Cl_2 \rightarrow S + H_2O$$

(c) Iron

The precipitate resulting from the reaction of chlorine and iron (i.e., $Fe(OH)_2$) in water serves two useful purposes. First, it helps remove iron; second, it helps produce a coagulant for the treatment of the water. The ultimate reaction is thus:

$$2FeCl_3 + 3Ca(HCO_3)2 \rightarrow 2Fe(OH)3$$
$$+ 6CO_2 + 3CaCl_2$$

(d) Cyanide

The destruction of cyanide by chlorine is most effective at high pH values (8.5–9.5), which can be created when necessary by the addition of NaOH. Sodium cyanate is produced, which decomposes, releasing nitrogen.

$$2Cl_2 + 4NaOH + 2NaCN \rightarrow 2NaCNO + 4NaCl$$
$$+ 2H_2O \qquad \text{(sodium cyanate)}$$

The complete reaction is thus:

$$5Cl_2 + 10NaOH + NaCN \rightarrow 2NaHCO_3$$
$$+ 10NaCl + N + 4H_2O$$

(e) Manganese

Chlorine reacts with manganese at pH 7–10, thus

$$MnSO_4 + Cl_2 + 4NaOH \rightarrow MnO_2 + 2NaCl$$
$$+ Na_2SO_4 + 2H_2O$$

(f) Methane

Methane is produced by anaerobic bacteria. Besides its unpleasant odor, methane could be explosive. It is best removed by aeration, but chlorine also reacts with it.

$$2CH_4 + Cl_2 + H_2O \rightarrow 4HCl + H_2O$$

9.3.7.2 The Present Practice of Water Chlorination

The current practice of chlorination is to add sufficient chlorine to oxidize all organic matter, iron, manganese, and other reducing substances in water as well as oxidize free ammonia in raw waters, and leave unused chlorine as free residual chlorine rather than the less active combined residual chlorine or chloramines. The process is known as "super-chlorination," "break-point chlorination," or "free residual chlorination" (see Fig. 9.2).

The reactions of chlorine with progressively large doses of chlorine are shown in Fig. 8.2. Free residual chlorine is not molecular chlorine reacting as a dissolved gas except at pH values below 5. When chlorine is dissolved in water, it dissociates to form two germicidal compounds, hypochlorous acid (HOCI) and the chlorite ion (OCI). The undissociated molecule of hypochlorite ion is 100 times more germicidal than the chlorite ion. At pH 5.6, the chlorine forms hypochlorous and hypochloric acid, thus:

$$Cl_2 + H_2O \rightarrow HCl + HClO$$

Free residual chlorine

$$HClO \rightarrow H + ClO$$

In earlier chlorination practice, only 0.1–0.2 ppm. of chlorine was added. The result was that though there

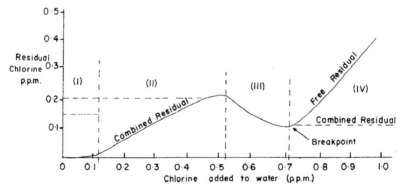

I = Destruction of chlorine by reducing compounds; no residual chlorine and no disinfection

II = Formation of chloro-organic compound and chloramines, used in earlier chlorination practice

III = Destruction of chloramines and chloro-organic compounds; breakpoint zone; chloramines oxidized.

IV = Formation of free chlorine and presence of presence of undestroyed chloro-organic compounds

Fig. 9.2 Modern (breakpoint) chlorination (Modified from Tebbutt 1992; http://water.me.vccs.edu/concepts/chlorchemistry.html) (Anonymous 2010b)

was residual chlorine, it was combined residual chorine in chloro-organic compounds and chloramines. These were not only less effective as disinfectants but also possessed odor and/or taste.

Undissociated hypochlorous acid is far more effective as an antimicrobial agent than the chlorite ion. Between pH 6 and pH 8.5–9, the hypochlorous acid is in equilibrium with the chlorite. Nearer the upper limit, nearly all (about 90%) of the hypochlorous acid is ionized to chlorite ion, whereas at about pH 6, it is mainly hypochlorous acid (see Table 9.2). Fortunately, most waters have a pH of 6.0–7.5, hence 50–95% of the free residual chlorine is present as hypochlorous acid. Many methods of chlorine determination in water unfortunately merely measure combined $HOCl + OCl$; the pH of the water must be known if the amount $HOCl$ is to be known.

9.3.7.3 Mode of Action of Chlorine Disinfection

It should be explained that chlorine in water does not *sterilize* (i.e., it does not remove all forms of living things). It merely disinfects (i.e., it destroys pathogenic organisms, much as the mild heat treatment of pasteurization does). Complete sterilization would be impractical because of its expense and is, in any case, not necessary.

The germicidal action of chlorine was first believed to be entirely due to the liberation of nascent oxygen by the reaction:

$$HOCl \rightarrow HCl + O$$

This was dispelled when it was shown that nascent oxygen was not released. It is now accepted that the disinfectant is hypochlorous acid, which has a small neutral molecule shaped like water and hence easily diffuses into the cell. The negative charge of the OCl ion, on the other hand, hampers its own penetration into the cell. The sites of action of the hypochlorous acid are the sulfhydryl (−SH) groups of enzymes.

9.3.7.4 Factors Affecting the Efficacy of Disinfection in Water by Chlorine (and the other Halogens)

Some of the factors affecting the efficacy of chlorine as a disinfectant in water are the type of organisms, the number of each organism, the concentration of chlorine, the contact time, the temperature, and the pH.

1. Type of organism
   Organisms vary in their resistance to killing by chlorine. In general, the order (of decreasing resistance) is: bacterial spores, protozoan cysts, viruses, and vegetative bacteria. Diseases produced by spore formers, e.g., anthrax *(B. anthracis)*, bolutilism *(Cl. botulimum)*, and tetanus *(Cl. tetanus)* are not normally transmitted via water although their spores may be transported therein. *Cl welchii*, an inhabitant of man's alimentary canal, is also sometimes used as an indicator of fecal contamination. It seems reasonable, therefore, for the spore formers to be used as test organisms for disinfection because of their greater resistance than *E. coli* and other non-sporulating bacteria. They have however not been used up till the present time.

   Cysts of the pathogenic protozoa *Entamoeba histolytica* and *Giardia lambia* are shed in feces of affected patients. Because the cysts are highly impermeable, they are said to be about 160 and 90 times more resistant than *E. coli* and hardier than enteroviruses, respectively, to hypochlorous acid.

   Chlorination as currently practiced does not remove all viruses from water and many of them persist after vegetative bacteria in water have been eliminated. Viruses, which are waterborne, have been discussed in Chap. 8. Table 9.3 compares the effectiveness of various disinfecting agents used in water.

2. The number of organisms, the concentration of chlorine, and the contact time
   As with other disinfecting agents, the greater the number of organisms the greater the concentration of chlorine required to kill a given number of organisms in a given time. In practice, the minimum contact time is 10–15 min.

3. Temperature
   The higher the temperature, the lower the rate of dissolution of chlorine or any gas. However, higher temperature affects the dissolved chlorine in two ways. First, it increases the rate of chlorine reactions with ammonia. Second, it affects the germicidal power of free residual chlorine. Thus, a 99.6–100% kill of the coxsackie A2 virus will require 4 min at pH 7 at 0–5°C, and 2 mg/l free residual chlorine; at 20–29°C it will require only 0.2 mg/l. residual chlorine – about tenfold reduction of chlorine. This emphasizes the need for recalculating the chlorine concentration

required for effective kills in outdoor disinfection in tropical countries, when the only data available are from temperate countries.

4. pH

The acidity of water affects the dissociation of HOCl. At lower pH values, HOCl predominates and at higher pH values the OCl (hyprochlorite) ion whose disinfecting ability is low, predominates. A well-defined equilibrium will be established at various pH values for each of the hypohalous acids, thus:

$$HOA \leftrightarrow H^+ + OX^-$$

When bromine is used, there is a preponderance of the more bactericidal hypobromous acid, HOBr, when the pH of the water is less than 8.7. With chlorine, HOCl predominates at pH less than about 7.45. Iodine is unusual because at the usual pH of water (5–8), there is hypoiodite ion, and the iodine exists mainly as molecular iodine I_2, and hypoiodous acid, HIO or H_2IO^+:

$$I_2 + H_2O \leftrightarrow H_2OI^+ + I^-$$

Iodine further differs in that whereas chlorine and bromine form chloramines and bromanines when ammonia is present in the water, iodine does not form iodamines with NH_3, nor does it react readily with organic matter. Iodine can however be discounted as a means of disinfecting water because it is physiologically active and may affect the thyroid. Apart from this, it is expensive. Iodine has however been shown to be quite suitable for use in swimming pools. In this connection, its lack of reactivity in water with ammonia and organic matter over a wide range of pH values is an advantage. It is thus able to remain in water as the molecular form I_2.

Bromine is also active, but it is expensive to produce; besides, it will require acclimatization by both water handlers and consumers to adapt to its use. There seems, therefore, to be no compelling reason to change to its use. Certain compounds of the halogens, chlorine, and bromine, may also be used. Notable among these are the compounds resulting from their reaction with ammonia, but these are generally less active than the hypohalous acids. Still, other compounds, bromine chlorine, and chlorine dioxide, have been tried mainly in wastewater.

The disinfecting activity of chlorine and its compounds, as well as ozone and uv are compared in Table 9.3. The table shows that while chlorine and its compounds are efficient disinfectants they are limited by their ineffectiveness, unlike ozone and uv, against protozoan cysts (Cryptosporidium).

9.3.7.5 Tests for Chlorine in Water

Chlorine in water reacts to form either free residual chlorine (HOCl + OCr) or combined residual chlorine (chloramines and chloro-organic compounds). These chlorine compounds are oxidizing to varying extents and this variability is the basis of chlorine tests in water. The tests are (a) the orthotolidine method (b) the starch – iodine method (c) amperometric method, and (d) titration with ferrous ammonium sulfate.

In the *orthotolidine* method, the free residual chlorine reacts in a matter of seconds with the colorless orthotolidine (a benzidine structure) to form a highly colored yellow holoquinone at low pH, thus:

The combined residual chlorine reacts much more slowly and can be removed with arsenite after the free chlorine has reacted. The yellow color can then be compared with a standard.

The *starch–iodine* method depends on the reaction:

$$HOCl + 2I + H+ \rightarrow I_2 + Cl + H_2O$$

This iodine is titrated with starch. The method is not very useful with drinking water and is more applicable with wastewater where all the chlorine is usually combined.

In the *amperometric* method, two platinum electrodes are immersed in a sample and a voltage of 10–15 mV applied. The reaction $HOCl + H + I_2 + Cl + H_2O$ also takes place. Iodine is oxidized at the anode and reduced at the cathode, thus causing current to flow. Changes in iodine concentration produce changes in current, which can be recorded. In water, it is useful in distinguishing free from residual chlorine. By altering the pH, the various combined residuals can be determined.

In the *titration with ferrous ammonium sulfate*, the titrating agents (e.g., sodium hexametaphosphate) react with chlorine at pH 7 to give a clear color when chlorine residual is present. Iodine is added and the addition of blue color is titrated and measured as monochloramine. Acidity after this neutralization to the original pH gives the dichloraning content on titration.

The most widely used of these methods is the ortho-tolidine–arsenite method, but it has been discontinued in some countries because the indicator is believed to be carcinogenic.

9.3.7.6 Alternative Methods of Disinfection Besides Chlorine and Other Halogens

Chlorine is an excellent water disinfectant whose advantages and disadvantages have already been discussed earlier in this chapter. Some work has gone into trying to find substitutes for chlorine. The reasons for this search are as follows. First, natural organic compounds in water, including humic acids, react with chlorine to produce volatile and nonvolatile halogenated compounds. The result is that the finished water contains a greater amount of these undesirable chloro-organic compounds than the raw water. For example, chloroform has been found to be less than 1 mg/l in raw water, but more than 300 mg/l after chlorination. Other compounds besides chloroform found in chlorinated water beside chloroform are carbon tetrachloride and 1,2-dichloroethane. These compounds confer unpleasant odors to water. The second reason for the search for other disinfection methods is that many viruses are known to be more resistant than *E. coli*. Hence, the absence of *E. coli* does not absolve a water sample from being a possible source of infection. The alternatives which have been considered outside the halogens are, in some cases being used, ozone and ultraviolet light (Weintraub et al 2005).

9.3.7.7 Ozone

Ozone, O_3, is produced on-site by introducing high voltage electric discharge (6–20 kV) across a dielectric discharge gap that contains oxygen-bearing gas.

Ozone is produced when oxygen (O_2) molecules are dissociated by an energy source into oxygen atoms; these subsequently collide with an oxygen molecule to form an unstable gas, ozone (O_3).

It is used in some European countries, notably France and Switzerland, for water disinfection. In the US, it is sometimes used to treat water discharged from wastewater treatment plants. Because of its powerful oxidizing properties, it is also used for removal of odor and taste, manganese, and organic compounds.

1. Advantages of ozone for water disinfection:
   The advantages of ozone are as follows:
   (a) Ozone is more effective than chlorine in destroying viruses and bacteria; it is highly virucidal while being at least as bactericidal as chlorine.
   (b) Ozone's germicidal action is extremely rapid – acting sometimes in a matter of seconds (see Fig. 8.4). However, in practice, the contact period allowed for efficient killing, depends on the turbidity of the water. Thus, for a 5–10 min period used in practice, 0.25–0.5 mg/l turbidity has been recommended for good-quality ground water. Owing to its rapidity of action, and to accommodate differences in the quality of the raw water, many plants adopt a general procedure of approximately 10–30 min for contact when treating waters with ozone.
   (c) There are no harmful residuals that need to be removed after ozonation because ozone decomposes rapidly.
   (d) After ozonation, there is no regrowth of microorganisms, except for those protected by the particulates in the wastewater stream.
   (e) Ozone is generated on-site, and thus, there are fewer safety problems associated with shipping and handling.
   (f) Ozonation elevates the dissolved oxygen (DO) concentration of the effluent. The increase in DO can eliminate the need for re-aeration and also raise the level of DO in the receiving stream.
   (g) Its efficacy is not affected by pH in the range (pH 5–8), and turbidity of up to 5 mg/l does not affect it.

2. Disadvantages of ozone: Ozone, however, has the following disadvantages:
   (a) Low dosages may not effectively inactivate some viruses, spores, and cysts.

(b) Ozonation is a more complex technology than is chlorine or UV disinfection, requiring complicated equipment and efficient contacting systems.

(c) Ozone is very reactive and corrosive, thus requiring corrosion-resistant material, such as stainless steel.

(d) Ozonation is not economical for water with high levels of suspended solids (SS), biochemical oxygen demand (BOD), chemical oxygen demand, or total organic carbon.

(e) Ozone is extremely irritating and possibly toxic, so off-gases from the contactor must be destroyed to prevent worker exposure.

(f) The cost of treatment can be relatively high, being both capital- and power-intensive.

(g) There is no measurable residual to indicate the efficacy of ozone disinfection. It is for this reason that post-ozonation chlorination is practiced in some countries.

Factors affecting the Efficacy of Ozonation in Water

1. *Temperature*

Temperature has an important influence on the half-life of ozone (i.e., time it takes for it to disintegrate). Table 9.4 shows the half-life of ozone in air and water. In water, the half-life of ozone is much shorter than in air; in other words, ozone decomposes faster in water. The solubility of ozone decreases at higher temperatures and is less stable. On the other hand, the reaction speed increases by a factor 2 or 3 per 10°C. Principally, ozone dissolved in water cannot be applied when temperatures are above 40°C, because at this temperature, the half-life of ozone is very short (Table 9.4).

2. *pH*

Ozone decomposes partly into OH⁻ radicals. When the pH value increases, the formation of OH⁻radicals increases. In a solution with a high pH value, there are more hydroxide ions present, see formulas below. These hydroxide ions act as an initiator for the decay of ozone:

$$O_3 + OH^- \rightarrow HO_2^- + O_2 \qquad (9.1)$$

$$O_3 + HO_2^- \rightarrow {}^{\bullet}OH + O_2^{\bullet-} + O_2 \qquad (9.2)$$

The radicals that are produced during reaction 2 can introduce other reactions with ozone, causing more

Table 9.4 Half-life of ozone in gas and water at different temperatures (From Lenntech Delft the Netherlands. http://www.lenntech.com/library/ozone/decomposition/ozone-decomposition.htm. With permission) (Anonymous 2009a)

Air		Dissolved in water, pH 7	
Temp (°C)	Half-life	Temp (°C)	Half-life
−50	3 months	15	30 min
−35	18 days	20	20 min
−25	8 days	25	15 min
20	3 days	30	12 min
120	1.5 h	35	8 min
250	1.5 s		

OH-radicals to be formed. The decay of ozone in a basic environment is much faster than in an acid environment.

3. *Dissolved solids concentration*

Dissolved ozone can react with a variety of matter, such as organic compounds, viruses, bacteria, etc. They react with ozone in different ways, causing ozone to break down to the OH- ion. Some dissolved compounds hasten the breakdown of ozone into the OH ion, while some delay the breakdown. Those which react with the OH ion and slow down the breakdown are referred to as scavengers. For example, carbonates are strong scavengers. The addition of the carbonate ion increases the half-life of ozone.

4. *Dissolved organic matter (DOM)*

Dissolved organic matter (DOM), also called dissolved organic carbon (DOC), is present in every kind of water. It is largely colloidal in nature; the materials, which confer color and odor in water, are colloidal and part of the DOM in water. Ozone is used to reduce DOC in water and hence color and odor. Ozone easily reacts with reactive chemical structures such as double bonds, activated aromatic compounds, amines, and sulfides. The OH- ion also directs with DOM.

Mode of Action of Ozone

When ozone decomposes in water, the free radicals hydrogen peroxy (HO_2) and hydroxyl (OH) that are formed have great oxidizing capacity and play an active role in the disinfection process. It is generally believed that the bacteria are destroyed because of protoplasmic oxidation resulting in cell wall disintegration (cell lysis). The effectiveness of disinfection depends on the susceptibility of the target organisms, the contact time, and the concentration of the ozone.

Ozone is a very strong oxidant microbicide and virucide. The mechanisms of disinfection using ozone include:

(a) Direct oxidation/destruction of the cell wall with leakage of cellular constituents outside of the cell.

(b) Reactions with radical by-products of ozone decomposition.

(c) Damage to the constituents of the nucleic acids (purines and pyrimidines).

(d) Breakage of carbon–nitrogen bonds in nucleic acids leading to depolymerization.

9.3.8 Ultraviolet Light

Special lamps are used to generate the radiation that creates UV light by striking an electric arc through low-pressure mercury vapor; a broad spectrum of radiation with intense peaks at UV wavelengths of 253.7 nm (nm) and a lesser peak at 184.9. The germicidal UV range exists between 250 and 270 nm. At shorter wavelengths (e.g., 185 nm), UV light is powerful enough to produce ozone, hydroxyl, and other free radicals that destroy bacteria.

Using ultraviolet (UV) light for drinking water disinfection dates back to 1916 in the U.S. UV costs have since declined as new UV methods to disinfect water and wastewater have been developed (Anonymous 2007). Currently, several states have developed regulations that allow systems to disinfect their drinking water supplies with UV light. The US Environmental Protection Agency (EPA) lists UV disinfection as an approved technology for small public water systems. The use of UV for disinfecting water and wastewater has advantages and disadvantages.

1. Advantages of UV for disinfecting water

   The advantages include:

   (a) UV has no known toxic or significant nontoxic byproducts.

   (b) It has no danger of overdosing.

   (c) With UV, there are no volatile organic compound (VOC) emissions or toxic air emissions.

   (d) No on-site smell and no smell in the final water product exist with UV.

   (e) Contact time with UV is very short; it occurs in seconds, but in minutes for chemical disinfectants.

   (f) UV requires very little space for equipment and contact chamber.

   (g) It improves the taste of water because some organic contaminants and nuisance

   (h) UV does not react

   (i) Has little or no impact on the environment.

2. Disadvantages of water disinfection using UV

   The use of UV has the following disadvantages:

   (a) UV radiation is not suitable for water with high levels of suspended solids, turbidity, color, or soluble organic matter; they react with UV radiation, and reduce disinfection

   (b) performance. In particular, high turbidity makes it difficult for radiation to penetrate water.

   (c) In comparison with chlorine, it has no disinfection residual.

   (d) Using ultraviolet light on a large scale is a more expensive process than the use of chemicals.

9.3.8.1 Mode of Action and Use of UV

Ultraviolet light kills microorganisms by causing the coalescing or dimerization of adjacent thymine bases in the DNA of organisms exposed to it. It has its maximum germicidal effect at 2,500–2,600 A wavelength (250.0–260.0 nm).

The major feature of the set-up for UV water sterilization is that water circulates round a chamber in which ultraviolet lamps are located. According to the U. S. Dept. of Health, Education & Welfare, Public Health Service, the retention time of water in the chamber should not be less than 15 s, at the maximum flow rate of the system. The flow rate must not exceed 0.0125 m^3/s per effective (arc length) inch of the lamp. The lamp must also emit light energy at 253.7 nm at an intensity of 4.85 UVW/sq. ft (0.005 UVW /sq. cm) at a distance of 2 m (see Fig. 10.23).

9.3.9 Iron and Manganese Control

Iron and manganese either together or alone are objectionable in water because beyond certain levels they stain clothes and they may give an unwelcome deposit in drinking water. They can make water appear red or yellow, create brown or black stains in the sink, and give off an easily detectable metallic taste. They can be aesthetically displeasing, but iron and manganese do not constitute health risks.

These minerals are to be found in waters obtained from certain deep waters, which contain CO_2 but not O_2. Under these conditions, the insoluble oxides of these metals are reduced and transformed to soluble carbonates. It is possible to determine the source of the staining or other observations resulting from the presence of iron in water, thus:

(a) When the water is clear but has rust or black-colored particles, then it is due to ferrous iron – Fe^{2+}.

(b) When the water from the tap is red, yellow, or rusty and sediments soon form on resting, then the culprit is ferric iron, Fe^{3+}.

(c) When in metallic pipes, slimy brown, red, or green masses on the pipes appear, this is diagnostic of iron bacteria.

Iron and manganese may be removed in any of the following ways:

1. Aeration:
   This introduces oxygen and hence causes oxidation and precipitation of oxides.

2. Use of contact bed containing iron and manganese oxides
   Water is passed over a bed of gravels coated with Fe and Mn oxides. These oxides enhance oxidation of Fe and Mn through catalytic action of preciously precipitated oxides. Contact beds are regenerated potassium permanganate $KMnO_4$.

3. Oxidation by Chlorine or $KMnO_4$
   Appropriate amounts of free residual chlorine will oxidize Fe and Mn. It is usually aided by a small amount at $CuSO_4$. Potassium permanganate may also be used.

9.3.10 Softening of Water

Hard water is water that has a high soap consuming power (i.e., water which will not produce lather unless a large amount of soap is used). Besides soap wastage, hard water also forms sediments or scales in kettles, industrial boilers, thereby requiring more heat for the same amount of water. Hard water also contributes to skin clogging and discoloration of porcelain, and shortening of the life of fabrics. Hardness in water is due primarily to the presence of Ca and Mg ions. The presence of the following may also cause slight increases in hardness: Fe, Mn, Cu, Ba, and Zn.

Total water "hardness" (including both Ca^{2+} and Mg^{2+} ions) is reported as ppm w/v (or mg/L) of $CaCO_3$. Water hardness usually measures the total concentration of Ca and Mg, the two most prevalent divalent metal ions, although in some geographical locations, iron and other metals may also be present at elevated levels. Calcium usually enters the water from either $CaCO_3$, as limestone or chalk, or from mineral deposits of $CaSO_4$. Water hardness may be temporary or permanent.

Temporary hardness is hardness that can be removed by boiling or by the addition of lime (calcium hydroxide). It is caused by a combination of calcium ions and bicarbonate ions in the water. Boiling, which promotes the formation of carbonate from the bicarbonate, will precipitate calcium carbonate out of solution, leaving water that is less hard on cooling. Upon heating, less CO_2 is able to dissolve into the water. Since there is not enough CO_2, the reaction cannot proceed and therefore the $CaCO_3$ will not "dissolve" as readily. Instead, the reaction is forced to go from right to left (i.e., products to reactants) to reestablish equilibrium, and solid $CaCO_3$ is formed. Heating water will remove hardness as long as the solid $CaCO_3$ that precipitates out is removed. After cooling, if enough time passes, the water will pick up CO_2 from the air and the reaction will again proceed from left to right, allowing the $CaCO_3$ to "redissolve" in the water.

$$2Ca\left(HCO_3\right)_2 \xleftrightarrow{\text{boiling}} Ca_2CO_3 + CO_2 \uparrow + H_2O$$

Permanent hardness is hardness that cannot be removed by boiling. It is usually caused by the presence of calcium and magnesium sulfates and/or chlorides in the water, which become more soluble as the temperature rises. Permanent hardness can be removed with ion exchange, or the lime-soda process.

In the ion exchange method, using *zeolite,* Ca and Mg ions are exchanged for Na.(see below for a detailed discussion of ion exchange treatment for water purification). In the lime-soda process, Ca and Mg ions are precipitated and removed by sedimentation and filtration. When water is softened with lime, it may be necessary sometimes to introduce CO_2, which reacts with any excess lime to form $CaCO_3$, which is precipitated before filtration.

9.3.11 Fluoridation

In some communities, small amounts of fluorides are added (about 1 ppm). It is believed that fluorides prevent dental decay. The fluorides used include sodium fluoride, sodium silicofluoride, and ammonium silicofluoride.

Many North American and Australian municipalities fluoridate their water supplies in the belief that this practice will reduce tooth decay at a low cost, and about 70% of the water drunk in the US is fluoridated. Since fluoridation began in 1945, there has been a drop in dental decay, but fluoridation is still controversial in some communities.

9.3.12 Algae Control (and Control of Taste and Odors)

Objectionable tastes and odors in water may be taken as evidence of pollution or unwholesomeness even when this is not necessarily so. But the effect is that consumers lose confidence in the water.

Source of odors and tastes include:

1. *The growth of living microorganisms*: Some living things release taste-producing materials into water. For example, the Protozoa, *Synura* and *Uroglena* impart fishy taste to water and actinomycetes impart a "soil" taste to water. Many algae impart a taste of grass, and some of them produce other particularly objectionable odors, e.g., *Nitella.* Some of the compounds producing odors in water have been identified as geosmin and mucidione. The former is produced by the bluegreen alga, *Anabaena circinalis.* The observation that *Bacillus* spp. (esp. *B. cereus*) can degrade geosmin has led to the suggestion that the bacillus be used in water as a form of biological control of odors.
2. *The decomposition products of dead microorganisms plants and animals*: When bacteria (e.g., *Beggiatoa, Crenothrix,* and *Sphaerotilus),* plants, and animal die, their decaying parts produce odors in water.
3. *The production of methane and the reduction of sulfates to sulfides*
   These activities take place under anaerobic conditions and give rise to odors.
4. *Odor and taste-producing compounds in sewage and industrial effluents*
   Odor and taste-producing compounds are sometimes present in industrial and sewage effluents and these maybe carried into water; reaction compounds of chlorine with organic compounds also give rise to odors.

9.3.12.1 Methods for the Control of Algae and Taste and Odor

1. Algae may be destroyed in water by the addition of $CuSO_4$. In small amounts, $CuSO_4$ has not been shown to be toxic to man, but where fish are present in natural reservoirs, some fish may be affected.
2. Aeration also helps remove some odor.
3. The combined residual chlorine may have an objectionable taste if it is present in the form of chloro-organic compounds rather than as chloramines. The dose of chlorine should then be increased to destroy the compounds and leave Cl_2 as the free residual type.
4. Water may be passed over activated charcoal to remove odors and tastes.
5. Ozone may be used to destroy odor-producing compounds.

9.3.13 Color and Turbidity Removal

Color in water is derived from the microbial degradation of organic materials and from the extraction of organic materials from soil. It has been suggested that a humic-acid-like exudate from the aquatic fungus *Aurebasidium pullulans* may contribute to the yellow color found in water. Color-producing chemicals are complex chemically. Some are aromatic polyhydric methoxy carboxylic acids sometimes similar to tannic acids, which are of plant origin. Materials conferring color are negatively charged and usually occur along with chelated iron and manganese. Colored waters are usually surface waters. Color removal can be accomplished by metallic salts. For $Al(OH)_3$, the optimal pH is 5.5–7.0, while for ferric salts it is 3.5–4.5. The correct pH should be worked out in laboratory in "jar" tests.

Turbidity occurs mainly in surface waters and is absent in ground waters because soil particles filter off the colloidal materials, which cause turbidity. The colloidal particles are absorbed in the coagulants used in water purification.

9.3.14 Miscellaneous Treatments Water Purification

9.3.14.1 Plumbosolvency Removal
Plumbosolvency is the ability of a solvent, notably water, to dissolve lead. In older premises where lead pipes were used, plumbosolvent water can attack lead

pipes and any lead in solder used to join copper pipes, leading to increased lead levels in the tap water. On account of this, lead becomes a common contaminant of drinking water. Plumbosolvency of water can be countered by increasing the pH with lime or sodium hydroxide (lye), or by the addition of phosphate at the water treatment works.

9.3.14.2 Radium (Radioactivity) Removal

Underground waters associated with certain rock formations such as the crystalline granite rock of north-central Wisconsin exhibit low-level radioactivity, primarily from radium, although it could also come from uranium. Radioactivity levels are measured in "picocuries" per liter of water, (pCi/L). Health risks are low in consuming waters with low-level radioactivity. However, consuming it over a lifetime increases such risks. The consumption of radionucleotides in water poses increasing risks of cancers, and the US water standards permit zero amounts of alpha particles, beta particles, and photon emitters, Radium 226 and Radium 228, and uranium. In areas where the raw water is high in radioactivity, the water may be treated with synthetic zeolite ion exchange resins, which could remove about 90% of the radium. The water may also be purified with reverse osmosis.

9.3.14.3 Reverse Osmosis

Reverse osmosis is a filtration process that uses pressure to force a solution through a membrane; the solute is retained on one side and allowing the solvent passes at the other. It is so called because it is reverse of normal osmosis in which the solvent moves into where the solute concentration is, passing through a membrane.

9.3.14.4 Ion Exchange

Ion exchange resins are insoluble matrix or support structure normally in the form of small (1–2 mm diameter) beads, colored white or yellowish, made from an organic polymer material. The solid ion exchange particles are either naturally occurring inorganic zeolites or synthetically produced organic resins. The synthetic organic resins are preferred as they can be designed for specific applications. A matrix of pores on the surface of the beads, easily trap and release ions. The trapping of ions takes place only with simultaneous releasing of other ions, and hence the process is called ion exchange. Different types of ion exchange resin are fabricated to selectively prefer one or several different types of ions.

Ion exchange resins are widely used in different separation, purification, and decontamination processes. Most commonly, they are used for water softening and water purification. Before the introduction of resins, zeolites, which are natural or artificial alumino-silicate minerals and which can accommodate a wide variety of cations such as Na^+, K^+, Ca^{2+}, Mg^{2+}, were used.

For water softening, ion exchange resins are used to replace the magnesium and calcium ions, found in hard water, with sodium ions. When in contact with a solution containing magnesium and calcium ions, but a low concentration of sodium ions, the magnesium and calcium ions preferentially migrate out of solution to the active sites on the resin, being replaced in solution by sodium ions. This process reaches equilibrium with a much lower concentration of magnesium and calcium ions in solution than was started with.

The resin can be recharged by washing it with a solution containing a high concentration of sodium ions (e.g., it has large amounts of common salt (NaCl) dissolved in it). The calcium and magnesium ions migrate off the resin, being replaced by sodium ions from the solution until a new equilibrium is reached. This is the method of operation used in dishwashers that require the use of "dishwasher salt." The salt is used to recharge an ion exchange resin, which itself is used to soften the water so that limescale deposits are not left on the cooking and eating utensils being washed.

For water purification, ion exchange resins are used to remove undesirable constituents, e.g., copper and lead ions from solution, replacing them with more innocuous ions, such as sodium and potassium

Ion exchange is widely used in the food and beverage, metals finishing, chemical and petrochemical, pharmaceutical, sugar and sweeteners, ground and potable water, nuclear, softening and industrial water, semiconductor, power, and a host of other industries. It is particularly employed for removing radioactive compounds in water purifying processes.

9.3.14.5 Electrodeionization

Electrodeionization (EDI) is a water treatment process that removes ionizable species from liquids using electrically active media and an electrical potential to effect ion transport. It differs from other water purification technologies such as conventional ion exchange in that it is does not require the use of chemicals such as acid and caustic soda. In traditional ion exchange units, after the contaminants are trapped onto the resin sites, the resin continues to exhaust and lose capacity. In EDI, the contaminants are continuously removed as they are attracted to one of the two electrical charges, and then migrate through the resin bed, through ion exchange membranes and into the concentrate stream where they are removed from the device.

Water splitting replaces the chemical regeneration process. For example, where R is impurity trapped on the mixed bed resin, the chemical regeneration process is

$$H_2SO_4 + R \rightarrow H^{2+} + RSO_4$$

$$NaOH + R \rightarrow OH^- + NaR$$

The hydrogen (H^+) ions and hydroxyl (OH^-) ions continuously regenerate the mixed resin in the EDI module. EDI is a polishing technology and requires reverse osmosis (RO) as pretreatment. The combination of RO–EDI provides the customer with a continuous, chemical-free system.

9.4 Purification of Bottled Water

Bottled water is very popular not only in the home, but it is also convenient for traveling, sports, and occasions where small quantities of drinking water with an assurance of good quality is expected. Many individuals do not however trust so-called purified water, in spite of what the manufacturers may write on the label. Many individuals believe that ordinary tap water is simply bottled. It would appear that in spite of this skepticism, some bottlers do make an effort to purify the water. Below is a flow chart of a system used by one bottler in the USA, which was made available to the author (Fig. 9.3).

9.5 Standards Required of Water

Water is used for many purposes, each of which requires that the water meets the standards, which in the main will ensure the health and safety of the users of the water. In this book, we have considered the use of water for drinking, recreation, and for the growth of shellfish. The standards required for each of these activities will be discussed in this section.

9.5.1 Standards Required for Drinking Water

Water is required by the human body constantly and an average adult probably consumes up to one liter or more per day. Since water must normally be consumed every day, unlike other food constituents, which may be eaten now and again, standards must be carefully set with the aim of protecting human health. Several considerations enter into the selection of standards for drinking water. These include:

(a) The public health statistics relating to morbidity and mortality due to a pathogen or chemical
(b) The population exposed
(c) The physical and chemical state of the substance
(d) The toxicity of the substance to man or to suitable experimental animals.
(e) The amount of the substance likely to be found in other sources

Water meant for human consumption must be free from chemical substances and microorganisms in types and amounts, which can be hazardous to health. Not only must it be safe but it must also be aesthetically acceptable. It is for this reason that the governments of various countries around the world set standards to be met in drinking water.

Ideally, the standards for drinking water should be uniform universally and used the world over. In practice, however, the standards depend on known and expected contaminants and the ability of the society or the government concerned to attain the standard; therefore, standards vary from country to country. The US standards will be discussed mainly, but standards of the European Union and the World Health Organization will also be mentioned for comparison.

In the United States, the US Environmental Protection Agency (EPA) sets the standards for drinking water. The Safe Drinking Water Act (SDWA), passed in 1974 and amended in 1986 and 1996, gives the EPA the authority to set drinking water standards. These standards are regulations that EPA sets to control the level of contaminants in the nation's drinking water. They are part of the Safe Drinking Water Act's "multiple barrier" approach to drinking water protection, which includes assessing and protecting drinking water sources; protecting wells and collection systems; making sure water is treated by qualified operators; ensuring the integrity of distribution systems; and making information available to the public on the quality of their drinking water. With

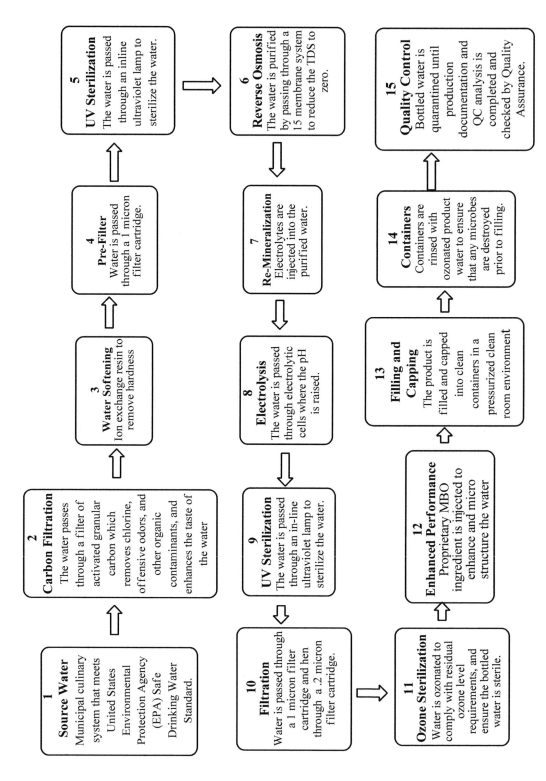

Fig. 9.3 Flowchart for the manufacture of bottled water by a manufacturer

the involvement of EPA, States, Tribes, drinking water utilities, communities and citizens, these multiple barriers ensure that tap water in the United States and its territories is safe to drink. In most cases, EPA delegates responsibility for implementing drinking water standards to States and Tribes.

There are two aspects to the EPA water standards:

(a) The National Primary Drinking Water Regulation (NPDWR) or the primary standard
(b) The National Secondary Drinking Water Regulation (NSDWR) or the secondary standard

The primary standard is a legally enforceable standard that applies to public water systems. The primary standards protect drinking water quality by limiting the levels of specific contaminants that can adversely affect public health and are known, or anticipated, to occur in water. They take the form of maximum contaminant levels or treatment techniques. The secondary standards are a set of nonenforceable guidelines regarding contaminants that may cause cosmetic effects (such as skin or tooth discoloration) or aesthetic effects (such as taste, odor, or color) in drinking water. EPA recommends secondary standards to water systems but does not require systems to comply. However, States may choose to adopt them as enforceable standards.

Table 9.5 gives the maximum contaminant levels (MCLs) of the various contaminants expected in the environment of the USA. It is arranged in terms of microorganisms, disinfectants, disinfection by-products, inorganic chemicals, organic chemicals, and radionuclides. The standards are the highest in the world.

The World Health Organization (WHO) has produced standards (1993) for drinking water. These are however merely recommendations about minimal standards, which the WHO itself recognizes will not necessarily be attained in all countries of the world or regions due to differences in the economic and technological capabilities of various countries. The latest standards of the European Union (EU) (1998) are compared with those of the WHO (1993) in Table 9.6. It must be emphasized all standards are revised from time to time. For the most part, the EU standards are higher than the WHO standards. For example, bromate (Br) is not mentioned at all in the WHO standards, whereas a standard of 0.01 mg/l is set in the EU standards; for Manganese (Mn), the WHO standard is set at 0.5 mg/l whereas that of the EU is 0.05 mg/l. The WHO standards are to be found in an earlier document,

WHO (1984). The standards set by the USEPA (2009) are higher than those of the EU and the WHO. The US standards updated on March 19, 2009 are given in Table 9.4. EU and WHO standards are given side by side in Table 9.6.

9.5.1.1 The Microbiological Standards

All the standards state that no sample should contain fecal coliforms. In addition, the US standards specify the absence of *Cryptosporidium, Giardia lambia, and Legionella.* It is recommended that, to be acceptable, drinking water should be free from any viruses which affect man. This objective may be achieved *(a)* by the use of a water supply from a source which is free from wastewater and is protected from fecal contamination; or *(b)* by adequate treatment of a water source that is subject to fecal pollution.

Adequacy of treatment cannot be assessed in an absolute sense because neither the available monitoring techniques nor the epidemiological evaluation is sufficiently sensitive to ensure the absence of viruses. However, it is considered at present that contaminated source water may be regarded as adequately treated when the following conditions are met:

1. A turbidity of 1 NTU or less is achieved.
2. Disinfection of the water with at least 0.5 mg/l of free residual chlorine after a contact period of at least 30 min at a pH below 8.0.

The turbidity condition must be fulfilled prior to disinfection if adequate treatment is to be achieved.

Disinfection other than by chlorination may be applied provided the efficacy is at least equal to that of chlorination as described above. Ozone has been shown to be an effective viral disinfectant, preferably for clean water, if residuals of 0.2–0.4 mg/l are maintained for 4 min. Ozone has advantages over chlorine for treating water containing ammonia but, unfortunately, it is not possible to maintain an ozone residual in the distribution system

9.5.1.2 Turbidity

Turbidity is a measure of the cloudiness of water. It is used to indicate water quality and filtration effectiveness (e.g., whether disease-causing organisms are present). Higher turbidity levels are often associated with higher levels of disease-causing microorganisms such as viruses, parasites, and some bacteria. These organisms can cause symptoms such as nausea, cramps, diarrhea, and associated headaches. According to EPA

Table 9.5 USEPA's national primary and secondary standards for drinking water (As of 19 March, 2009) (http://www.epa.gov/safewater/contaminants/index.html) (Anonymous 2010b)

Contaminant	MCLG[a] (mg/L)[b]	MCL or TT[a] (mg/L)[b]	Potential health effects from ingestion of water	Sources of contaminant in drinking water
Microorganisms				
Cryptosporidium	Zero	TT[c]	Gastrointestinal illness (e.g., diarrhea, vomiting, cramps)	Human and animal fecal waste
Giardia lamblia	Zero	TT[c]	Gastrointestinal illness (e.g., diarrhea, vomiting, cramps)	Human and animal fecal waste
Heterotrophic plate count	N/a	TT[c]	HPC has no health effects; it is an analytic method used to measure the variety of bacteria that are common in water. The lower the concentration of bacteria in drinking water, the better maintained the water system is.	HPC measures a range of bacteria that are naturally present in the environment
Legionella	Zero	TT[c]	Legionnaire's Disease, a type of pneumonia	Found naturally in water; multiplies in heating systems
Total coliforms (including fecal coliform and *E. Coli*)	Zero	5.0%[d]	Not a health threat in itself; it is used to indicate whether other potentially harmful bacteria may be present[e]	Coliforms are naturally present in the environment; as well as feces; fecal coliforms and *E. coli* only come from human and animal fecal waste.
Turbidity	N/a	TT[c]	Turbidity is a measure of the cloudiness of water. It is used to indicate water quality and filtration effectiveness (e.g., whether disease-causing organisms are present). Higher turbidity levels are often associated with higher levels of disease-causing microorganisms such as viruses, parasites, and some bacteria. These organisms can cause symptoms such as nausea, cramps, diarrhea, and associated headaches.	Soil runoff
Viruses (enteric)	Zero	TT[c]	Gastrointestinal illness (e.g., diarrhea, vomiting, cramps)	Human and animal fecal waste
Contaminant	MCLG[a] (mg/L)[b]	MCL or TT[a] (mg/L)[b]	Potential health effects from ingestion of water	Sources of contaminant in drinking water
Disinfection by-products				
Bromate	Zero	0.010	Increased risk of cancer	By-product of drinking water disinfection
Chlorite	0.8	1.0	Anemia; infants & young children: nervous system effects	By-product of drinking water disinfection
Haloacetic acids (HAA5)	N/a[f]	0.060[g]	Increased risk of cancer	By-product of drinking water disinfection
Total Trihalomethanes (TTHMs)	N/a[f]	0.080[g]	Liver, kidney, or central nervous system problems; increased risk of cancer	By-product of drinking water disinfection

(continued)

Table 9.5 (continued)

Contaminant	MRDLG[a] (mg/L)[b]	MRDL[a] (mg/L)[b]	Potential health effects from ingestion of water	Sources of contaminant in drinking water
Disinfectants				
Chloramines (as Cl_2)	MRDLG = 4[a]	MRDL = 4.0[a]	Eye/nose irritation; stomach discomfort, anemia	Water additive used to control microbes
Chlorine (as Cl_2)	MRDLG = 4[a]	MRDL = 4.0[a]	Eye/nose irritation; stomach discomfort	Water additive used to control microbes
Chlorine dioxide (as ClO_2)	MRDLG = 0.8[a]	MRDL = 0.8[a]	Anemia; infants & young children: nervous system effects	Water additive used to control microbes

Contaminant	MCLG[a] (mg/L)[b]	MCL or TT[a] (mg/L)[b]	Potential health effects from ingestion of water	Sources of contaminant in drinking water
Inorganic chemicals				
Antimony	0.006	0.006	Increase in blood cholesterol; decrease in blood sugar	Discharge from petroleum refineries; fire retardants; ceramics; electronics; solder
Arsenic	0[g]	0.010 as of 01/23/06	Skin damage or problems with circulatory systems, and may have increased risk of getting cancer	Erosion of natural deposits; runoff from orchards, runoff from glass and electronics production wastes
Asbestos (fiber >10 µm)	7 million fibers per liter	7 MFL	Increased risk of developing benign intestinal polyps	Decay of asbestos cement in water mains; erosion of natural deposits
Barium	2	2	Increase in blood pressure	Discharge of drilling wastes; discharge from metal refineries; erosion of natural deposits
Beryllium	0.004	0.004	Intestinal lesions	Discharge from metal refineries and coal-burning factories; discharge from electrical, aerospace, and defense industries
Cadmium	0.005	0.005	Kidney damage	Corrosion of galvanized pipes; erosion of natural deposits; discharge from metal refineries; runoff from waste batteries and paints
Chromium (total)	0.1	0.1	Allergic dermatitis	Discharge from steel and pulp mills; erosion of natural deposits
Copper	1.3	TT[h]; action level = 1.3	Short-term exposure: Gastrointestinal distress Long-term exposure: Liver or kidney damage People with Wilson's Disease should consult their personal doctor if the amount of copper in their water exceeds the action level	Corrosion of household plumbing systems; erosion of natural deposits
Cyanide (as free cyanide)	0.2	0.2	Nerve damage or thyroid problems	Discharge from steel/metal factories; discharge from plastic and fertilizer factories

(continued)

Table 9.5 (continued)

Contaminant	MCLG[a] (mg/L)[b]	MCL or TT[a] (mg/L)[b]	Potential health effects from ingestion of water	Sources of contaminant in drinking water
Fluoride	4.0	4.0	Bone disease (pain and tenderness of the bones); Children may get mottled teeth	Water additive, which promotes strong teeth; erosion of natural deposits; discharge from fertilizer and aluminum factories
Lead	Zero	TT[h]; action level = 0.015	Infants and children: Delays in physical or mental development; children could show slight deficits in attention span and learning abilitiesAdults: Kidney problems; high blood pressure	Corrosion of household plumbing systems; erosion of natural deposits
Mercury (inorganic)	0.002	0.002	Kidney damage	Erosion of natural deposits; discharge from refineries and factories; runoff from landfills and croplands
Nitrate (measured as nitrogen)	10	10	Infants below the age of 6 months who drink water containing nitrate in excess of the MCL could become seriously ill and, if untreated, may die. Symptoms include shortness of breath and blue-baby syndrome.	Runoff from fertilizer use; leaching from septic tanks, sewage; erosion of natural deposits
Nitrite (measured as nitrogen)	1	1	Infants below the age of 6 months who drink water containing nitrite in excess of the MCL could become seriously ill and, if untreated, may die. Symptoms include shortness of breath and blue-baby syndrome.	Runoff from fertilizer use; leaching from septic tanks, sewage; erosion of natural deposits
Selenium	0.05	0.05	Hair or fingernail loss; numbness in fingers or toes; circulatory problems	Discharge from petroleum refineries; erosion of natural deposits; discharge from mines
Thallium	0.0005	0.002	Hair loss; changes in blood; kidney, intestine, or liver problems	Leaching from ore-processing sites; discharge from electronics, glass, and drug factories

Contaminant	MCLG[a] (mg/L)[b]	MCL or TT[a] (mg/L)[b]	Potential health effects from ingestion of water	Sources of contaminant in drinking water
Organic chemicals				
Acrylamide	Zero	TT[i]	Nervous system or blood problems; increased risk of cancer	Added to water during sewage/wastewater treatment
Alachlor	Zero	0.002	Eye, liver, kidney, or spleen problems; anemia; increased risk of cancer	Runoff from herbicide used on row crops
Atrazine	0.003	0.003	Cardiovascular system or reproductive problems	Runoff from herbicide used on row crops
Benzene	Zero	0.005	Anemia; decrease in blood platelets; increased risk of cancer	Discharge from factories; leaching from gas storage tanks and landfills

(continued)

Table 9.5 (continued)

Contaminant	MCLG[a] (mg/L)[b]	MCL or TT[a] (mg/L)[b]	Potential health effects from ingestion of water	Sources of contaminant in drinking water
Benzo(a)pyrene (PAHs)	Zero	0.0002	Reproductive difficulties; increased risk of cancer	Leaching from linings of water storage tanks and distribution lines
Carbofuran	0.04	0.04	Problems with blood, nervous system, or reproductive system	Leaching of soil fumigant used on rice and alfalfa
Carbontetrachloride	Zero	0.005	Liver problems; increased risk of cancer	Discharge from chemical plants and other industrial activities
Chlordane	Zero	0.002	Liver or nervous system problems; increased risk of cancer	Residue of banned termiticide
Chlorobenzene	0.1	0.1	Liver or kidney problems	Discharge from chemical and agricultural chemical factories
2,4-D	0.07	0.07	Kidney, liver, or adrenal gland problems	Runoff from herbicide used on row crops
Dalapon	0.2	0.2	Minor kidney changes	Runoff from herbicide used on rights of way
1,2-Dibromo-3-chloropropane (DBCP)	Zero	0.0002	Reproductive difficulties; increased risk of cancer	Runoff/leaching from soil fumigant used on soybeans, cotton, pineapples, and orchards
o-Dichlorobenzene	0.6	0.6	Liver, kidney, or circulatory system problems	Discharge from industrial chemical factories
p-Dichlorobenzene	0.075	0.075	Anemia; liver, kidney, or spleen damage; changes in blood	Discharge from industrial chemical factories
1,2-Dichloroethane	Zero	0.005	Increased risk of cancer	Discharge from industrial chemical factories
1,1-Dichloroethylene	0.007	0.007	Liver problems	Discharge from industrial chemical factories
cis-1,2-Dichloroethylene	0.07	0.07	Liver problems	Discharge from industrial chemical factories
trans-1,2-Dichloroethylene	0.1	0.1	Liver problems	Discharge from industrial chemical factories
Dichloromethane	Zero	0.005	Liver problems; increased risk of cancer	Discharge from drug and chemical factories
1,2-Dichloropropane	Zero	0.005	Increased risk of cancer	Discharge from industrial chemical factories
Di(2-ethylhexyl) adipate	0.4	0.4	Weight loss, liver problems, or possible reproductive difficulties.	Discharge from chemical factories
Di(2-ethylhexyl) phthalate	Zero	0.006	Reproductive difficulties; liver problems; increased risk of cancer	Discharge from rubber and chemical factories
Dinoseb	0.007	0.007	Reproductive difficulties	Runoff from herbicide used on soybeans and vegetables
Dioxin (2,3,7,8-TCDD)	Zero	0.00000003	Reproductive difficulties; increased risk of cancer	Emissions from waste incineration and other combustion; discharge from chemical factories

(continued)

Table 9.5 (continued)

Contaminant	MCLG[a] (mg/L)[b]	MCL or TT[a] (mg/L)[b]	Potential health effects from ingestion of water	Sources of contaminant in drinking water
Diquat	0.02	0.02	Cataracts	Runoff from herbicide use
Endothall	0.1	0.1	Stomach and intestinal problems	Runoff from herbicide use
Endrin	0.002	0.002	Liver problems	Residue of banned insecticide
Epichlorohydrin	Zero	TT[i]	Increased cancer risk, and over a long period of time, stomach problems	Discharge from industrial chemical factories; an impurity of some water treatment chemicals
Ethylbenzene	0.7	0.7	Liver or kidneys problems	Discharge from petroleum refineries
Ethylene dibromide	Zero	0.00005	Problems with liver, stomach, reproductive system, or kidneys; increased risk of cancer	Discharge from petroleum refineries
Glyphosate	0.7	0.7	Kidney problems; reproductive difficulties	Runoff from herbicide use
Heptachlor	Zero	0.0004	Liver damage; increased risk of cancer	Residue of banned termiticide
Heptachlor epoxide	Zero	0.0002	Liver damage; increased risk of cancer	Breakdown of heptachlor
Hexachlorobenzene	Zero	0.001	Liver or kidney problems; reproductive difficulties; increased risk of cancer	Discharge from metal refineries and agricultural chemical factories
Hexachlorocyclo-pentadiene	0.05	0.05	Kidney or stomach problems	Discharge from chemical factories
Lindane	0.0002	0.0002	Liver or kidney problems	Runoff/leaching from insecticide used on cattle, lumber, gardens
Methoxychlor	0.04	0.04	Reproductive difficulties	Runoff/leaching from insecticide used on fruits, vegetables, alfalfa, livestock
Oxamyl (Vydate)	0.2	0.2	Slight nervous system effects	Runoff/leaching from insecticide used on apples, potatoes, and tomatoes
Polychlorinated-biphenyls (PCBs)	Zero	0.0005	Skin changes; thymus gland problems; immune deficiencies; reproductive or nervous system difficulties; increased risk of cancer	Runoff from landfills; discharge of waste chemicals
Pentachlorophenol	Zero	0.001	Liver or kidney problems; increased cancer risk	Discharge from wood preserving factories
Picloram	0.5	0.5	Liver problems	Herbicide runoff
Simazine	0.004	0.004	Problems with blood	Herbicide runoff
Styrene	0.1	0.1	Liver, kidney, or circulatory system problems	Discharge from rubber and plastic factories; leaching from landfills
Tetrachloroethylene	Zero	0.005	Liver problems; increased risk of cancer	Discharge from factories and dry cleaners

(continued)

Table 9.5 (continued)

Contaminant	MCLG[a] (mg/L)[b]	MCL or TT[a] (mg/L)[b]	Potential health effects from ingestion of water	Sources of contaminant in drinking water
Toluene	1	1	Nervous system, kidney, or liver problems	Discharge from petroleum factories
Toxaphene	Zero	0.003	Kidney, liver, or thyroid problems; increased risk of cancer	Runoff/leaching from insecticide used on cotton and cattle
2,4,5-TP (Silvex)	0.05	0.05	Liver problems	Residue of banned herbicide
1,2,4-Trichlorobenzene	0.07	0.07	Changes in adrenal glands	Discharge from textile finishing factories
1,1,1-Trichloroethane	0.20	0.2	Liver, nervous system, or circulatory problems	Discharge from metal degreasing sites and other factories
1,1,2-Trichloroethane	0.003	0.005	Liver, kidney, or immune system problems	Discharge from industrial chemical factories
Trichloroethylene	Zero	0.005	Liver problems; increased risk of cancer	Discharge from metal degreasing sites and other factories
Vinyl chloride	Zero	0.002	Increased risk of cancer	Leaching from PVC pipes; discharge from plastic factories
Xylenes (total)	10	10	Nervous system damage	Discharge from petroleum factories; discharge from chemical factories

Contaminant	MCLG[a] (mg/L)[b]	MCL or TT[a] (mg/L)[b]	Potential health effects from ingestion of water	Sources of contaminant in drinking water
Radionuclides				
Alpha particles	None[g] ——— Zero	15 picocuries per Liter (pCi/L)	Increased risk of cancer	Erosion of natural deposits of certain minerals that are radioactive and may emit a form of radiation known as alpha radiation
Beta particles and photon emitters	None[g] ——— Zero	4 millirems per year	Increased risk of cancer	Decay of natural and man-made deposits of certain minerals that are radioactive and may emit forms of radiation known as photons and beta radiation
Radium 226 and Radium 228 (combined)	None[g] ——— Zero	5 pCi/L	Increased risk of cancer	Erosion of natural deposits
Uranium	Zero	30 ug/L as of 12/08/03	Increased risk of cancer, kidney toxicity	Erosion of natural deposits

Notes

[a]**Definitions**: Maximum Contaminant Level (MCL) – The highest level of a contaminant that is allowed in drinking water. MCLs are set as close to MCLGs as feasible using the best available treatment technology and taking cost into consideration. MCLs are enforceable standards.

Maximum Contaminant Level Goal (MCLG) – The level of a contaminant in drinking water below which there is no known or expected risk to health. MCLGs allow for a margin of safety and are non-enforceable public health goals.

Maximum Residual Disinfectant Level (MRDL) – The highest level of a disinfectant allowed in drinking water. There is convincing evidence that addition of a disinfectant is necessary for control of microbial contaminants.

(continued)

Table 9.5 (continued)

Maximum Residual Disinfectant Level Goal (MRDLG) – The level of a drinking water disinfectant below which there is no known or expected risk to health. MRDLGs do not reflect the benefits of the use of disinfectants to control microbial contaminants. Treatment Technique – A required process intended to reduce the level of a contaminant in drinking water.

[b]Units are in milligrams per liter (mg/L) unless otherwise noted. Milligrams per liter are equivalent to parts per million.

[c]EPA's surface water treatment rules require systems using surface water or ground water under the direct influence of surface water to (1) disinfect their water, and (2) filter their water or meet criteria for avoiding filtration so that the following contaminants are controlled at the following levels:

- Cryptosporidium: (as of 1/1/02 for systems serving >10,000 and 1/14/05 for systems serving <10,000) 99% removal.
- Giardia lamblia: 99.9% removal/inactivation
- Viruses: 99.99% removal/inactivation
- Legionella: No limit, but EPA believes that if *Giardia* and viruses are removed/inactivated, *Legionella* will also be controlled.
- Turbidity: At no time can turbidity (cloudiness of water) go above 5 nephelometric turbidity units (NTU); systems that filter must ensure that the turbidity go no higher than 1 NTU (0.5 NTU for conventional or direct filtration) in at least 95% of the daily samples in any month. As of January 1, 2002, turbidity may never exceed 1 NTU, and must not exceed 0.3 NTU in 95% of daily samples in any month.
- HPC: No more than 500 bacterial colonies per milliliter.
- Long-Term 1 Enhanced Surface Water Treatment (Effective Date: January 14, 2005); Surface water systems or (GWUDI) systems serving fewer than 10,000 people must comply with the applicable Long-Term 1 Enhanced Surface Water Treatment Rule provisions (e.g., turbidity standards, individual filter monitoring, Cryptosporidium removal requirements, updated watershed control requirements for unfiltered systems).
- Long-Term 2 Enhanced Surface Water Treatment Rule (Effective Date: January 4, 2006) – Surface water systems or GWUDI systems must comply with the additional treatment for Cryptosporidium specified in this rule based on their Cryptosporidium bin classification calculated after the completion of source water monitoring.
- Filter Backwash Recycling; The Filter Backwash Recycling Rule requires systems that recycle to return specific recycle flows through all processes of the system's existing conventional or direct filtration system or at an alternate location approved by the state.

[d]More than 0.5% samples are tested positive for coliform in a month. (For water systems that collect fewer than 40 routine samples per month, no more than one sample can be total coliform-positive per month.) Every sample that has total coliform must be analyzed for either fecal coliforms or E. coli if two consecutive TC-positive samples, and one is also positive for *E. coli* fecal coliforms, system has an acute MCL violation.

[e]Fecal coliform and E. coli are bacteria whose presence indicates that the water may be contaminated with human or animal wastes. Disease-causing microbes (pathogens) in these wastes can cause diarrhea, cramps, nausea, headaches, or other symptoms. These pathogens may pose a special health risk for infants, young children, and people with severely compromised immune systems.

[f]Although there is no collective MCLG for this contaminant group, there are individual MCLGs for some of the individual contaminants:

- Trihalomethanes: bromodichloromethane (zero); bromoform (zero); dibromochloromethane (0.06 mg/L): chloroform (0.07 mg/L).
- Haloacetic acids: dichloroacetic acid (zero); trichloroacetic acid (0.02 mg/L); monochloroacetic acid (0.07 mg/L). Bromoacetic acid and dibromoacetic acid are regulated with this group but have no MCLGs.

[g]The MCL values are the same in the Stage 2 DBPR as they were in the Stage 1 DBPR, but compliance with the MCL is based on different calculations. Under Stage 1, compliance is based on a running annual average (RAA). Under Stage 2, compliance is based on a locational running annual average (LRAA), where the annual average at each sampling location in the distribution system is used to determine compliance with the MCLs. The LRAA requirement will become effective April 1, 2012 for systems on schedule 1; October 1, 2012 for systems on schedule 2; and October 1, 2013 for all remaining systems.

[h]Lead and copper are regulated by a treatment technique that requires systems to control the corrosiveness of their water. If more than 10% of tap water samples exceed the action level, water systems must take additional steps. For copper, the action level is 1.3 mg/L, and for lead is 0.015 mg/L.

[i]Each water system must certify, in writing, to the state (using third-party or manufacturer's certification) that when acrylamide and epichlorohydrin are used in drinking water systems, the combination (or product) of dose and monomer level does not exceed the levels specified, as follows:

- Acrylamide = 0.05% dosed at 1 mg/L (or equivalent)
- Epichlorohydrin = 0.01% dosed at 20 mg/L (or equivalent)

National Secondary Drinking Water Regulations

National Secondary Drinking Water Regulations (NSDWRs or secondary standards) are non-enforceable guidelines regulating contaminants that may cause cosmetic effects (such as skin or tooth discoloration) or aesthetic effects (such as taste, odor, or color) in drinking water. EPA recommends secondary standards to water systems but does not require systems to comply. However, states may choose to adopt them as enforceable standards.

(continued)

Table 9.5   (continued)

List of National Secondary Drinking Water Regulations

Contaminant	Secondary standard
Aluminum	0.05 to 0.2 mg/L
Chloride	250 mg/L
Color	15 (color units)
Copper	1.0 mg/L
Corrosivity	noncorrosive
Fluoride	2.0 mg/L
Foaming agents	0.5 mg/L
Iron	0.3 mg/L
Manganese	0.05 mg/L
Odor	3 threshold odor number
pH	6.5-8.5
Silver	0.10 mg/L
Sulfate	250 mg/L
Total dissolved solids	500 mg/L
Zinc	5 mg/L

Last updated on Wednesday, March 18th, 2009. http://www.epa.gov/safewater/contaminants/index.html

regulations, at no time can turbidity (cloudiness of water) go above 5 nephelometric turbidity units (NTU); systems that filter must ensure that the turbidity go no higher than 1 NTU (0.5 NTU for conventional or direct filtration) in at least 95% of the daily samples in any month. As of January 1, 2002, turbidity may never exceed 1 NTU, and must not exceed 0.3 NTU in 95% of daily samples in any month.

9.5.1.3 Chemical Standards

The EPA standards are by far the most comprehensive and include maximum quantities, expected in water, of disinfectants, by-products of disinfectants, inorganic chemicals such as antimony, arsenic, lead and mercury, organic chemicals such as acrylamide and benzene, radionucleotides such as alpha particles.

9.5.2   Standards Required for Recreational Waters

Standards set for drinking water include microbiological, chemical, radiological, turbidity, etc. For recreational waters, the standards appear to be mainly microbiological. Furthermore, since diseases, which can be contacted in recreational waters, are not enteric ones that enter by the oral–fecal route, it has been suggested that the standards should include the content of water of other indicators, organisms besides those used for drinking water. Microorganisms that are used to assess the microbial quality of swimming pool and similar environments include heterotrophic plate count – HPC (a general measure of nonspecific microbial levels), fecal indicators (such as thermotolerant coliforms, *E. coli*), *Pseudomonas aeruginosa*, *Staphylococcus aureus* and *Legionella* spp. HPC, thermotolerant coliforms, and *E. coli* are indicators in the strict sense of the definition.

As health risks in pools and similar environments may be fecal or non-fecal in origin, both fecal indicators and non-fecally derived microorganisms (e.g., *P. aeruginosa*, *S. aureus*, and *Legionella* spp.) should therefore be examined. Fecal indicators are used to monitor for the possible presence of fecal contamination; HPC, *Pseudomonas aeruginosa*, and *Legionella* spp. can be used to examine growth, and *Staphylococcus aureus* can be used to determine non-fecal shedding. The absence of these organisms, however, does not guarantee safety, as some pathogens are more resistant to treatment than the indicators, and there is no perfect indicator organism.

In practice, enteric organisms have mainly been used. Thus, to protect human health from waterborne pathogens, the EPA recommends monitoring marine recreational waters for *enterococci* (recommended threshold: geometric mean of 35/100 mL), and in

Table 9.6 Comparative assessment of EU and WHO water standards (From Lenntech Delft the Netherlands; http://www.lenntech.com/WHO-EU-water-standards.htm. With permission) (Anonymous 2009c)

Items	WHO standards 1993	EU standards 1998
Suspended solids	No guideline	Not mentioned
COD	No guideline	Not mentioned
BOD	No guideline	Not mentioned
Oxidisability		5.0 mg/l O_2
Grease/oil	No guideline	Not mentioned
Turbidity	No guideline[a]	Not mentioned
pH	No guideline[b]	Not mentioned
Conductivity	250 μ/cm	250 μ/cm
Color	No guideline[c]	Not mentioned
Dissolved oxygen	No guideline[d]	Not mentioned
Hardness	No guideline[e]	Not mentioned
TDS	No guideline	Not mentioned
Cations (positive ions)		
Aluminum (Al)	0.2 mg/l	0.2 mg/l
Ammonia (NH4)	No guideline	0.50 mg/l
Antimony (Sb)	0.005 mg/l	0.005 mg/l
Arsenic (As)	0.01 mg/l	0.01 mg/l
Barium (Ba)	0.3 mg/l	Not mentioned
Beryllium (Be)	No guideline	Not mentioned
Boron (B)	0.3 mg/l	1.00 mg/l
Bromate (Br)	Not mentioned	0.01 mg/l
Cadmium (Cd)	0.003 mg/l	0.005 mg/l
Chromium (Cr)	0.05 mg/l	0.05 mg/l
Copper (Cu)	2 mg/l	2.0 mg/l
Iron (Fe)	No guideline[f]	0.2
Lead (Pb)	0.01 mg/l	0.01 mg/l
Manganese (Mn)	0.5 mg/l	0.05 mg/l
Mercury (Hg)	0.001 mg/l	0.001 mg/l
Molibdenum (Mo)	0.07 mg/l	Not mentioned
Nickel (Ni)	0.02 mg/l	0.02 mg/l
Nitrogen (total N)	50 mg/l	Not mentioned
Selenium (Se)	0.01 mg/l	0.01 mg/l
Silver (Ag)	No guideline	Not mentioned
Sodium (Na)	200 mg/l	200 mg/l
Tin (Sn) inorganic	No guideline	Not mentioned
Uranium (U)	1.4 mg/l	Not mentioned
Zinc (Zn)	3 mg/l	Not mentioned
Anions (negative ions)		
Chloride (Cl)	250 mg/l	250 mg/l
Cyanide (CN)	0.07 mg/l	0.05 mg/l
Fluoride (F)	1.5 mg/l	1.5 mg/l
Sulfate (SO4)	500 mg/l	250 mg/l
Nitrate (NO3)	(See Nitrogen)	50 mg/l
Nitrite (NO2)	(See Nitrogen)	0.50 mg/l

Microbiological parameters		
Escherichia coli	Not mentioned	0 in 250 ml
Enterococci	Not mentioned	0 in 250 ml
Pseudomonas aeruginosa	Not mentioned	0 in 250 ml
Clostridium perfringens	Not mentioned	0 in 100 ml
Coliform bacteria	Not mentioned	0 in 100 ml
Colony count 22°C	Not mentioned	100/ml
Colony count 37°C	Not mentioned	20/ml
Other parameters		
Acrylamide	Not mentioned	0.0001 mg/l
Benzene (C_6H_6)	Not mentioned	0.001 mg/l
Benzo(a)pyrene	Not mentioned	0.00001 mg/l
Chlorine dioxide (ClO_2)	0.4 mg/l	
1,2-Dichloroethane	Not mentioned	0.003 mg/l
Epichlorohydrin	Not mentioned	0.0001 mg/l
Pesticides	Not mentioned	0.0001 mg/l
Pesticides – total	Not mentioned	0.0005 mg/l
PAHs	Not mentioned	0.0001 mg/l
Tetrachloroethene	Not mentioned	0.01 mg/l
Trichloroethene	Not mentioned	0.01 mg/l
Trihalomethanes	Not mentioned	0.1 mg/l
Tritium (H_3)	Not mentioned	100 Bq/l
Vinyl chloride	Not mentioned	0.0005 mg/l

[a]Desirable: less than 5 NTU
[b]Desirable: 6.5–8.5
[c]Desirable: 15 mg/l Pt-Co
[d]Desirable: less than 75% of the saturation concentration
[e]Desirable: 150–500 mg/l
[f]Desirable: 0.3 mg/l

freshwater, *E. coli* (126/100 mL threshold) or *enterococci* (33/100 mL). EPA requires that at least five samples be taken at equally spaced interval over a 30-day period.

Drinking water regulations of the EPA specify the absence of *Legionella*, while the EU regulations specify the absence of *Pseudomonas*; these two organisms are more usually linked to recreational waters.

9.5.3 Standards Required for Shellfish Harvesting Waters

For shellfish harvesting, total coliform (not exceeding a geometric mean of 70 MPN per 100 mL, with not more than 10% of the samples taken during any 30-day period exceeding 230 MPN per 100 mL) and fecal coliform monitoring (median concentration should not exceed 14 MPN per 100 mL, with not more than 10 % of the samples taken during any 30-day period exceeding 43 MPN per 100 mL) is recommended.

References

Anonymous (1984) *Guidelines for drinking-water quality.* Vol. 1, Recommendations. Geneva: World Health Organization.

Anonymous (2006a) *Setting Standards for Safe Drinking Water.* US Environmental Protection Agency. http://www.epa.gov/safewater/standard/setting.html. Accessed on 5 June 2007.

Anonymous (2006b) *Guidelines for drinking-water quality [electronic resource] : Incorporating first addendum.* Vol. 1, Recommendations. – 3rd Ed. 1.Potable water – standards. 2.Water – standards. 3.Water quality – standards. 4. Guidelines. Geneva: World Health Organization.

Anonymous (2007) Different Water Filtration Methods Explained. http://www.freedrinkingwater.com/water-education/quality-water-filtration-method.htm. Accessed on 28 Aug 2007.

Anonymous (2009a) Ozone decomposition. http://www.lenntech.com/library/ozone/decomposition/ozone-decomposition.htm. Accessed 15 Sept 2010.

Anonymous (2009b) Drinking Water Standards: Maximum Allowable Levels of Various Contaminants. http://www.epa.gov/safewater/contaminants/index.html. Accessed 12 Sept 2010.

Anonymous (2009c) Comparative Assessment of EU and WHO Water Standards. http://www.lenntech.com/WHO-EU-water-standards.htm. Accessed 12 Sept 2010.

Anonymous (2010a) Water Disinfection: Evaluating Alternative Methods In Light of Heightened Security Concerns. http://www.americanchemistry.com/s_chlorine/sec_content.asp?CID=1133&DID=4530&CTYPEID=109. Accessed 30 Aug 2010.

Anonymous (2010b) Water disinfection. Chlorination Chemistry. http://water.me.vccs.edu/concepts/chlorchemistry.html Accessed 30 Sept 2010.

Anonymous (2010c) More UV Sterilizer info. http://www.ocean-reeflections.com/uv_school.htm. Accessed 2 June 2010.

Bitton, G. (2005). *Wastewater microbiology.* Hoboken: Wiley.

Singley, J. E., Robinson, J., & Dearborn, B. (2006). *Water Treatment. Kirk-Othmer Encyclopedia of Chemical Technology.* New York: Wiley.

Tebbutt, T. H. Y. (1992). *Principles of Water Quality Control* (4th ed.). Oxford: Butterworth-Heinemann.

Weintraub, J. M., Wright, J. M., & Venkatapathy, R. (2005). Alternative disinfection practices and future directions for disinfection by-product minimization. *Water Enyclopedia, 2,* 90–94.

Part V

Waste Disposal in Aquatic and Solid Media

Waste Disposal in the Aqueous Medium: Sewage Disposal

10

Abstract

Wastes are discarded or unwanted materials resulting from the domestic and industrial activities of humans. Wastes carried in water are described as sewage. It is important to determine the quantity of carbon in sewage so as to know the most appropriate method to use for treating it. One of the most common methods is the Biochemical Oxygen Demand (BOD) which measures the oxygen consumed by microorganisms to degrade the carbon over a 5-day period. Other methods are chemical in nature and related to the potential of the microbial breakdown. These chemical methods are the Permanganate Value (PV) Test, the Chemical Oxygen Demand (COD), Total Organic Carbon (TOC), Total Suspended Solids (TSS), and Volatile Suspended Solids (VSS).

Aerobic methods for disposing of sewage are: the Activated Sludge System, the Trickling Filter, the Rotating Disks, and the Oxidation Pond. Anaerobic methods include the Septic Tank, the Imhoff Tank, and Cesspools. Advanced wastewater treatment (AWT) methods are expensive and used for producing water for specialized use. They include: Reverse Osmosis, Nanofiltration, Ultrafiltration and Microfiltration, Electrodialysis, Activated Charcoal, Ion exchange, UV Oxidation, and Precipitation.

Keywords

Activated sludge • Trickling filter • Cesspools • Imhoff tank • Biochemical Oxygen Demand (BOD) • Chemical Oxygen Demand (COD) • Total Organic Carbon (TOC) • Advanced Wastewater treatment (AWT)

10.1 Nature of Wastes

Wastes, discarded and unwanted materials, result inevitably from human activities, whether domestic or industrial. If wastes are allowed to accumulate on the ground, or if dumped indiscriminately into rivers and other bodies of water, unacceptable environmental problems would result (Eckenfelder 2000). Wastes are therefore disposed of in order to reduce their harmful effects on the environment. Three approaches are used:

(a) The wastes may be concentrated and isolated from the normal environment. For example, small amounts of highly toxic wastes may be concentrated and dumped into the depths of the ocean; human excreta may be isolated by being buried underground.

(b) The wastes may be diluted in the environment to a concentration at which they are not harmful. An example of this occurs in some developing countries where sewage is not treated but is diluted by introduction through pipes far into the sea.

(c) Finally, the wastes may be treated mechanically, biologically, or chemically to render them harmless. The materials resulting from such treatments are expected to meet standards laid down by authorities set up in various countries to monitor such activities. For instance, the US Environmental Protection Agency (USEPA) has set standards which sewage treatment effluents must meet regarding bacterial numbers in domestic sewage and chemicals in case of industrial sewage (see below).

Governments the world over institute legislation which regulates the handling of wastes, including those resulting from industry. In the US, the USEPA works to develop and enforce regulations that implement environmental laws enacted by the Congress. EPA is responsible for researching and setting national standards for a variety of environmental programs, and delegates to states the responsibility for issuing permits, and for monitoring and enforcing compliance (Anonymous 2004).

Wastes may be carried in water (sewage) or may be solid (municipal solid wastes, MSW). This chapter will discuss some of the methods used in treating sewage. Chapter 11 will discuss the treatment of MSW.

10.2 Methods for the Determination of Organic Matter Content in Sewage and Wastewaters

Sewage is the water borne wastes of human activities, both domestic and industrial. The water borne wastes generated by some 60% of the US population are collected in sewer systems and carried by some 14 billion gallons of water a day before being treated. The rest receives some form of treatment to improve the quality of the water before it is released for reuse. Of this volume, about 10% is allowed to pass untreated into rivers, streams, and the ocean.

Wastewaters are sampled and analyzed in order to determine the efficiency of the treatment system in use. This is particularly important at the point of the discharge of the treated wastewater into rivers, streams, and other natural bodies of water. If wastewater discharged into natural water is rich in degradable organic matter, large numbers of aerobic micro-organisms will develop to break down the organic matter. They will use up the oxygen and as a consequence, fish and other aquatic life which require oxygen will die. Furthermore,

anaerobic bacteria will develop following the exhaustion of oxygen; the activities of the latter will result in foul odors. Some of the methods for analyzing the organic matter content of wastewaters are given below (Andrew 1996; Anonymous 2006).

10.2.1 Determination of Dissolved Oxygen

Dissolved oxygen (DO) is one of the most important, though indirect, means of determining the organic matter content of waters. The heavier the amount of degradable material present in water, the greater the growth of aerobic organisms and hence the lesser the oxygen content. The Winkler method is widely used for determining the oxygen in water. In this method, dissolved oxygen reacts with manganous oxide to form manganic oxide. On acidification in an iodide solution, iodine is released in an amount equivalent to the oxygen reacting to form the manganic oxide. The iodine may then be titrated using thiosulphate. Membrane electrodes are now available for the same purpose. In these electrodes, oxygen diffuses through the electrode and reacts with a metal to produce a current proportional to the amount of oxygen reacting with the metal.

10.2.1.1 The Biological or Biochemical Oxygen Demand Test

1. *The Dilution Method*

   In the Biochemical Oxygen Demand (BOD) procedure (El-Rehaili 1994), two bottles of 250–300 ml. capacity and with ground-glass stoppers are thoroughly washed with chromic acid mixture and dried before use. They are filled with a suitable dilution of the sewage or water whose BOD is to be determined (Anonymous 2006). The dissolved oxygen in one is determined immediately and in the other after five 5 days' incubation at 20°C. The difference, the loss of oxygen during incubation, is the BOD, i.e., a measure of oxygen consumed during the stabilization of the organic materials present. As a precaution to prevent the entry of air, special BOD bottles having flared mouths are used; water is added to the flared mouths.

   Alternately, the tightly sealed bottles may be inverted in a water bath. The bottle is incubated in a water-bath set at 20°C or preferably in a cooled incubator. Light is prevented from reaching it so that algae present in it may not produce O_2 during photosynthesis and thence give wrong results.

The water or sewage whose BOD is to be determined is usually so diluted that the BOD is not more than 5.0 mg/l. The amount of dilution is worked out of experience. The dilution water is usually pure distilled water, but de-ionized water is preferable. The water is aerated by bubbling O_2 through it for a few hours before use or by leaving it in a half-filled jar loosely stoppered with cotton wool for about 3–5 days.

It is often necessary to introduce a pure or mixed culture of micro-organisms to (i.e. "seed") the diluted water, when, as in the case of drinking water, the microbial population is low. The seed used in determining the BOD of drinking water may consist of 1–10 ml of settled domestic sewage or a drop of lactose broth containing an active culture which is known to produce a positive, presumptive coliform test. The amount of seed chosen should give at least 0.6 mg/l after 5 days' incubation. Seeding may also be done when the material is an industrial waste with no known population capable of breaking it down. The seed for such an industrial waste may be developed in the laboratory by bubbling O_2 through a culture, obtained from soil or sewage and grown in the industrial waste. The seed may then be added to the diluted water.

The chemical procedure of the test is a determination of dissolved oxygen (DO) using the Winkler method employing the Rideal–Stewart modification or the Alkaline-Azide modification (also known as Alsterbury [azide] modification). The last term is usually used.

The principle of the test is that manganous hydroxide is oxidized to manganic hydroxide in a highly alkaline solution. When the set-up is acidified in the presence of iodine, the manganic hydroxide dissolves and the free iodine is liberated in an amount equivalent to the O_2 originally dissolved in the sample. The free iodine is titrated with a sodium thiosulphate standard solution using starch as, an indicator.

$$2Mn(OH)_2 + O_2 \rightarrow 2Mn(OH)_3 + H_2O \quad (10.1)$$

$$Mn(OH)_3 + H_2SO_4 \rightarrow Mn(OH)_2 + MnSO_4 + O_2 \quad (10.2)$$

$$O_2 + KI + H_2O \rightarrow . KOH + I_2 \quad (10.3)$$

The normality of the thiosulphate solution is so adjusted that 1 ml is equivalent to 1 mg/l dissolved O_2 when 200 ml of the original sample is titrated, i.e., an N/160 standard solution.

The aim of this discussion has been to examine the concept and principles of BOD. The practical details of the determination will be found in *Standard Methods for the Examination of Waters* and *Waste Waters. 21st edition* (Anonymous 2006) and *International standards for Drinking Water (WHO). Microbial Activity and the BOD Test*

The breakdown of organic matter in sewage and other waters takes place slowly and is not usually complete in the standard 5-day period of incubation, except for easily oxidized materials such as glucose. Domestic sewage is only about 65% oxidized at this period and industrial sewage might be 40% or less. The BOD required for the breakdown of all stabilizable organic matter is the ultimate BOD (UOD or UBOD). In sewage and industrial wastes, it appears, as has been shown earlier, that the first-order reaction rate is followed at the early stages. In a first-order reaction, the rate is directly proportional to the concentration of the reacting substances.

Before stabilization is complete, the BOD is known to be exerted at two stages. The first demand is used to break down the carbonaceous matter. The second is that in which nitrification of nitrogen-containing compounds occurs.

$$2NH_3 + 3O_2 \xrightarrow[\text{bacteria}]{\text{nitrifying}} 2NO_2 + 2H^+ 2H_2O$$
$$2NO_2 + O_2 + 2H^+ \xrightarrow[\text{bacteria}]{\text{Nitrifying}} 2NO_3 + 2H^+$$

With raw sewage, nitrification is not important until the 8th to 10th day, because of the slow growth of the bacteria. In practice, the shape of the 5-day BOD curve is not affected by nitrification.

In effluents from sewage treatment plants, nitrification can be evident in 1 or 2 days due to the large number of nitrifying bacteria already present. When therefore the BOD of treated effluents is to be determined, thiourea or TCMP (2-chloro-6 (trichloromethyl) pyridine is added to inhibit nitrification, so that only BOD due to carbonaceous breakdown is determined.

Nitrification is the process in which ammonia is converted first into nitrites and later into nitrates as shown in the equations above. It occurs not only in water but also in soil. Although it can be brought about by a few heterotrophs, nitrification is brought about mainly by autotrophic bacteria. The bacteria derive energy needed for transforming (reducing) CO_2 into organic material not with energy from light as in photosynthetic organisms, but by the transfer of electrons from ammonia and nitrites. The organisms involved are small Gram-negative, polarly flagellated rods. The best known of these are *Nitrosomonas* which oxidizes ammonia to nitrites and *Nitrobacter* which oxidizes nitrites to nitrates. Others about which much less is known include *Nitrosococcus, Nitrospira,* and *Nitrosocystics.* They grow slowly and are cultivable on silica gel, but not on agar unless it is specially purified.

Some Criticisms of the BOD Test

Despite its wide use, the BOD test has come under serious criticisms:

(a) When used for sewage disposal studies (especially the aerobic system), the conditions of the BOD are unnatural: The microbial population and the dissolved oxygen are far less than that actually exist during the activated sludge treatment, a major sewage treatment procedure.

(b) The standard 5-day BOD method gives no idea of the *rate* of O_2 uptake, i.e., it gives only the oxygen consumed after 5 days but not from day to day. This criticism has been circumvented by the use of the respirometer. It has been found, using an electrolytic respirometer, that a 3-day BOD at 20°C gave results equivalent to those obtained by the usual method after 5 days.

(c) BOD is only very loosely related to the actual organic matter content of a water, since it represents an overall value of the respiration of a numerically and taxonomically unknown population of micro-organisms, in a medium whose composition is usually unknown.

(d) The assumption that the oxygen consumed is that used by the aerobic bacteria is incorrect. Under natural condition, protozoa use up oxygen within the 5-day period, when at the end of the first 2–3 days bacteria have developed (see Fig. 10.1).

(e) The BOD test is a static test in a batch culture, whereas the condition in a stream is more like a continuous culture (Eckenfelder 2000).

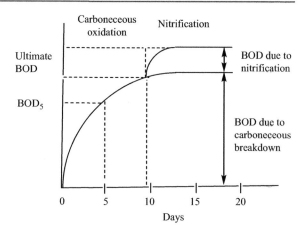

Fig. 10.1 Typical BOD curve for raw domestic sewage

2. *Respirometric Determination of BOD*

The standard BOD method described above provides information only at the end of the 5-day period. The single information provided does not offer enough grounds for the computation of the reaction rate and the ultimate BOD.

Since the BOD test is really a measure of oxygen consumption, a respirometer which measures gas volumes is an ideal instrument for its determination. Commercial respirometers study only small volumes of liquid. Attempts have therefore been made to design simple respirometers which will measure relatively large volumes of liquid, with dilute suspensions of organic matter (*Standard Methods for the Examination of Water and Waste Water*) (Anonymous 2006). Some are available commercially.

Figure 10.2 describes a simple respirometer, which can be constructed in a laboratory workshop with good facilities. It consists of three parts.

(a) A reaction vessel with a suitable stirrer – e.g., a magnetic stirrer.

(b) An adaptor-unit or container which holds potassium hydroxide or some other suitable CO_2-absorbing material.

(c) An electrolysis unit which contains a weak electrolyte, e.g., H_2SO_4. This unit also serves as a manometer to detect pressure changes and also an oxygen generator to maintain a constant partial pressure in the atmosphere within the sample container.

The respirometer functions by providing semi-continuous and automatic registering, and

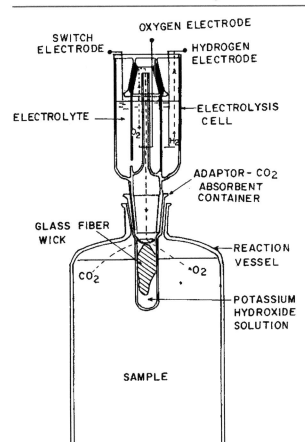

SWITCH ELECTRODE

OXYGEN ELECTRODE

HYDROGEN ELECTRODE

ELECTROLYTE

ELECTROLYSIS CELL

ADAPTOR - CO_2 ABSORBENT CONTAINER

GLASS FIBER WICK

REACTION VESSEL

CO_2

O_2

POTASSIUM HYDROXIDE SOLUTION

SAMPLE

STIRRING MAGNET

Fig. 10.2 Respirometric method of BOD determination (From Young and Baumann 1976. With permission)

adjustment of the pressure changes brought about within the reaction vessel because of O_2 consumption by microorganisms.

3. *Substitutes for the BOD Test*

Because of the shortcomings of the BOD test, which is only an indirect measure of the organic matter in water and sewage, many pollution scientists wish to discard it entirely. Indeed the real nature of the BOD test is so often forgotten that one often reads in sanitary engineering textbooks of "BOD removal" – as if BOD were synonymous with organic water.

Two tests which have been suggested as substitutes are the Chemical Oxygen Demand (COD) and a direct measurement of the organic carbon using

methods such as the Walkley Black method. The COD is the total oxygen consumed by the chemical oxidation of that portion of organic materials in water which can be oxidized by a strong chemical oxidant. The strong oxidant selected is a mixture of potassium dichromate and sulfuric acid which is refluxed with the sample of water being studied. The excess dichromate is titrated with ferrous ammonium sulfate. The amount of oxidizable organic matter measured as oxygen equivalent is proportional to the dichromate consumed. It is a more rapid test than the BOD and when the water contains materials toxic to micro-organisms, such as in some industrial effluents, it is (along with the determination of total carbon) the only method available to determine the organic load. Unless a catalyst is present, however, it cannot measure a biologically degradable compound such as acetic acid. Furthermore, it measures cellulose which is not easily broken down in water, and certainly not during the 5-day BOD. Where wastes contain readily degradable organic materials and nothing toxic, it can be taken as the UBOD or 20-day BOD.

The direct carbon determination determines the total carbon content of the water including those not ordinarily sampled in the COD determination. A strong chemical oxidant oxidizes all the carbon in the water. The method, therefore, has similar uses to the COD.

Both methods can really not substitute for the BOD determination which measures the easily degradable materials (Anonymous 2006; El-Rehaili 1994).

10.2.1.2 Permanganate Value Test

This Permanganate Value (PV) method determines the amount of oxygen used up by a sample in 4 h from a solution of potassium permanganate in dilute H_2SO_4 in a stoppered bottle at 27°C. It gives an idea of the oxidizable materials present in water, although the actual oxidation is only 30–50% of the theoretical value. The method records the oxidation of organic materials such as phenol and aniline as well as those of sulfide, thiosulfate, and thiocyanate and would be useful in some industries. However, because oxidation is incomplete it is not favored by some workers.

10.2.1.3 Chemical Oxygen Demand

The chemical oxygen demand is the total oxygen consumed by the chemical oxidation of that portion of organic materials in water which can be oxidized by a strong chemical oxidant. The oxidant used is a mixture of potassium dichromate and sulfuric acid and is refluxed with the sample of water being studied. The excess dichromate is titrated with ferrous ammonium sulfate. The amount of oxidizable material measured in oxygen equivalent is proportional to the dichromate used up. It is a more rapid test than BOD and since the oxidizing agents are stronger than that used in the PV test, the method can be used for a wider variety of wastes. Furthermore, when materials toxic to bacteria are present, it is perhaps the best method available. Its major disadvantage is that bulky equipment and hot concentrated sulfuric acid are used (Anonymous 2006; El-Rehaili 1994).

10.2.1.4 Total Organic Carbon

Total organic carbon (TOC) provides a speedy and convenient way of determining the degree of organic contamination. A carbon analyzer using an infrared detection system is used to measure total organic carbon. Organic carbon is oxidized to carbon dioxide.

The CO_2 produced is carried by a "carrier gas" into an infrared analyzer that measures the absorption wavelength of CO_2. The instrument utilizes a microprocessor that will calculate the concentration of carbon based on the absorption of light in the CO_2. The amount of carbon will be expressed in mg/l. TOC provides a more direct expression of the organic chemical content of water than BOD or COD.

10.2.1.5 Total Suspended Solids

The term "total solids" refers to matter suspended or dissolved in water or wastewater, and is related to both specific conductance and turbidity. "Total solids" (also referred to as total residue) is the term used for material left in a container after evaporation and drying of a water sample. Total Solids include both total suspended solids (TSS), the portion of total solids retained by a filter and total dissolved solids (TDS), the portion that passes through a filter. Total solids can be measured by evaporating a water sample in a weighed dish, and then drying the residue in an oven at 103–105°C. The increase in weight of the dish represents the total solids. Instead of total solids, laboratories often measure total suspended solids and/or total dissolved solids. To measure total suspended solids (TSS), the water sample is filtered through a pre-weighed filter. The residue retained on the filter is

dried in an oven at 103–105°C until the weight of the filter no longer changes. The increase in weight of the filter represents the total suspended solids. TSS can also be measured by analyzing for total solids and subtracting total dissolved solids.

10.2.1.6 Volatile Suspended Solids

Volatile suspended solids (VSS) are those solids (mg/l) which can be oxidized to gas at 550 ° C. Most organic compounds are oxidized to CO_2 and H_2O at that temperature; inorganic compounds remain as ash.

10.3 Systems for the Treatment of Sewage

Several methods exist for the treatment of sewage and they may be grouped into aerobic as shown below:
Aerobic methods:
1. The activated sludge system
2. The Trickling filter
3. The oxidation pond
Anaerobic methods
1. The septic tank
2. The Imhoff tank
3. The cesspool

10.3.1 Aerobic Breakdown of Raw Waste Waters

The basic microbiological phenomenon in the aerobic treatment of wastes in aqueous environments is as follows:
1. The degradable organic compounds in the wastewater (carbohydrates, proteins, fats, etc.) are broken down by aerobic micro-organisms mainly bacteria and to some extent, fungi. The result is an effluent with a drastically reduced organic matter content.
2. The materials difficult to digest form a sludge which must be removed from time to time and also treated separately.

The discussion will therefore be under two headings: aerobic breakdown of raw wastewater (dealing with the activated sludge system, the trickling filter system, and the oxidation), and followed by the anaerobic breakdown of sludge resulting from the aerobic breakdown.

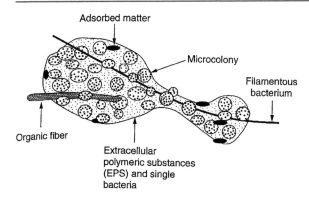

Fig. 10.3 Diagram of a floc (From Nielsen 2002. With permission)

10.3.1.1 The Activated Sludge System

The activated sludge method (Nielsen 2002; Lindera 2002; Kosric and Blaszczyk 1992) is the most widely used method for treating wastewaters. Its main features are as follows:

(a) It uses a complex population of microorganisms of bacteria and protozoa.

(b) This community of microorganisms has to cope with an uncontrollably diverse range of organic and inorganic compounds some of which may be toxic to the organisms.

(c) The microorganisms occur in discreet aggregates known as flocs which are maintained in suspension in the aeration tank by mechanical agitation or during aeration or by the mixing action of bubbles from submerged aeration systems. Flocs consist of bacterial cells, extracellular polymeric substances, adsorbed organic matter, and inorganic matter. Flocs are highly variable in morphology, typically 40–400 µm and not easy to break apart (see Fig. 10.3).

(d) When floc particles first develop in the activated sludge process, the particles are small and spherical. As the sludge age increases and the short filamentous organism within the floc particles began to elongate, the floc forming bacteria now flocculate along the lengths of the filamentous organisms. The presence of long filamentous organisms results in a change in the size and shape of floc particles. These organisms provide increased resistance to shearing action and permit a significant increase in the number of floc-forming bacteria in the floc particles. The floc particles increase in size to medium and large and change from spherical to irregular.

(e) The flocs must have good settling properties so that separation of the biomass of microorganisms and liquid phases can occur efficiently and rapidly in the clarifier. Sometimes, proper separation is not achieved giving rise to problems of bulking and foaming.

(f) Some of the settled biomass is recycled as "returned activated sludge" (RAS) to inoculate the incoming raw sewage because it contains a community of organisms adapted to the incoming sewage.

(g) The solid undigested sludge may be further treated into economically valuable products.

The advantages of the activated sludge system over the other methods to be discussed are its efficiency, economy of space, and versatility. The flow diagrams of the conventional set-up and various modifications thereof are given in Fig. 10.4.

Microbiology of the Activated Sludge Process

The activated-sludge process is a biological method of wastewater treatment brought about by a variable and mixed community of microorganisms in an aerobic aquatic environment. These microorganisms derive nourishment from organic matter in aerated wastewater for the production of new cells. Some of the bacteria carry out nitrification and convert ammonia nitrogen to nitrate nitrogen. The consortium of microorganisms, the biological component of the process, is known collectively as activated sludge.

There are three essential requirements for the activated sludge process:

(a) A mixed population of micro-organisms able to degrade the components of the sewage, must be present.

(b) The population must be able to grow in the environment of the aeration tank.

(c) The organisms must grow in such a way that they will form flocs and settle out in the sedimentation tank.

The activated sludge microbial community is specialized and is less diverse than of the biological or trickling filter (see below). Bacteria make up about 95% of the activated sludge and biomass, and aerobic bacteria are the dominant bacteria. Bacteria are followed by protozoa, fungi, and rotifers; nematodes are few and algae are usually absent.

The organisms involved are bacteria and ciliates (protozoa). It was once thought that the formation of

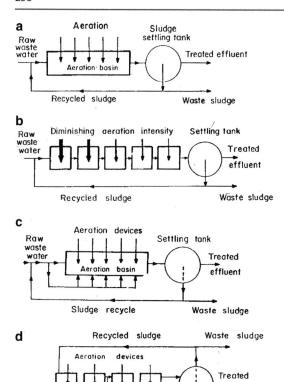

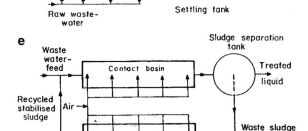

Fig. 10.4 Schematic representation of the activated sludge set-up. (**a**) Conventional aeration; (**b**) tapered aeration with direct introduction of raw sewage; (**c**) tapered aeration with tank introduction of raw sewage; (**d**) step aeration; (**e**) contact stabilization (From Okafor 2007. With permission)

flocs which are essential for sludge formation was brought about by the slime-forming organism, *Zooglea ramigera*. It is now known that a wide range of bacteria are involved, including *Pseudomonas, Achromobacter, Flavobacterium* to name a few.

Various bacteria have been reported in the activated sludge setup depending on the chemical nature of the sludge. Perhaps, the most important one is the rod shaped bacterium, *Zoogloea ramigera*, which produces a large quantity of extracellular slime matrix. This

organism is believed to be the main agent for flocculation, although other organisms can also form flocs. Other bacteria which have been reported are *Pseudomonas. Achromobacter. Flavobacterium. Microbacterium. Micrococci, Bacillus, Comomonas, Azotobacter, Staphylococcus, Bdellovibrio, Nitrobacter,Nitrosomonas,Alcaligenes* spp.,*Brevibacterium* spp., *Beggiatoa* spp., *Corynebacterium* spp., and *Sphaerotilus* spp.

Fungi encountered are *Zoophagus* spp.,*Arthrobotrys, Geotrichum candidum, Pullularia* spp., *Alternaria* spp., *Penicillium* spp., and *Cephalosporium* spp.

Among the protozoa, ciliates dominate and among ciliates the following are most common: *Aspidisca costata, Carchesium polypinum, Chilodonella uncinata, Opercularia coarcta* and *O. microdiscum, Trachelophyllum pusillum, Vorticella convallaria* and *V. microstoma*. Most of these are sessile and attached to the sludge flocs. In addition, amoebae and flagellates are also seen. Yeasts and algae are of rare occurrence.

A succession of protozoa has been observed. Flagellates (e.g., *Heteronema, Bodo*) occur only in the immature sewage, being replaced in about 3 weeks first by free swimming ciliates (e.g., *Paramecium caudatum*), then by crawling ciliates (e.g., *Aspidica costata*), and finally by attached ciliates (e.g., *Vorticella* and *Epistylis*). Flagellates are found in poor quality systems or in "young" as opposed to mature activated sludge systems. This succession is illustrated in Fig. 10.5).

Rotifers are rarely found in large numbers in wastewater treatment processes. The principal role of rotifers is the removal of bacteria and the development of floc. Rotifers contribute to the removal of effluent turbidity by removing non-flocculated bacteria. Mucous secreted by rotifers at either the mouth opening or the foot aids in floc formation. Rotifers require a longer time to become established in the treatment process. Rotifers indicate increasing stabilization of organic wastes.

Bulking in Activated Sludge Systems

"Bulking" is a growth condition in which the sludge has poor setting properties, because of loose cotton wool-like growth of filamentous organisms. Bulking may also create problems of aeration by the trapping of oxygen; the net effect maybe inadequate stabilization. Some authors have distinguished bulking into two types. The first type, in which *Sphaerotilus* is always present, is according to these authors, caused by overload. The second type of bulking in which *Sphaerotilus*

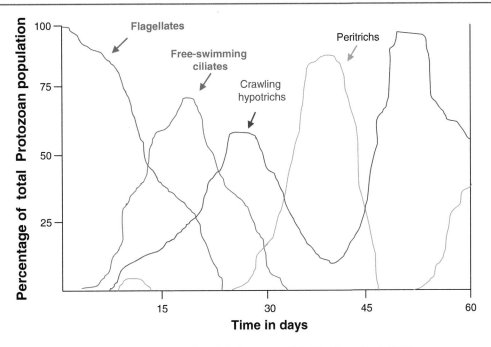

Fig. 10.5 Succession of dominant protozoa in an activated sludge system (Modified from Curds 1965)

is absent, is due to underloading in the aeration chamber. *Streptothrix* and non-sheath-forming bacteria are implicated in this type. Organisms which bring about bulking are regarded as nuisance organisms and include, apart from *Sphaerotilus* and *Streptothrix*, *Geolrichum, Beggiatoa, Bacillus,* and *Thiothrix*.

Nutrition of Organisms in the Activated Sludge Process
Provided, the wastes are nutritionally satisfactory sludge will be formed whether the sewage is rich in colloids as in domestic sewage or in soluble materials as in some industrial sewage. While domestic sewage is usually rich in organisms and in nutrients, some industrial sewages may be deficient in the key nutrients of nitrogen and phosphorous. These must therefore be added. Sometimes, the industrial activated sewage must also be seeded with soil in order to develop a population capable of stabilizing it. In some systems, the aerobic nitrogen fixing bacteria, *Azotobacter* are added to provide nitrogen.

Modifications of the Activated Sludge System
Several modifications of the activated sludge procedure exist.
1. *The conventional activated sludge set-up*: The basic components of the conventional system are an aeration tank and a sedimentation tank. Before raw wastewater enters the aeration tank, it is mixed with a portion of the sludge from the sedimentation tank. The raw water is therefore broken down by organisms already adapted to the environment of the aeration tank. The incoming organisms from the sludge exist in small flocs which are maintained in suspension by the vigor of mixing in the aeration tank. It is the introduction of already adapted flocs of organisms that gave rise to the name activated sludge. Usually 25–50% of the flow through the plant is drawn off the sedimentation tank. Other modifications of the activated sludge system are given below.
2. *Tapered aeration*: This system takes cognizance of the heavier concentration of organic matter and hence of oxygen usage at the point where the mixture of raw sewage and the returned sludge enters the aeration tank. For this reason, the aeration is heaviest at the point of entry of wastewaters and diminishes toward the distal end. The diminishing aeration may be made directly into the main aeration tank (Fig. 10.3b, c) or a series of tanks with diminishing aeration may set up.
3. *Step aeration*: In step aeration, the feed is introduced at several equally spaced points along with length of the tank, thus creating a more uniform demand in the tank. As with tapered aeration, the aeration may be done in a series of tanks.

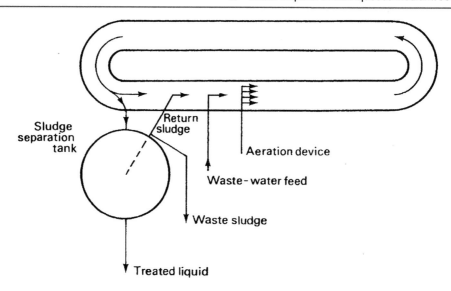

Fig. 10.6 Schematic representation of the Pasveer Ditch activated sludge set-up (From Okafor 2007. With permission)

4. *Contact stabilization*: This is used when the waste-water has a high proportion of colloidal material. The colloid-rich wastewater is allowed contact with sludge for a short period of 1–1½ h, in a contact basin which is aerated. After settlement in a sludge separation tank, part of the sludge is removed and part is recycled into an aeration tank from where it is mixed with the in-coming wastewater.

5. *The Pasveer ditch*: This consists of a stadium-shaped shallow (about 3 ft) ditch in which continuous flow and oxygenation are provided by mechanical devices. It is essentially the conventional activated sludge system in which materials are circulated in ditch rather than in pipes (Fig. 10.6).

6. *The deep shaft process*: The deep shaft system for wastewater treatment was developed by Agricultural Division of Imperial Chemical Industries (ICI) in the UK, from their air-lift fermentor used for the production of single cell protein from methanol. It consists of an outer steel-lined concrete shaft measuring 300 ft or more installed into the ground. Wastewater, and sludge re-cycle are injected down an inner steel tube. Compressed air is injected at a position along the center shaft deep enough to ensure that the hydrostatic weight of the water above the point of injection is high enough to force air bubbles downward and prevent them coming upward. The air dissolves lower down the shaft providing oxygen for the aerobic breakdown of the wastes. The water rises in the outer section of the shaft (Fig. 10.7). The system has the advantage of great rapidity in reducing the BOD and about 50% reduction in the sludge. Space is also saved.

7. *Enclosed tank systems and other compact systems*: Since the breakdown of waste in aerobic biological treatment is brought about by aerobic organisms, efficiency is sometimes increased by the use of oxygen or oxygen enriched air. Enclosed tanks, in which the wastewater is completely mixed with the help of agitators, are used for aeration of this type. Sludge from a sedimentation tank is returned to the enclosed tank along with raw water as in the case with other systems. The advantage of the system is the absence, (or greatly reduced) obnoxious smell from the exhaust gases, and increased efficiency of waste stabilization. This system is widely used in industries the world over.

8. Compact activated sludge systems: These do not have a separate sedimentation tank. Instead, sludge separation and aerobic breakdown occur in a single tank. The great advantage of such systems is the economy of space (Fig. 10.8).

Efficiency of Activated Sludge Treatments

The efficiency of any system is usually determined by a reduction in the BOD of the wastewater before and after treatment. Efficiency depends on the amount of aeration, and the contact time between the sludge and the raw wastewater. Thus, in conventional activated sludge plants the contact time is about 10 h, after which

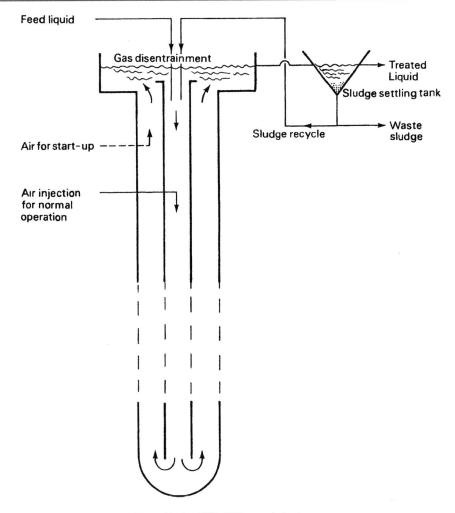

Fig. 10.7 The deep shaft aeration system (From Okafor 2007. With permission)

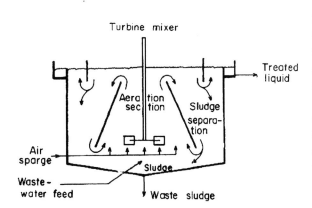

Fig. 10.8 Compact activated sludge system (From Okafor 2007. With permission)

90–95% of the BOD is removed. When the contact time is less (in the high-rate treatment), BOD removal is 60–70% and the sludge produced is more. With longer contact time, say several days, BOD reduction is over 95% and sludge extremely low.

With systems where oxygen is introduced as in the closed tank system, or where there is great oxygen solubility as in the deep shaft system, contact time could be as short as 1 h but with up to 90% BOD reduction along with substantially reduced sludge.

10.3.1.2 The Trickling Filter

The trickling filter consists of round rocks 1–4 in. diameter arranged in a bed 6–10 ft deep (Fig. 10.9). Sewage

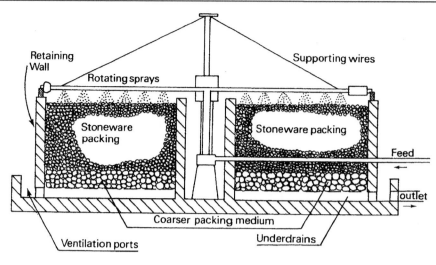

Fig. 10.9 Section through trickling filter bed (From Okafor 2007. With permission)

Fig. 10.10 Scheme illustrating two arrangements of trickling filter: conventional and single stage (From Okafor 2007. With permission)

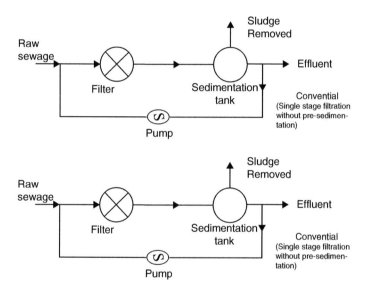

is uniformly spread on the bed by a rotating distributor which is powered by an electric motor or driven by hydraulic impulse where a hydraulic head is available. The sewage percolates by gravity over the rocks and through the spaces between the rocks and the effluent is collected in an under-drain. From the under-drain, the liquid is allowed to settle in a sedimentation tank which is an integral part of the system.

The sludge consisting of microorganisms and undecomposed matter is removed from time to time and may be used as manure. Various modifications may be made to this basic design. In one modification (see

Fig. 10.10), the sewage may be pre-sedimented before filtration; in another, two filters may operate; and in yet others, the effluent may be re-cycled.

Trickling filters may be low-rate or high-rate. Low rate filters handle 2–4 million gallons per acre per day (mgad). They are very efficient in BOD reduction and may attain 85–90% BOD. The effluent is usually still rich in oxygen; hence, nitrates are produced in abundance by nitrifying bacteria. High rate filters handle 10–40 mgad and BOD reduction is only 65–75%. The effluent is usually low in nitrates since much of the O_2 is used in breaking down the organic matter load.

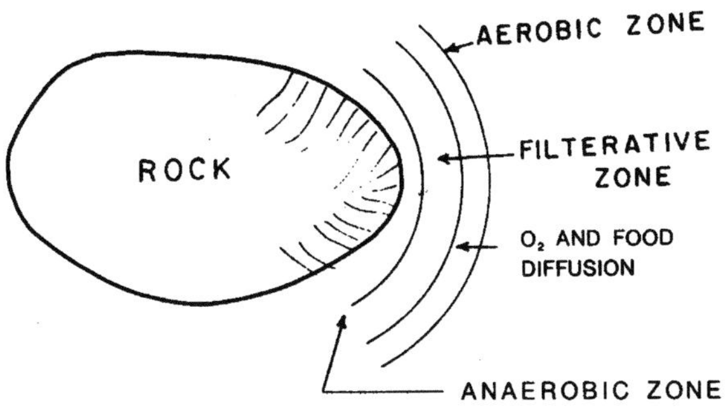

Fig. 10.11 Oxygen and food zones in the slime coating of rocks in the trickling filter (From Okafor 2007. With permission)

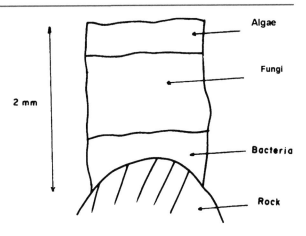

Fig. 10.12 Diagrammatic transverse section of the slime layers in mature trickling filter (From Okafor 2007. With permission)

Microbiology of the Trickling Filter

As the sewage percolates over the stones, micro-organisms break down the organic matter in it. Although the process is described as aerobic, most of the bacteria are in fact facultative. Aerobic conditions operate mainly when the filter is fresh and in the outer areas of the coating of micro-organisms on the rocks in mature filters (see Fig. 10.11). The innermost portion of the coating of micro-organisms may in fact be anaerobic. Bacteria are the most important organisms in the trickling filter, but other organisms are also present.

Fungi, for instance, are to be found in the aerobic zone. The fungi, encountered include mainly *Fusarium* and *Geotrichum*. Others are *Trichosporon, Sepedonium, Saccharomyces,* and *Ascoidea*. Fungi are more common in low pH sewages and in some industrial sewages.

Protozoa are present: the flagellates, free-swimming ciliates and, to some extent, amoeba at the upper part of the filter rocks, and the stalked ciliates in the bottom portion. It has been suggested that this stratification is a result of the availability of soluble food at the various levels, the greater portion of such food being available in the surface region. Some protozoa encountered in the surface region are *Trepomonas agilis* and *Vorticella microstoma*; in the middle region, *Paramecium candatum* and *Opercularia;* and in the innermost layer, *Arcella vulgaris* and *Aspidica costata.*

Algae, because of their need for sunlight, are found In the upper layers of the filter and may clog the filter (see Fig. 10.12). The only groups of algae found in both trickling filters and oxidation ponds (see below) are *Chlorophyceae, Euglenaphyceae, Cyanophyceae,* and *Chrysophyceae*. In the outer portions of the filter, particularly in older filters, algae are to be found throughout the coating of the rocks especially in the middle

layer as shown in Fig. 10.1. The algae involved include the unicellular *Chlorella, Phormidium, Oscillatoria,* or the sheet-forming multicellular *Stigeoclonium* and *Ulothrix*. Algal photosynthesis provides only some of the oxygen required by the aerobic bacteria which contribute to the breakdown of organic matter. Worms, snails, and larvae are also to be encountered, but these contribute little to the process of filtration.

The micro-organisms adhere to the rock by weak (van der Waals) forces and grow in one direction only as liquid flows over the film of microbial coating. As the microbial layer increases in thickness, the innermost organisms die and the microbial layer drops off from the rock. A new growth starts thereafter. Since the breakdown is brought about by aerobes, a filter is most efficient when the microbial layer is thinnest or when the filter stones regularly shed their slime coatings.

10.3.1.3 Rotating Discs

Also known as rotating biological contactors, these consist of closely packed discs about 10 ft in diameter and 1 in. apart. Discs made of plastic or metal may number up to 50 or more and are mounted on a horizontal shaft which rotates slowly, at a rate of about 0.5–15 revolutions/min. During the rotation, 40–50% of the area of the discs is immersed in liquid at a time. A slime of micro-organisms, which decompose the wastes in the water, builds up on the discs. When the slime is too heavy, it sloughs off and is separated from the liquid in a clarifier. It has a short contact time and produces little sludge (Fig. 10.13).

Fig. 10.13 Structure of
rotating discs (rotating
biological contactor)
(From Okafor 2007. With
permission)

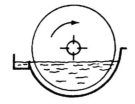 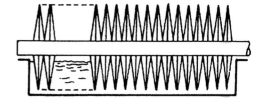

Left: Transverse section   Right: Side view

10.3.1.4 Oxidation Ponds

Oxidation ponds (also called or Stabilization Ponds) are shallow lagoons about three feet deep into which sewage is discharged at a single point, usually at the center but also occasionally at the side. After suitable periods of holding, the effluent which usually has a low BOD is discharged at a single point. The effluent is usually low in coliforms and may be discharged into a river or used as raw water source. Oxidation ponds are especially appropriate in warm, sunny climates.

Oxidation Ponds are a common sewage treatment method for small communities because of their low construction and operating costs. Oxidation ponds represent 12% of all sewage treatment plants in the US. New oxidation ponds can treat sewage fairly efficiently but require maintenance and periodic de-sludging in order to maintain this standard.

They may be made of one or up to four shallow ponds in series. The natural processes of algal and bacteria growth exist in a mutually dependent relationship.

Oxygen is supplied from natural surface aeration and by algal photosynthesis. Bacteria present in the wastewater use the oxygen and feed on organic material, breaking it down into nutrients and carbon dioxide. These are in turn used by the algae. Other microbes in the pond such as protozoa remove additional organic and nutrients to polish the effluent.

There are normally at least two ponds constructed. The first pond reduces the organic material using aerobic digestion, while the second pond polishes the effluent and reduces the pathogens present in sewage. Sewage enters a large pond after passing through a settling and screening chamber. After retention for several days, the flow is often passed into a second pond for further treatment before it is discharged into a drain. Bacteria already present in sewage acts to break down organic matter using oxygen from the surface of the pond. Oxidation ponds need to be de-sludged periodically in order to work effectively.

Oxidation ponds require large amounts of land and the degree of treatment is weather dependent. The only operation necessary, if at all, is to alter by appropriate valves, the point of the discharge of the raw sewage.

A number of problems may however be associated with oxidation ponds. First, they permit the growth of mosquitoes in countries where malaria and other mosquito-borne diseases abound. The mosquito larvae can be controlled by spraying the ponds with oil. Second, aquatic weeds may clog them but if they are up to 3 ft deep weeds do not grow except at the edges. Depending upon the design. oxidation ponds must be freed of sludge approximately every 10 years. They are sometimes used for the primary stabilization of wastes from dairies; often, however, they are employed as secondary or tertiary treatment facilities.

Oxidation ponds (Gerhardt and Oswald 1990) are usually employed as secondary or tertiary treatment. Occasionally, they are used as primary treatment plants in which case anaerobic conditions tend to occur because of the heavy load. This anaerobic condition is particularly apt to occur in the dark when, as is seen later, the photosynthetic organisms largely responsible for the provision of oxygen to the aerobic bacteria, are inactive.

Oxidation ponds are most often used as secondary treatment for wastewaters or for waters such as shed wastes which contain heavy loads of organic matter. Oxidation pond systems may be a two-pond system (see Fig. 10.14) or a multi-pond one (see Fig. 10.15). Multi-pond systems are usually described as an Advanced Integrated Pond System (AIPS), in which different ponds fulfill different functions. It consists of a series of ponds in the ground, which use heterotrophic bacteria and algae to treat the wastewater. Wastewater first passes into deep pits in the Advanced Facultative Pond. Here, solids are broken down by fermentation anaerobically in a fermentation pit to produce methane and achieve the removal of many pathogens. The water

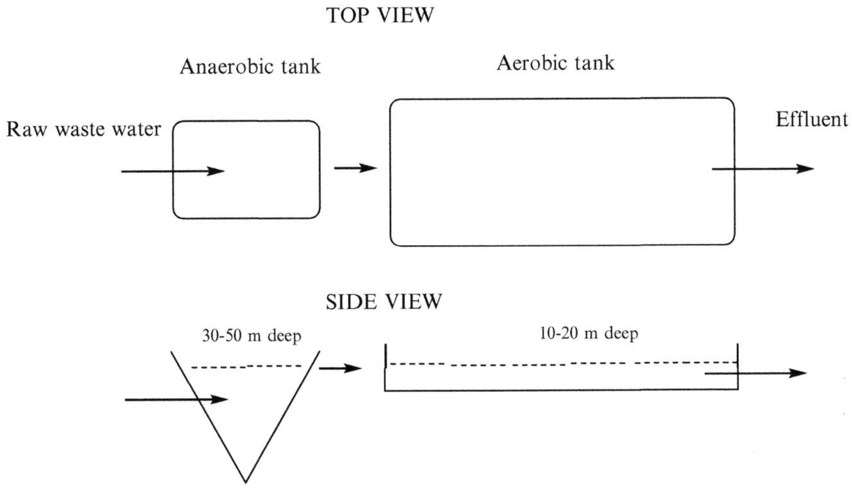

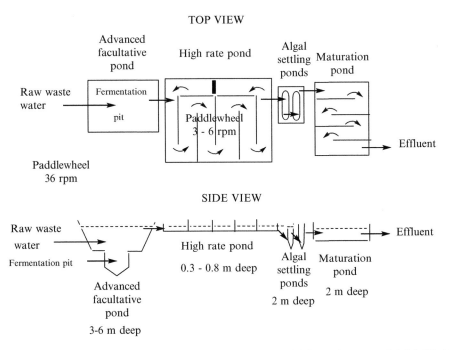

Fig. 10.14 Setup in a two-pond oxidation pond system (Not drawn to scale) (Modified from Gerhardt and Oswald 1990)

Fig. 10.15 Setup in a multiple pond system (Advanced Integrated pond System, AIPS; not drawn to scale) (Modified from Gerhardt and Oswald 1990)

then passes into a High Rate Pond (HRP) for rapid growth of algae and concomitant production of oxygen; in this pond, organic materials are oxidized. In some systems, paddlewheels which revolve slowly help to increase aeration in the HRP. Small ponds immediately down stream remove algae. The water next passes into Maturation Pond where it is held for a period before being discharged. Maturation tanks pro-vide opportunities for pathogen reduction, and when the aim is to reduce pathogens the water may be passed through a series of maturation tponds.

Effluent from AIPS is typically suitable for aqua-culture because the fish feed on the algae in it. Under such conditions, the algal settling and maturation ponds may be eliminated or the water may held for shorter periods in them.

The side view of the AIPS (Fig. 10.15) shows that the deepest portion is the fermentation pit where anaerobic breakdown takes place; the HRP is the shallowest being less than a meter deep, so as to encourage aeration.

The Microbiology of the Oxidation Pond

The bulk of the stabilization in an oxidation pond is brought about by aerobic bacteria, but zones of growth by anaerobic and facultative bacteria may also exist depending on the depth of the pond. Oxygen is supplied to the aerobic bacteria by two means: (a) Oxygen released by algae during photosynthesis, and (b) Diffusion of oxygen into the water assisted by natural winds and sometimes by floating turbine aerators.

In large oxidation ponds with retention periods of 3 weeks to 3 months, algal aeration by photosynthesis is not very effective; but in smaller ponds with retention time of less than a week, algal photosynthetic oxygen effectively supplies the oxygen required by the aerobic bacteria. Turbine aerators which float on the ponds are often used in large oxidation ponds to encourage O_2 diffusion into the pond. The CO_2 released by the aerobes is used by the algae (see Figs. 10.15 and 10.16).

Bacteria, protozoa, algae, and rotifers are all to be found in the oxidation pond. The predominant bacteria are *Pseudomonas, Flavobacterium* and *Alcaligines*, but this depends to some extent on the nature of the sewage. Coliforms die off rapidly because they cannot compete for food and because of the predatory activity of ciliates. Some authors suggest that antibiotics released by algae are effective in killing off bacteria.

The algae commonly involved include *Chlorella, Spirogyra, Vaucheria* and *Ulothrix*. Some algae are confined to the surface, e.g., *Oscillatoria* while others are benthic, e.g. *Scendemus*.

Some of the free-swimming ciliates include *Paramecium* and *Colpidium* whereas the stalked ciliates include *Vorticella*.

In some oxidation pond designs, only one pond is used, especially when it is used for treating effluents from other treatments such as activated sludge or the Imhoff tank (see below). When however it is used on its own, say in dairy or abattoir wastes where the organic matter content is high, at least two ponds are involved. The first is usually deep, and decomposition in it is usually anaerobic. The subsequent ponds are usually aerobic and involve algae (see Fig. 10.15).

10.3.2 Anaerobic Sewage Systems

10.3.2.1 Treatment of the Sludge from Aerobic Sewage Treatment Systems: Anaerobic Breakdown of Sludge

As has been seen above, sludge always accompanies the aerobic breakdown of wastes in water. Its disposal is a major problem of waste treatment. Sludge consists of micro-organisms and those materials which are not readily degradable particularly cellulose. The solids in sludge form only a small percentage by weight and generally do not exceed 5% (Fig. 10.17).

The goals of sludge treatment are to stabilize the sludge and reduce odors, remove some of the water and reduce volume, decompose some of the organic matter and reduce volume, kill disease causing organisms, and disinfect the sludge. Untreated sludges are about 97% water. Settling the sludge and decanting off the separated liquid removes some of the water and reduces the sludge volume. Settling can result in a sludge with about 96–92% water. More water can be removed from sludge by using sand drying beds, vacuum filters, filter presses, and centrifuges resulting in sludges with between 80% and 50% water. This dried sludge is called a sludge cake. Anaerobic digestion is used to decompose organic matter to reduce volume. Digestion also stabilizes the sludge to reduce odors. Caustic chemicals can be added to sludge or it may be heat treated to kill disease-causing organisms. Following treatment, liquid and cake sludges are usually spread on fields, returning organic matter and nutrients to the soil.

The most common method of treating sludge, however, is by anaerobic digestion and this will be discussed below.

Anaerobic digestion consists of allowing the sludge to decompose in digesters under controlled conditions for several weeks. Digesters themselves are closed tanks with provision for mild agitation, and the introduction of sludge and release of gases. About 50% of the organic matter is broken down to gas, mostly methane. Amino acids, sugars alcohols are also produced. The broken-down sludge may then be de-watered and disposed of by any of the methods described above. Sludge so treated is less offensive and consequently easier to handle. Organisms responsible for sludge breakdown are sensitive to pH values outside 7–8, heavy metals, and detergents and these should not be introduced into digesters. Methane gas is also pro-

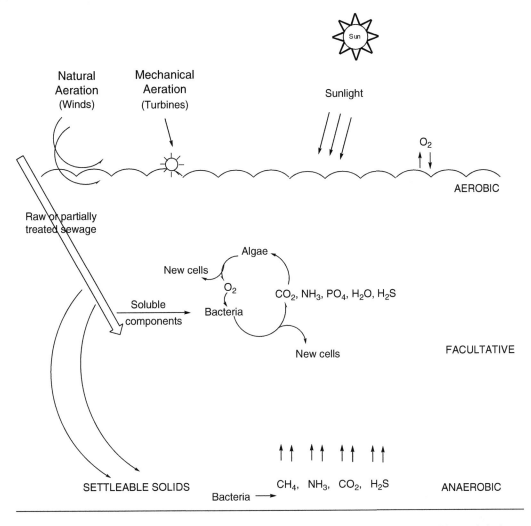

Fig. 10.16 Scheme depicting the microbiological activity in an oxidation pond (From Okafor 2007. With permission)

duced and this may sometimes be collected and used as a source of energy. Figure 10.17 shows the various processes to which sludge may be subjected and some anaerobic sludge digester designs.

10.3.2.2 The Septic Tank

Septic tanks are small scale sewage systems not connected to main sewage systems. About 26% use this system in North America. Both in North America and in Europe, they are limited to the rural areas. In many parts of the developing world, however, they are the main method of disposing of domestic sewage (Fig. 10.18).

A septic tank generally consists of a tank of between 1,000 and 1,500 gal (4,000–5,500 l) which is con-

nected to an inlet wastewater pipe at one end and to a septic drain field or soak away or soakage pit at the other. These pipe connections are generally made via T pipes which allow liquid entry and outflow without disturbing any crust on the surface; i.e., direct current between the inlet and outlet is prevented using baffles and pipe tees.

In the US, the size of the septic tank to be constructed depends on the local authority approving the construction of the residence. Ultimately, it is based on the number of persons expected to use the building. In practice, some authorities base it on the expected flow of wastewater per day, while others base it on the number of rooms Table 10.1 shows tank sizes and the number of persons they should serve.

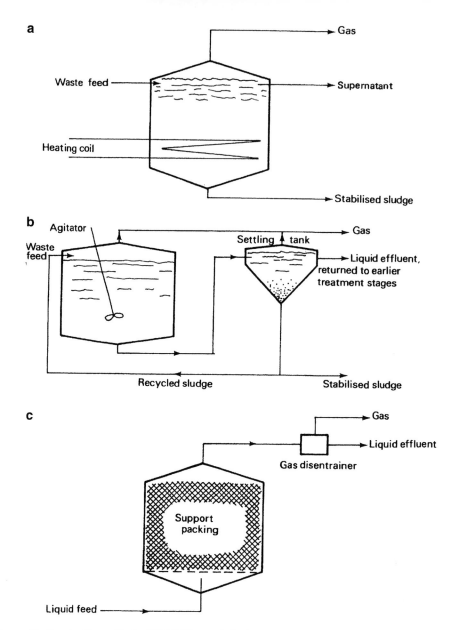

a

Gas

Waste feed

Supernatant

Heating coil

Stabilised sludge

b

Agitator

Gas

Waste feed

Settling tank

Liquid effluent, returned to earlier treatment stages

Recycled sludge

Stabilised sludge

c

Gas

Liquid effluent

Gas disentrainer

Support packing

Liquid feed

Fig. 10.17 Anaerobic digester (From Okafor 2007. With permission)

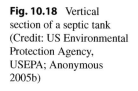

Fig. 10.18 Vertical section of a septic tank (Credit: US Environmental Protection Agency, USEPA; Anonymous 2005b)

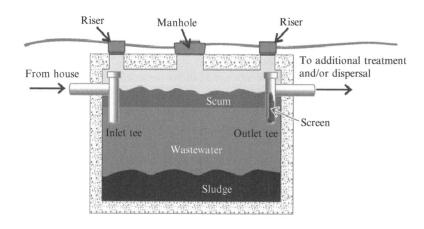

Riser

Manhole

Riser

From house

To additional treatment and/or dispersal

Scum

Inlet tee

Outlet tee

Screen

Wastewater

Sludge

Table 10.1 Septic tanks sizes and number of persons to be served (using bedroom numbers) (Compiled from requirements of the Eldorado County, California, http://www.edcgov.us/emd/envhealth/septic_tank_sizes.html. With permission)

| Number of bedrooms | Size (gal) | Septic tank dimensions | | | Actual gallons |
		Length	Height	Breadth	
3	1,000 (minimum)	8'	4'8"	6'	987
4	1,200	8'	5'8"	6'	1,234
5	1,500	9'6"	5'8"	6'	1,486
6	2,000	16'	4'8"	6'	2,064

Wastewater enters through the inlet pipe and the baffle directs it to the bottom. The heavier materials remain at the bottom where they are broken down by anaerobic bacteria, thereby reducing the volume of solids. The oils and fats are not usually broken down as easily as the other materials and they float to the top of scum. As new sewage is added, the accompanying added liquid displaces the top portion of the liquid in the tank into the pipe leading to the soakage pit, taking with it dissolved broken down materials.

Anaerobic decomposition is not as efficient as aerobic and the unbroken down materials form sludge which must be removed from time to time along with the fat scum. Septic tanks have some problems which are the result of the added sewage and its breakdown products remaining in the tank (as opposed to sewers which move the added sewage to the processing location). Thus, excessive addition of fats and oils or flushing down non-biodegradable materials such as sanitary towels may lead to early filling up of the tank. Similarly, the addition of chemicals, such as acid or sodium hydroxide, which can kill off the microorganisms carrying out the decomposition, may also lead to an early filling up of the tank. Roots from trees growing nearby may rupture the tank and shrubbery growing above the tank, or the drain field may clog and or rupture them.

10.3.2.3 The Imhoff Tank

This is named after its inventor, the German Karl Imhoff. It is widely used. It has an upper flow or sedimentation chamber through which the sewage passes at low velocity, and a lower or digestion chamber into which the heavier sewage particles sediment and are broken down. Methane, which is produced by anaerobic breakdown, is discharged through a pipe and may be collected for use as fuel, while the scum (see Fig. 10.19) and the undigested sediment are transferred from the sedimentation tank by mechanical means to drying beds. The dried sludge is then used as manure.

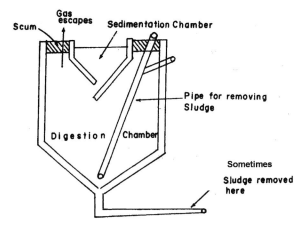

Fig. 10.19 Diagrammatic cross-section through an Imhoff tank. (From Okafor 2007. With permission)

The effluent from the tanks may be further treated either with a trickling filter or in an oxidation pond. Unpleasant odors which accompany black foams sometimes appear at the gas vents but the odors may be reduced some-what by introducing lime into the vent. Animal manure (i.e., dung and trash) which may also help in odor reduction is added to the gas vents.

The Imhoff tank has no mechanical parts and is relatively easy and economical to operate. It provides sedimentation and sludge digestion in one unit and when operating properly, produces a satisfactory primary effluent with a suspended solids removal of 40–60% and a BOD reduction of 15–35%. The two-storey design requires a deep over-all tank. The Imhoff tanks is best suited to small municipalities and large institutions where the population is 5,000 or less, and a greater degree of treatment is not needed.

10.3.2.4 Cesspools

Cesspools are shallow disposal systems that are generally constructed as a concrete cylinder with an open bottom and/or perforated sides (drywell). They are

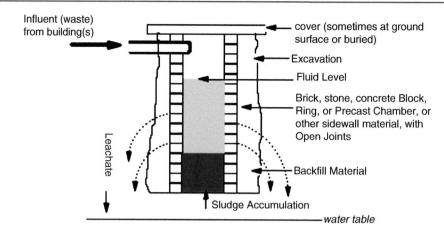

Fig. 10.20 Diagram of a cesspool (From http://www.epa.gov/region9/water/groundwater/uic-hicesspools.html; Anonymous 2010a. With permission)

used by multi-family residential units, churches, schools, and public meeting facilities, office buildings, industrial and commercial buildings, shopping malls, hotels and restaurants, highway rest stops, state parks, and camp grounds (Fig. 10.20).

This untreated sanitary waste can enter shallow groundwater and contaminate drinking water resources because they are designed to isolate but not to treat sanitary waste. They may introduce into ground water the following undesirable items: Nitrates, total suspended solids, and coliform bacteria exceeding the quantities recommended for drinking water, as well as other constituents of concern such as phosphates, chlorides, grease, viruses, and chemicals used to clean cesspools (e.g., trichloroethane and methylene chloride). On account of this, the EPA has from 2,000 banned new large-capacity cesspools and existing ones must be closed by 2005.

The EPA defines a cesspool as typically a "dry-well" which sometimes has an open bottom and/or perforated sides, and receives untreated sanitary waste. The EPA considers a cesspool large capacity when used by

(a) Multiple dwelling, community, or regional system for the injection of waste (e.g., a townhouse complex or apartment building), or

(b) Any nonresidential cesspool that is used solely for the disposal of sanitary waste and has the capacity to serve 20 or more people per day (e.g., a rest stop or church)

Breakdown of organic matter is anaerobic as no aeration is introduced. The cesspool resembles a septic tank in which the broken down materials seep into the ground without going through a T-pipe.

10.4 Advanced Wastewater Treatment

Ordinarily, effluents from waste treatment plants such as the activated sludge or the Imhoff systems are discharged into rivers or streams. Such water is sometimes insufficiently treated to provide reusable water for industrial and/or domestic recycle; in addition, it may cause an excessive increase in nutrients such as nitrogen and phosphorus leading to eutrophication. Thus, additional treatment steps have been added to wastewater treatment plants to provide for further organic and solids removals or to provide for removal of nutrients and/or toxic materials. The need for the further treatment of effluents from conventional or secondary treatments has been accentuated by the interest in the control of environmental pollution. This further treatment of effluents is known as Advanced Wastewater Treatment (AWT) or tertiary wastewater treatment. The EPA defines AWT as "any treatment of sewage that goes beyond the secondary or biological water treatment stage, and includes the removal of nutrients such as phosphorus and nitrogen, and a high percentage of suspended solids."

Apart from the possibility of drinking, water resulting from AWT can be used for the following: recreational water, e.g. in swimming pools; irrigation in agricultural practice; industrial processes such as cooling, and in replenishing underground water.

Several components appear to be included in the term AWT, and what is included varies from one locality to another.

Some constituents of effluents which may give rise to concern are:

1. Inorganics – metals, nitrate, phosphorous, and total dissolved solids (TDS)
2. Organics – trace organics, pesticides, color
3. Micro-organisms – viruses, bacteria
4. Physical – suspended solids, turbidity

10.4.1 Methods Used in Advanced Wastewater Treatment

Some methods used in AWT are discussed below. They are usually expensive and in the USA, are usually used for purified bottled water and for specialized uses. Where AWT methods are adopted for municipal large-scale treatment, the users usually agree to pay more for the product. The AWT methods to be discussed are: (1) Reverse osmosis, nanofiltration, ultrafiltration and microfiltration; (2) Electrodialysis; (3) Activated charcoal; (4) Ion exchange; (5) UV oxidation; and (6) Precipitation.

1. *Reverse Osmosis, Nanofiltration, Ultrafiltration and Microfiltration*

In *reverse osmosis*, wastewater is forced through cellulose acetate membranes at high pressure (about 500 lb/in.²). Salts and organic materials are rejected by the membrane, but water is allowed to pass. This is the reverse of the normal osmosis process, which is the natural movement of solvent from an area of low solute concentration, through a membrane, to an area of high solute concentration when no external pressure is applied. The membrane here is semi-permeable, and allows the passage of solvent but not of solute. Ordinary osmosis is the diffusion of a solvent through a selective membrane from a solution of higher solvent concentration to one of lower concentration. However, in reverse osmosis, because of the higher pressure applied, solvent will flow from the solution that is lower in concentration to that which is higher. The rejected organic macro-molecules tend to block the membrane and have to be removed. This process requires that a high pressure be applied on the high concentration side of the membrane, usually 2–17 bar (30–250 psi) for fresh and brackish water, and 40–70 bar (600–1,000 psi)

for seawater, which has around 24 bar (350 psi) natural osmotic pressure which must be overcome. This process is best known for its use in the desalination of sea water to fresh water, but it has also been used to produce water for medical and industrial applications.

Nanofiltration is very similar to reverse osmosis. The key difference is the degree of removal of monovalent ions such as chlorides. Reverse osmosis removes the monovalent ions at 98–99% level at 200 psi. Nanofiltration membranes' removal of monovalent ions is slightly less, and varies between 50% and 90% depending on the material and manufacture of the membrane (Fig. 10.21).

Both methods are applied in drinking water purification process steps, such as water softening, decoloring, and micro pollutant removal such as coloring agents. Nanofiltration is a pressure related process, during which separation takes place, based on molecule size. Membranes bring about the separation. The technique is mainly applied for the removal of organic substances, such as micro pollutants and multivalent ions. In water treatment, specifically they are used for the removal of pesticides from groundwater, the removal of heavy metals from wastewater, wastewater recycling in laundries, water softening, and nitrates removal.

In *ultrafiltration*, the membrane functions as a molecular sieve. It separates dissolved molecules on the basis of size by passing a solution through a very fine filter. The ultrafilter is a thin, selectively permeable membrane that retains most macromolecules above a certain size including colloids, microorganisms, and pyrogens (fever-causing compounds, mainly derived from the cell walls of Gram-negative bacteria). Smaller molecules, such as solvents and ionized contaminants, are allowed to pass into the filtrate. Thus the ultrafilter retains a fraction that is rich in large molecules and a filtrate that contains few, if any, of these molecules.

Ultrafiltration removes most particles including pyrogens, microorganisms, and colloids above their rated size, and produces the highest quality water for the least amount of energy. It is also regenerable. However, it will not remove dissolved inorganics.

Microfiltration will not remove dissolved inorganics, chemicals, pyrogens or all colloidals; it is not regenerable and it is expensive. However, it requires minimal maintenance and removes all

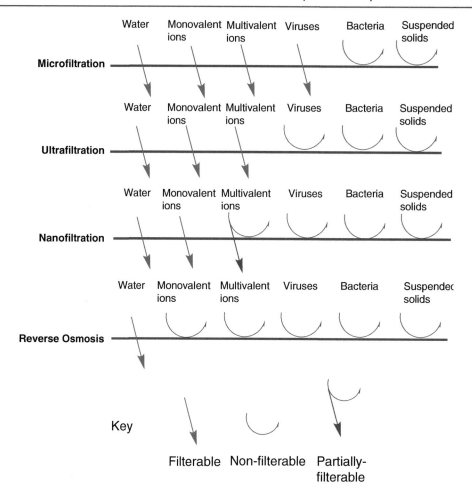

Fig. 10.21   Comparison of the capabilities of different filtration methods (Modified from Pinnau 2008)

particles and microorganisms greater than the pore size (Anonymous (2007a).

2. *Electrodialysis*

Electrodialysis is a membrane process, during which ions are transported through semi permeable membrane, under the influence of an electric potential (Anonymous 2005a). The membranes are arranged in a stack and are cation- or anion-selective, and thus either positive ions or negative ions will flow through. Cation-selective membranes are polyelectrolytes with negatively charged matter, which rejects negatively charged ions and allows positively charged ions to flow through. By placing multiple membranes in a row, which alternately allow positively or negatively charged ions to flow through, the ions can be removed from wastewater. The membranes are separated from each other in the stack by non-conductive spacers (see Fig. 10.22).

In some columns, concentration of ions will take place and in other columns ions will be removed. The concentrated saltwater flow is circulated until it has reached a value that enables precipitation. At this point, the flow is discharged. This technique can be applied to remove ions from water. Particles that do not carry an electrical charge are not removed. Cation-selective membranes consist of sulfonated polystyrene, while anion-selective membranes consist of polystyrene with quaternary ammonia.

Sometimes pre-treatment is necessary before the electro dialysis can take place. Suspended solids with a diameter that exceeds 10 μm need to be removed, or else they will plug the membrane pores. There are also substances that are able to neutralize a membrane, such as large organic anions, colloids, iron oxides, and manganese oxide.

Fig. 10.22 Diagram illustrating the principle of electrodialysis (Modified from (Anonymous 2005a)

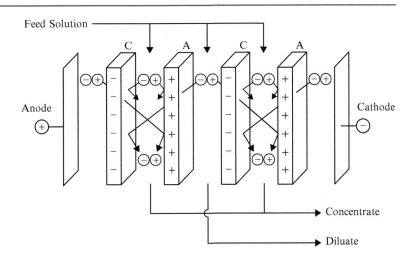

These disturb the selective effect of the membrane. Pre-treatment methods, which aid the prevention of these effects, are active carbon filtration (for organic matter), flocculation (for colloids), and filtration techniques.

Electrodialysis is cost competitive compared to reverse osmosis when producing drinkable water, but it has a number of shortcomings for the production of water of higher purity than drinking, say for laboratory or medical work. Electrodialysis does not remove organics, pyrogens (materials which give fever in injection water, e.g., walls of Gram negative bacteria) and metallic elements which have weak or nonexistent surface charges because they are attached to the membranes. Some colloids and detergents, can plug the membranes' pores reducing their ionic transport ability, and requiring frequent cleaning. Furthermore, electrodialysis releases hydrogen gas which is potentially explosive. It is relatively expensive to build because it uses expensive materials such as stainless steel and platinum, and its operation is equally expensive because of the high amount of electricity it consumes. Finally, the system requires a skilled operator and routine maintenance. Electrodialysis is used to desalt sea water, brackish ground water and water from estuaries, or river mouths.

3. *Activated Charcoal*
Activated charcoal or activated carbon, is a material with an exceptionally high surface area. Just 1 g of activated carbon has a surface area of approximately 500 m^2, determined by nitrogen gas adsorption. It has a wide range of applications including gas purification, metal extraction, water purification, medicine, sewage treatment, air filters in gas masks and filter masks, filters in compressed air, and many other applications.

It is used in homes to remove taste and odor from drinking water, as well as organic compounds such as volatile organic compounds, pesticides, benzene, some metals, chlorine, and radon. It must be changed regularly; bacteria may grow in it using the organic materials accumulated on the filter.

4. *Ion exchange*
Ion exchange resins are insoluble matrix or support structure normally in the form of small (1–2 mm diameter) beads, colored white or yellowish, made from an organic polymer material. A matrix of pores on the surface of the beads easily trap and release ions. The trapping of ions takes place only with simultaneous releasing of other ions, and hence the process is called ion exchange. Different types of ion exchange resin are fabricated to selectively prefer one or several different types of ions.

Ion exchange resins are widely used in different separation, purification, and decontamination processes. Most common, they are used for water softening and water purification. Before the introduction of resins, zeolites, which are natural or artificial alumino-silicate minerals, which can accommodate a wide variety of cations, such as Na$^+$, K$^+$, Ca^{2+}, Mg^{2+}, were used.

For water softening, ion-exchange resins are used to replace the magnesium and calcium ions found in hard water with sodium ions. When in contact with a solution containing magnesium and

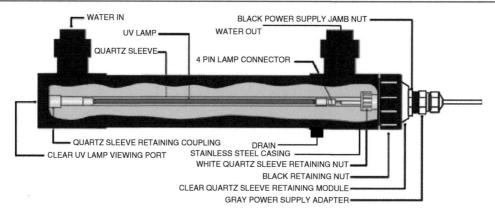

Fig. 10.23 Schematic diagram of a generalized apparatus for sterilizing water with UV light (Modified from http://www.oceanreeflections.com/uv_school.htm; Anonymous 2010b)

calcium ions, but a low concentration of sodium ions, the magnesium and calcium ions preferentially migrate out of solution to the active sites on the resin, being replaced in solution by sodium ions. This process reaches equilibrium with a much lower concentration of magnesium and calcium ions in solution than was started with.

The resin can be recharged by washing it with a solution containing a high concentration of sodium ions (e.g., it has large amounts of common salt (NaCl) dissolved in it. The calcium and magnesium ions migrate off the resin, being replaced by sodium ions from the solution until a new equilibrium is reached. This is the method of operation used in dishwashers that require the use of "dishwasher salt." The salt is used to recharge an ion exchange resin which itself is used to soften the water so that limescale deposits are not left on the cooking and eating utensils being washed.

For water purification, ion exchange resins are used to remove undesirable constituents e.g., copper and lead ions from solution, replacing them with more innocuous ions, such as sodium and potassium.

5. *UV Oxidation*

Ultraviolet (UV) oxidation is an important purification technology used in the production of high-purity water for the chemical, food and beverage, pharmaceutical and semiconductor industries and also for laboratory work. When combined with other purification technologies in a complete water system, UV oxidation provides unique benefits in the reduction of dissolved organics and microor-

ganisms. In order to oxidize organics and reduce TOC in purified water, a UV lamp which emits both UV185 and UV254 must be used. UV185 not only breaks organic bonds, but also generates free radicals, which are short-lived, highly reactive molecules or atoms which can rapidly oxidize many organic and some inorganic molecules, including one of the most powerful oxidizing species, the hydroxyl radical (OH•). The OH• created may freely react with organic molecules to partially ionize or fully oxidize them to CO_2 and water.

The hardware used to generate UV radiation in a water purification system includes a low-pressure, mercury vapor lamp, a ballast, and a power supply (see Fig. 10.23). The lamp consists of a sealed quartz tube with electrodes (cathodes) on each end. The lamp tube contains a small amount of mercury and an inert gas, such as argon or neon, at a very low pressure. The power supply energizes the ballast which regulates the current to the lamp. Electrical current from the ballast pre-heats the lamp cathodes.

6. *Precipitation*

The following treatment methods have been approved by the USEPA for removing nitrates/nitrites: Ion exchange, Reverse Osmosis, Electrodialysis. However precipitation is the main method for the removal of phosphorous in effluents of wastewater treatment.

Mineral addition and lime addition are the principal methods for in-plant removal of phosphorus from wastewater. The most commonly used of these metal salts are: Alum, a hydrated aluminum sulfate $(Al_2SO_3).18\ H_2O$); sodium aluminate $(Na_2O.Al_2O_3)$; ferric sulfate $(Fe_2(SO_4)_3)$; ferrous sulfate $(FeSO_4)$; ferric chloride $(FeCl_3)$; and ferrous chloride $(FeCl_2)$. Mineral addition is usually followed by anionic polymer addition, which aids flocculation; the pH may require adjustment depending on the particular process. In lime addition, phosphorus removal is attained through the chemical precipitation of hydroxyapatite, $Ca_5OH\ (PO_4)_3$.

References*

Andrew, W. (1996). *Biotechnology for waste and wastewater treatment*. Westwood: Noyes.

Anonymous. (2004). *Local limits development guidance* (EPA 833-R-04-002A). Washington, DC: Office of Wastewater Management, USEPA.

Anonymous. (2005a). Electrodialysis: A method to deionize water and to recover the salt. http://www.pca-gmbh.com/appli/ed.htm. Accessed 25 July 2010.

Anonymous. (2005b). *A homeowner's guide to septic systems* (EPA-832-B-02-005). Washington, DC: United States Environmental Protection Agency.

Anonymous. (2006). *Standard methods for the examination of water and wastewater*. Washington, DC: American Public Health Association, American Water Works Association, Water Environment Federation.

Anonymous. (2007). Different water filtration methods explained. http://www.freedrinkingwater.com/water-education/quality-water-filtration-method.htm. Accessed 28 Aug 2007.

Anonymous. (2010a). Underground Injection Control (UIC). http://www.epa.gov/region9/water/groundwater/uic-hicesspools.html. Accessed 10 Sept 2010.

Anonymous. (2010b). More UV sterilizer info. http://www.oceanreeflections.com/uv_school.htm. Accessed 12 Sept 2010.

Curds, C. (1965). An Ecological Study of the Ciliated Protozoa in Activated Sludge. *Oikos, 15*, 282–289.

Eckenfelder, W. W. (2000). *Industrial water pollution control*. Boston: McGraw-Hill.

El-Rehaili, A. M. (1994). Implications of activated sludge kinetics based on total or soluble BOD, COD and TOC. *Environmental Technology, 15*, 1161–1172.

Gerhardt, M.B., & Oswald, W.J. (1990). Advanced ingrated ponding system in sewage reuse. In P. Edwards, & R.S.V. Pullin (Eds.), *Wastewater-fed aquaculture. Proceedings of the international seminar on wastewater reclamation and re-use for aquaculture, Calcutta, India, 6–9 December, 1988*. Bangkok, Thailand: Environmental Sanitation Information Centre, Asian Institute of Technology.

Kosric, N., & Blaszczyk, R. (1992). Industrial effluent processing. In *Encyclopedia of microbiology* (Vol. 2, pp. 473–491). San Diego: Academic.

Lindera, K. C. (2002). Activated sludge – the process. In *Encyclopedia of environmental microbiology* (Vol. 1, pp. 74–81). New York: Wiley-Interscience.

Nielsen, P. H. (2002). Activated sludge – the floc. In *Encyclopedia of environmental microbiology* (Vol. 1, pp. 54–61). New York: Wiley-Interscience.

Okafor, N. (2007). *Modern industrial microbiology and biotechnology*. Enfield: Science Publishers.

Pinnau, I. (2008). Membranes for water treatment: Properties and characterization. http://www.stanford.edu/group/ees/rows/presentations/Pinnau.pdf. Accessed 21 Aug 2010.

Young, J. C., & Baumann, E. R. (1976). The electrolytic respirometer – I factors affecting oxygen uptake measurements. *Water Research, 10*, 1031–1040.

* References and further reading may be available for this article. To view references and further reading, you must purchase this article

The Disposal of Municipal Solid Wastes

11

Abstract

The publication of Rachel Carson's *Silent Spring* (1962) stimulated eventual worldwide interest in the environment, leading to the founding of the United Nations Environmental Programme (UNEP) and the inauguration of national environmental ministries, agencies or laws world-wide.

Solid wastes management involves waste reduction, reuse and recycling, composting, incineration with or without energy recovery, and landfilling.

Modern incinerators scrub the flue gases of incineration to reduce the harmful components. Newer methods of treating wastes include plasma arc gasification, in which waste is treated at very high temperatures and pressures, melting the waste into a nontoxic dross and yielding a fuel, syngas, for generating electricity. Pyrolysis operates at about 430°C. Supercritical water oxidation (SCWO) is the destruction technology for organic compounds and toxic materials at very high temperature and pressure, converting them to carbon dioxide, hydrogen to water, and chlorine atoms to chloride ion.

Keywords

Municipal solid wastes • Environment and development • Waste management • Incineration • Landfills • Plasma gasification • Composts • Energy from waste • Recycling • Reuse • Supercritical water oxidation • Pyrolysis

11.1 The Nature of Wastes in General

Wastes are unwanted materials or outcomes resulting from *a particular* human activity. The fact that a material or outcome is a waste in a particular activity does not render that material or outcome totally unwanted for all other activities. Indeed what is waste under one condition may be the corner stone of activity in another. Thus molasses is a waste material in sugar manufacture, but it is a major input in the manufacture of many fermentation products such as antibiotics. Similarly, farm trash such as corn cobs, wastes from corn growing, may be the basis for improving soil qualities through composting.

Wastes differ in nature and weight according to the activities which generate them. Thus wastes by weight were highest in construction and manufacturing, followed by mining and by municipal activities, agriculture in the European Union (EU) in 2002. (Anonymous 2005).

The nature of the wastes generated in any society or country clearly relates to the economic activity of that society. Thus, although wastes from construction and manufacturing are expected to be heavier by their

N. Okafor, *Environmental Microbiology of Aquatic and Waste Systems*,
DOI 10.1007/978-94-007-1460-1_11, © Springer Science+Business Media B.V. 2011

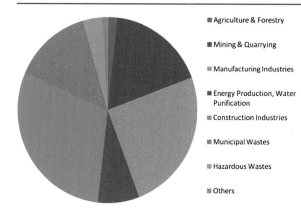

■ Agriculture & Forestry

■ Mining & Quarrying

■ Manufacturing Industries

■ Energy Production, Water Purification

■ Construction Industries

■ Municipal Wastes

■ Hazardous Wastes

■ Others

Fig. 11.1 Wastes generated in the European Union (EU), by activity, in 2002 (Pie chart generated from data in Anonymous 2005)

nature than those from agriculture, Fig. 11.1 can be interpreted to mean that more activities go on in construction and industry than in agriculture; since agricultural wastes are at the bottom of the chart, while manufacturing and construction are at the top. It can also be interpreted to mean that Western Europe is more of a manufacturing region and that its agriculture is very efficient, yielding little wastes, since, comparatively, so little agricultural waste is produced.

Municipal Solid Waste (MSW) is waste collected by, or on behalf of, a local authority. It comprises mostly household waste, but it may include some commercial and industrial wastes. European Union legislation require the pretreatment, including presorting, of waste before it is sent to landfill (Anonymous 2009a).

11.2   The World-Wide Development of Interest in the Environment

Although the environment has influenced human affairs from as early as the seventh century AD (see Table 11.1), the earliest country in modern times to develop a concerted government (please also see Section 7.3.2 on oil spills) interest on the environment is the USA in the 1970s. This followed the public interest generated by *Silent Spring*, a book by Rachel Carson (1962), which documented the detrimental effects of pesticides on the environment, particularly on birds; the book stated that the insecticide DDT had been the cause of thinner egg shells, resulting in reproductive problems for the birds. It also criticized the chemical industry for spreading disinformation and

government officials for uncritically accepting industry claims. *Silent Spring*, published in 1962, led to widespread public concerns on pesticide use and the pollution of the environment.

The immense public interest in these concerns regarding environmental pollution eventually led to the creation of the US Environmental Protection Agency (US EPA) on December 2, 1970 to consolidate in one agency a variety of federal research, monitoring, standard-setting, and enforcement activities to ensure environmental protection. From regulating auto emissions to banning the use of DDT; from cleaning up toxic waste to protecting the ozone layer; from increasing recycling to revitalizing inner-city brown fields, the US EPA's achievements have resulted in cleaner air, purer water, and better protected land (Anonymous 2010a).

The EPA submitted the Agency's *2006–2011 Strategic Plan* to Congress on September 29, 2006 as required under the Government Performance and Results Act (GPRA) of 1993. This revised *Strategic Plan* maintains the five goals that were described in the *2003–2008 Strategic Plan*, but reflects a sharpened focus on achieving more measurable environmental results.

The five goals are Clean Air and Global Climate Change, Clean and Safe Water, Land Preservation and Restoration, Healthy Communities and Ecosystems, and Compliance and Environmental Stewardship. The *Strategic Plan* serves as the Agency's road map and guides it in establishing the annual goals that need to be met along the way. It helps provide a basis from which the organization can focus on the highest priority environmental issues and ensure that taxpayer dollars are used effectively. The EPA produces numerous data on environmental affairs in the US.

11.2.1 The Stockholm Conference, 1972: Beginning of World-Wide Interest in the Protection of the Environment

Today, almost all (if not all) countries of the world have legislation, a Department, a Ministry, an Agency or such-like body whose function is the protection of the environment. This is as a result of the Conference on the Human Environment, held in Stockholm, Sweden, June 5–16, 1972 (also called the Stockholm Conference), the first of a series of world environmental conferences, to be held on a 10-yearly basis.

Table 11.1   Timeline of history of environmentalism (Modified from www.public.iastate.edu/~sws/enviro%20and%20society and http://en.wikipedia.org/wiki/Timeline_of_history_of_environmentalism, Anonymous 2010a, b). A listing of events that have shaped humanity's perspective on the environment, human induced disasters, environmentalists that have had a positive influence, and environmental legislation

Pre-twentieth century	
Seventh century: AD 676	Cuthbert of Lindisfarne enacts protection legislation for birds on the Farne islands (Northumberland, UK)
Fourteenth century: AD 1366	City of Paris forces butchers to dispose of animal wastes outside the city (Ponting); similar laws would be disputed in Philadelphia and New York nearly 400 years later
Fourteenth century: AD 1388	British Parliament passes an act forbidding the throwing of filth and garbage into ditches, rivers and waters. City of Cambridge also passes the first urban sanitary laws in England
Fifteenth century: AD 1420–1427	Madeira islands : destruction of the laurisilva forest, by fire that burnt for 7 years.
Seventeenth century: AD 1609	Hugo Grotius publishes *Mare Liberum* (The Free sea) with arguments for the new principle that the sea was international territory and all nations were free to use it for seafaring trade. The ensuing debate had the British empire and France claim sovereignty over territorial waters to the distance within which cannon range could effectively protect it, the three mile (5 km) limit
Seventeenth century: AD 1690	US Colonial Governor, William Penn requires Pennsylvania settlers to preserve 1-acre (4,000 m²) of trees for every 5 acres cleared
Eighteenth century: AD 1720	In India, hundreds of Bishnois Hindus of Khejadali go to their deaths trying to protect trees from the Maharaja of Jodhpur, who needed wood to fuel the lime kilns for cement to build his palace
Eighteenth century: AD 1739	Benjamin Franklin and neighbors petition Pennsylvania Assembly to stop waste dumping and remove tanneries from Philadelphia's commercial district
Eighteenth century: AD 1748	Jared Eliot, clergyman and physician, writes *Essays on Field Husbandry in New England* promoting soil conservation
Eighteenth century: AD 1762–1769	Philadelphia committee led by Benjamin Franklin attempts to regulate waste disposal and water pollution
Nineteenth century: AD 1820	World human population reached one billion
Nineteenth century: AD 1828	Carl Sprengel formulates the Law of the Minimum stating that economic growth is limited not by the total of resources available, but by the scarcest resource
Nineteenth century: AD 1845	First use of the term "carrying capacity" in a report by the US Secretary of State to the Senate
Nineteenth century: AD 1860	*Henry David Thoreau* delivers an address to the Middlesex (Massachusetts) Agricultural Society, entitled "The Succession of Forest Trees," in which he analyzes aspects of what later came to be understood as forest ecology and urges farmers to plant trees in natural patterns of succession
Nineteenth century: AD 1862	John Ruskin publishes *Unto This Last*, which contains a proto-environmental indictment of the effects of unrestricted industrial expansion on both human beings and the natural world. The book influences Mahatma Gandhi, William Morris and Patrick Geddes
Nineteenth century: AD 1866	The term Ecology is coined (in German as Oekologie by Ernst Heinrich Philipp August Haeckel (*1834–1919*) in his Generelle Morphologie der Organismen. Haeckel was an anatomist, zoologist, and field naturalist appointed professor of zoology at the Zoological Institute, Jena, in 1865
Nineteenth century: AD 1872	The term acid rain is coined by Robert Angus Smith in the book *Air and Rain*
Nineteenth century: AD 1872	German graduate student Othmar Zeidler first synthesizes DDT, later to be used as an insecticide
Nineteenth century: AD 1876	British River Pollution Control Act makes it illegal to dump sewage into a stream
Nineteenth century: AD 1895	Sewage cleanup in London means the return of some fish species (grilse, whitebait, flounder, eel, smelt) to the River Thames
Twentieth century	
1900s	
AD 1902	George Washington Carver writes *How to Build Up Worn Out Soils*
	7300 ha of land in the Lake District of the Andes foothills in Patagonia are donated by Francisco Moreno as the first park, Nahuel Huapi National Park, in what eventually becomes the National Park System of Argentina

(continued)

Table 11.1   (continued)

AD 1905	The term smog is coined by Henry Antoine Des Voeux in a London meeting to express concern over air pollution
	San Francisco earthquake and subsequent fires destroy much of the city
AD 1909	US President Theodore Roosevelt convenes the North American Conservation Conference, held in Washington, D.C. and attended by representatives of Canada, Newfoundland, Mexico, and the United States
1910s	
AD 1910	Scientific American reports alcohol-gasoline anti-knock blend is "universally" expected to be the fuel of the future
	Spanish Flu kills between 50 and 100 million people worldwide
1920s	
1921	Thomas Midgley discovers lead components to be an efficient antiknock agent in gasoline engines. In spite of the well known toxic effects, lead was in ubiquitous use. First banned from use in Japan 1986
1928	Thomas Midgley develops chlorofluorocarbons (CFC's) as a nontoxic refrigerant. The first warnings of damage to stratospheric ozone were published by Molina and Rowland 1974. They shared the 1995 Nobel Prize for Chemistry for their work
1929	The Swann Chemical Company develops polychlorinated biphenyls (PCBs) for transformer coolant use. Research in the 1960s revealed PCBs to be potent carcinogens. Banned from production in the US 1976, probably one million tonnes of PCBs were manufactured in total globally
1930s	
1930	World human population reached two billion
1933	First legislation on Animal rights adopted, Germany
1939	The insecticidal properties of DDT discovered by Paul Hermann Müller, who was awarded the 1948 Nobel Prize in Physiology and Medicine for his efforts. The first ban on its use came in 1970
1940s	
1949	First known dioxin exposure incident, in a Nitro, West Virginia herbicide production plant. Extensively used during the Vietnam War 1961–1971 as Agent Orange. Production ban in the US on some component from 1970
1950s	
1951	World Meteorological Organization (WMO) established by the United Nations
1954	The first nuclear power plant to generate electricity for a power grid started operations at Obninsk, Soviet union on 27 June
1956	Minamata disease, a neurological syndrome caused by severe mercury poisoning
1957	The first substantial nuclear accident happened on 10 October 1957 in Windscale, England
1960s	
1960	World human population reached three billion
1961	World Wildlife Fund (WWF) registered as a charitable trust in Morges, Switzerland, an international organization for the conservation, research and restoration of the natural environment
1962	Rachel Carson publishes *Silent Spring*
1969	National Environmental Policy Act including the first requirements on Environmental impact assessment
1970s	
1970	Earth Day – April 22, millions of people gather in the United States for the first Earth day organized by Gaylord Nelson, former senator of Wisconsin, and Denis Hayes, Harvard graduate student
	US Environmental Protection Agency (USEPA) established

(continued)

Table 11.1   (continued)

1971	The international environmental organization Greenpeace founded in Vancouver, Canada. Greenpeace has later developed national and regional offices in 41 countries worldwide
	Friends of the earth founded
	International Institute for Environment and Development established in London, UK. One offshoot is the World Resources Institute with its biannual report *World resources* since 1984
1972	The conference on the human environment, held in Stockholm, Sweden 5–16 June, the first of a series of world environmental conferences, to be held on a 10-yearly basis
	United Nations Environment Programme founded as a result of the Stockholm conference
	The Club of Rome publishes its report Limits to Growth, which has sold 30 million copies in more than 30 translations, making it the best selling environmental book in world history
	In the US Clean Water Act
	First photograph of the whole illuminated Earth taken from space, Apollo 17, resulting in the famous "Blue Marble" photograph, said to have been at least partly responsible for launching the modern environmental movement
1973	OPEC announces oil embargo against United States
	World Conservation Union (IUCN) meeting drafts the Convention on International Trade in Endangered Species of Wild Fauna and Flora (CITES)
1974	Chlorofluorocarbons are first hypothesized to cause ozone thinning.
	World human population reached four billion
1976	Dioxin accidental release in Seveso, Italy on 10 July, killing animals and traumatizing the population
1977	Surface Mining Control and Reclamation Act (US)
1978	Brominated flame-retardants replaces PCBs as the major chemical flame retardant. Swedish scientists noticed these substances to be accumulating in human breast milk 1998. First ban on use in the EU 2004
1979	Three Mile Island, worst nuclear power accident in US history
1980s	
1980	Superfund (Comprehensive Environmental Response, Compensation, and Liability Act or CERCLA)
1982	United Nations Convention on the Law of the Sea (UNCLOS) is signed on December the 10th at Montego Bay. Part XII of which significantly developed port-state control of pollution from ships
1983	Dr. Gro Harlem Brundtland, the first woman prime minister of Norway, as chairperson of UN World Commission on Environment and Development; UNCED coined the 'sustainable development'.
1984	Bhopal disaster in the Indian state of Madhya Pradesh (Methyl isocyanate leakage
1986	Chernobyl, world's worst nuclear power accident occurs at a plant in Ukraine.
1987	World human population reached five billion
	The report of the Brundtland Commission, our common future on sustainable development, is published
1988	Ocean Dumping Ban Act (US)
	Intergovernmental Panel on Climate Change (IPCC) was established by two United Nations organizations, the World Meteorological Organization (WMO) and the United Nations Environment Programme (UNEP) to assess the "risk of human-induced climate change"
1989	Exxon Valdez creates largest oil spill in US history (see also Table 7.11)
	Montreal Protocol on substances that deplete the ozone layer entered into force on January 1. Since then, it has undergone five revisions, in 1990 (London)
1990s	
1990	European Environment Agency was established by EEC Regulation 1210/1990 and became operational in 1994. It is headquartered in Copenhagen, Denmark
1991	World's worst oil spill (820,000 tones) occurs January 23, 1991 in the Persian Gulf (Kuwait) during war with Iraq

(continued)

Table 11.1 (continued)

1992	The Earth Summit, held in Rio de Janeiro from June 3 to June 14, was unprecedented for a United Nations conference, in terms of both its size and the scope of its concerns; Global warming major issue in Rio
	United Nations Framework Convention on Climate Change opened for signature on 9 May ahead of the Earth Summit in Rio de Janeiro
	The international convention on biological diversity opened for signature on 5 June in connection with the Earth Summit in Rio de Janeiro
	World Ocean Day began on 8 June at the Earth summit in Rio de Janeiro
	The metaphor ecological footprint is coined by William Rees
1993	The Great Flood of 1993 was one of the most destructive floods in United States history involving the Missouri and Mississippi river valleys
1994	United Nations Convention to Combat Desertification
	The first genetically modified food crop released to the market. It remains a strongly controversial environmental issue.
1996	Western Shield, a wildlife conservation project is started in Western Australia, and through successful work has taken several species off of the state, national, and international (IUCN) Endangered species lists
1997	Kyoto Protocol was negotiated in Kyoto, Japan in December. It is actually an amendment to the United Nations Framework Convention on Climate Change (UNFCCC). Countries that ratify this protocol commit to reduce their emissions of carbon dioxide and five other greenhouse gases
1999	World human population reached six billion
Twenty-first century	
2001	USA rejects the Kyoto Protocol
2002	Earth Summit, held in Johannesburg a United Nations conference; Focused for first time on problems of developing countries; a commitment to halve the number of people in the world who lack basic sanitation by 2015
2003	The world's largest reservoir, the Three Gorges Dam begins filling 1 June
	European Heat Wave resulting in the premature deaths of at least 35,000 people
2004	Earthquake causes large tsunamis in the Indian Ocean, killing nearly a quarter of a million people
	The Kyoto Protocol came into force on February 16 following ratification by Russia on November 18, 2004
2005	Hurricanes Katrina, Rita, and Wilma cause widespread destruction and environmental harm to coastal communities in the US Gulf Coast region, including numerous oil spills
2006	Former U.S. vice president Al Gore releases *An Inconvenient Truth*, a documentary that describes global warming. The next year, Gore is awarded the Nobel Peace Prize (jointly with the Intergovernmental Panel for Climate Change) for this and related effort
	World human population reached 6.5 billion
2007	The IPCC release the IPCC Fourth Assessment Report.
2009	The Energy Action Coalition hosted the second national youth climate conference to be held at the Washington Convention Center from February 27 to March 2, 2009. The conference aims to attract more than 10,000 students and young people and will include a Lobby Day
2010	The Gulf of Mexico oil spill (also known as Deepwater Horizon oil spill and the BP oil spill) resulted from a sea-floor oil gusher and flowed for 3 months before being capped on July 15, 2010, but not before 4.9 million barrels or 205.8 million gallons of crude oil had gushed from it. With 627,000 tonnes of crude oil, this would appear to be the largest spill in history, second only to the Lakeview Gusher, California, USA, March 1910 – September, 1911 which spilled 1,230,000 tonnes

This first major environmental (1972) conference of the United Nations was a watershed in the development of international environmental politics and was held at the initiative of the government of Sweden. Attended by the representatives of 113 countries, 19 inter-governmental agencies, and more than 400 inter-governmental and nongovernmental organizations, it is widely recognized as the beginning of modern political and public awareness of global environmental problems. It has been suggested that at least three achievements are the lasting legacy of the Stockholm United Nations Conference on the Environment and Development (UNCED): The creation of the UN Environment Programme (UNEP), the call for cooperation to reduce marine pollution, and the establishment of a global monitoring network have been cited as of especially lasting significance. The third aspect, the establishment of a global monitoring network can be seen, notwithstanding other more specific resolutions, to be the impetus for the world-wide establishment of environmental protection agencies by national governments around the world (Anonymous 2010c). The initiative with the EPA in the US, a well-developed economy, has been discussed above. To illustrate this world-wide interest in the environment at governmental level, brief discussions of the development of governmental interest in environmental protection in two economically developed areas of the world, namely the European Union and Japan, and two developing countries, Ghana and Egypt will be undertaken.

11.2.2 Environmental Regulation in the European Union

The 32-member European Union has as its environment-regulating body, the European Environment Agency (EEA), which is headquartered in Copenhagen, Denmark. The regulation establishing the EEA was adopted by the European Union in 1990 and it came into force in late 1993 with the following as its functions:

- To help the [EU] Community and its member countries make informed decisions about improving the environment, integrating environmental considerations into economic policies and moving toward sustainability
- To coordinate the European environment information and observation network (Anonymous 2010d)

11.2.3 Environmental Regulations in Japan

In Japan, with a population of about 130 million, environmental pollution has accompanied industrialization since the Meiji period (1868–1912). One of the earliest cases was the copper poisoning caused by drainage beginning as early as 1878. During the 20 years since the establishment of the Environment Agency in 1971, the environmental situation at the national and global levels has undergone substantial changes. At the national level, notable achievements have been made in combating severe pollution during the period of high economic growth.

Current Japanese environmental policy and regulations were the consequence of a number of environmental disasters in the 1950s and 1960s. One of the most famous was the *Minamata disease* episode in which there were many casualties from eating fish which had been contaminated by methyl mercury.

In the 1990s, Japan's environmental legislation was further tightened and in 1993, the government reorganized the environment law system and passed the *Basic Environment Law* and related laws. The law includes restriction of industrial emissions, restriction of products, restriction of wastes, improvement of energy conservation, promotion of recycling, restriction of land utilization, arrangement of environmental pollution control programs, relief of victims, and provision for sanctions. The Environment Agency was elevated to a full-fledged Ministry of the Environment in 2001.

Japan has of recent taken a much more proactive approach to waste management. As a signatory of the *Kyoto Protocol*, and host of the 1997 conference which created it, Japan is under treaty obligations to reduce its carbon dioxide emissions level by 6% less than the level in 1990, and to take other steps related to curbing climate change (Anonymous 2010e).

11.2.4 Governmental Regulation of the Environment in Ghana

In Ghana, which is in west Africa with a population of about 24 million, the central governmental environmental body is the Ghana Environmental Protection Agency (EPA). Its history began at a time of growing world-wide concern on the dangers posed to the environment through human activities and which prompted the United Nations to convene the conference in Stockholm on the Human Environment, in June 1972. An outcome of this meeting was the establishment of the United Nations Environmental Program (UNEP). The decision of the Ghana Government to establish an Environmental Protection Council was a direct outcome of recommendations of the Stockholm Conference. Prior to that, the National Committee on the Human Environment was formed by the Ministry of Foreign Affairs in 1971 as a result of concern expressed by the Economic Commission for Africa and the Organization of African Unity regarding the need to conserve and protect Africa's natural resources.

On 23rd May 1973, the government of Ghana announced its decision to establish an Environmental Protection Commission; the government soon renamed it the Environmental Protection Council. In 1994, it was transformed into an Agency by the Environmental Protection Agency Act, 1994 (Act 490). The EPA became a corporate body with powers to sue and be sued. It was also given the responsibility of regulating the environment and ensuring the implementation of government policies on the environment.

The Ghana EPA's objectives are to:
- Create awareness to mainstream environment into the development process at the national, regional, district, and community levels;
- Ensure that the implementation of environmental policy and planning are integrated and consistent with the country's desire for effective, long-term maintenance of environmental quality;
- Ensure environmentally sound and efficient use of both renewable and nonrenewable resources in the process of national development;
- Guide development to prevent, reduce, and as far as possible, eliminate pollution and actions that lower the quality of life;
- To apply the legal processes in a fair, equitable manner to ensure responsible environmental behavior in the country;
- Continuously improve EPA's performance to meet changing environmental trends and community aspirations;
- Encourage and reward a commitment by all EPA staff to a culture based on continuous improvement and on working in partnership with all members of the Ghanaian community (Anonymous 2010f).

11.2.5 Egypt's Environmental Affairs Agency

The other developing country, Egypt, is an Arab country in the Mediterranean region of North Africa with a population of about 80 million. Egypt had participated in the Conference on the Human Environment, Stockholm, 1972 and had indeed proposed with the overwhelming support of participating countries including the USA, Sweden, Germany, and many others for a Second United Nations Conference on the Human Environment within 5 years in Egypt. (Anonymous 2010g). Although this was not implemented, it showed that Egypt was fully in the spirit of the conference.

Environmental matters are handled in Egypt by the Egyptian Environmental Affairs Agency (EEAA) officially launched in 1982 by a Presidential Decree No. 631 of the year 1982. Its existence was formalized in 1994 by a decree issued by the Agency's Executive Head.

In June 1997, a Ministry of State for Environmental Affairs (MSEA) was created. The MSEA has focused, in close collaboration with the national and international development partners, on defining environmental policies, setting priorities and implementing initiatives within a context of sustainable development. MSEA and EEAA are the highest authority in Egypt responsible for promoting and protecting the environment, and coordinating adequate responses to these issues.

The functions of the EEAA are to:
(a) Prepare the draft laws concerning the Environment
(b) Implement the experimental projects
(c) Prepare the Environmental Training and Planning Policy
(d) Draft the necessary norms and standards to ensure that the environment is not polluted
(e) Formulate the basis and procedures for the assessment of environmental impacts of projects
(f) Supervise the Environmental Protection and Development Fund (Anonymous 2010g).

11.3 Nature of Municipal Solid Wastes

Municipal Sold Wastes (MSW), also known as trash or garbage, include everyday items such as product packaging, grass clippings, furniture, clothing, bottles, food scraps, newspapers, appliances, paint, and batteries.

Items such as construction and demolition debris, municipal wastewater treatment sludges, and nonhazardous industrial wastes are not generally included in those described as MSW.

According to the US EPA, in 2005, United States residents, businesses, and institutions produced more than 245 million tons of MSW, which is approximately 4.5 lb of waste per person per day. About one third of MSW in that year was made up of paper and paper board (34%), followed by yard trimmings (13%), and food scraps and plastics (about 12% each). (Fig. 11.1). (Anonymous 2004, 2005).

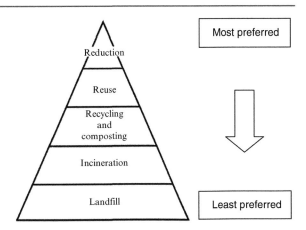

Fig. 11.2 Diagram illustrating alternative methods of treating wastes in an integrated waste management system

11.3.1 Integrated Solid Waste Management

Integrated waste management involves a series of activities dealing with solid wastes including storage, sorting, collection, transportation, and alternative treatment procedures designed to minimize the adverse effect of solid wastes on the environment and/or on human health (Anonymous (2004). The alternative procedures, in the generally accepted hierarchy of preferences, beginning with the most desirable, include the following:

- Making every effort to reduce wastes in the first place
- Reusing materials
- Recycling them
- Making composts of degradable waste components
- Incinerating the materials with or without recovery of energy or burial of the materials in landfills (see Fig. 11.2)

Which of the above alternatives is adopted in terms of the actual procedure and the order of the procedure, in any society, municipality, state, country, or regional organization of countries depends on the following:

- *Economic affordability*, i.e., the costs of waste management systems are acceptable to all sectors of the community served, including householders, commerce, industry, institutions, and government;
- *Social acceptability* in which the waste management system adopted not only meets the needs of the local community, but also reflects the values and priorities of that society; and
- *Environmental effectiveness* which ensures that the overall environmental burdens of managing waste are reduced, both in terms of consumption of resources (including energy) and the reduction

of emissions to air, water, and land (Anonymous 2004).

The management of wastes includes its collection, transportation, processing, recycling or disposal, and monitoring of waste materials. Waste management is undertaken for one or more of the following reasons: To reduce the adverse effect of their accumulation on human health, avoid the physical deterioration of the environment for aesthetic considerations, or to recover resources.

Waste management may involve solid, liquid, gaseous, or radioactive substances, each of which requires a different approach and techniques. What method or approach is adopted depends on the economic capability of the country or community, within the same country whether it is urban or rural and whether the wastes relate domestic or industrial activities.

It now seems generally accepted by the US EPA, the EU's EEA, and environmental regulatory bodies around the world that in the concept of a hierarchy of waste management options, the most desirable option is to prevent waste in the first place (see Fig. 11.2) and the least desirable option is to dispose of the waste with no recovery of either materials and/or energy. Between these two extremes, there is a variety of waste treatment options that may be used as part of a waste management strategy to recover materials (e.g., furniture reuse, glass recycling, or organic waste composting) or generate energy from the wastes (e.g., through incineration, or digesting biodegradable wastes to produce usable gases). In order to reduce the uncontrollable amounts of green house gases emitted to the atmosphere, some countries and regional

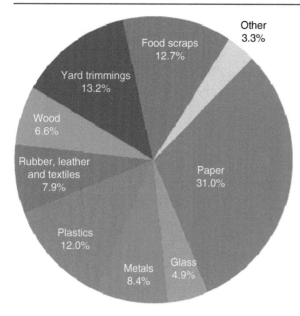

Fig. 11.3 Total municipal solid waste (by materials) generated in the US in 2008 (250) million tons) before recycling (From USEPA: http://www.epa.gov/wastes/nonhaz/municipal/pubs/msw2008rpt.pdf, Anonymous 2009a)

governmental organizations, for example, the EU, now limit the amount of biodegradable municipal waste (BMW) sent for disposal in landfill.

The generally accepted integrated waste management hierarchy includes the following four components, listed in order of preference:

1. Source reduction
2. Recycling
3. Incineration with energy recovery
4. Disposal through:
   (a) Composting
   (b) Landfilling
   (c) Incineration without energy recovery

Although EPA encourages the use of the top of the hierarchy whenever possible, all four components remain important within an integrated waste management system (Fig. 11.3).

11.3.1.1 Source Reduction

Perhaps the best slogan to describe source reduction is "Eliminate waste before it is created". Source reduction (also sometimes termed "waste prevention") deals with the design, manufacture, purchase, or use of materials or products designed to reduce the amount of waste generated; it includes reuse of second-hand products, waste elimination, repairing broken-down

items instead of buying new ones, encouraging consumers to avoid using disposable products (such as disposable cutlery), designing products to be refillable or reusable (such as cotton instead of plastic shopping bags), package reduction and substitution, and designing products that use less material to achieve the same purpose. Source reduction is at the top of the solid waste management hierarchy because it is generally superior to both recycling and disposal from an environmental and economic perspective. Source reduction is a pro-active, practical way to preempt the need to collect, process, and/or dispose of trash and recyclables by preventing their generation. Practices such as grass cycling, backyard composting, two-sided copying of paper, and transport packaging reduction by industry have yielded substantial benefits through source reduction. It has many environmental benefits: It prevents emissions of many greenhouse gases, reduces pollutants, saves energy, conserves resources, and reduces the need for new landfills and incinerators (Anonymous 2004).

11.3.1.2 Recycling

Recycling is passing a substance through a system that enables that substance to be reused; it enables the reprocessing of materials into new products. In waste recycling, waste materials are collected, sorted, separated, and those to be reused are cleaned-up.

Recycling waste means that fewer new products and consumables need to be produced, and thus raw materials are saved and energy consumption is reduced. Recycling generally prevents the waste of potentially useful materials, reduces the consumption of raw materials and reduces energy usage, and hence greenhouse gas emissions, compared to new production. Recycling, including composting, diverted 79 million tons of material away from disposal in 2005, up from 15 million tons in 1980, when the recycle rate was just 10% and 90% of MSW was being combusted with energy recovery or disposed of by landfilling. Typical materials that are recycled include batteries, recycled at a rate of 99%, paper and paperboard at 50%, and yard trimmings at 62%. These materials and others may be recycled through curbside programs, drop-off centers, buy-back programs, and deposit systems.

Recycling prevents the emission of many greenhouse gases and water pollutants, saves energy, supplies valuable raw materials to industry, creates jobs, stimulates the development of greener technologies,

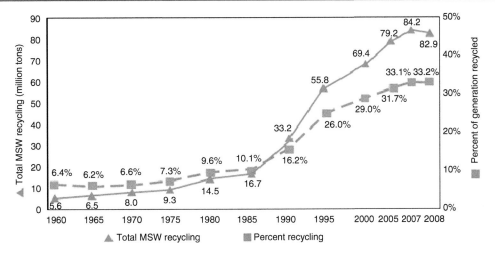

Fig. 11.4 Recycling rates in the US, 1960–2008 (From *municipal solid wastes in the US* http://www.epa.gov/wastes/nonhaz/municipal/pubs/msw2008rpt.pdf, Anonymous 2009a)

conserves resources for our children's future, and reduces the need for new landfills and combustors.

Recycling also helps reduce greenhouse gas emissions that affect global climate. In 1996, recycling of solid waste in the United States prevented the release of 33 million tons of carbon into the air, roughly the amount emitted annually by 25 million cars. The recycling rate has been rising steadily in the US and in 2005 had reached about 32% (see Fig. 11.4).

A breakdown of the rates of recycling of various items is given in Table 11.2, from where it will be seen that the most widely recovered items were nonferrous metals (69%), rubber and leather (11%), paper and paper board (55%), steel (38%), plastics (7%) and wood (10%).

Around the world, many of the waste items recycled are paper, glass jars and bottles (especially from hotels since most drinks are put in bottles), plastics and metals such as aluminum cans. There are usually three bottle banks, one for each color of glass: clear, green, and brown.

With regard to bottles in many parts of the world, they are collected in bottle banks from where they are sent for recycling.

Plastics make up a large amount of waste, since they are available in numerous forms especially for packaging. There are two main types of plastic: thermoplastics, which are the most common; and thermosets. Thermoplastics melt when heated and can therefore be remolded. This enables thermoplastics to be recycled relatively easily. Plastic waste tends to be sorted by hand, either at a materials recycling facility or the householder can separate it.

The metals recycled come from iron and steel and aluminum. Most of this waste comes from scrap vehicles, cookers, fridges, and other kitchen appliances. It is mainly made up of aluminum drinks cans and tin-plated steel food cans. Aluminum is an expensive metal and can therefore produce high incomes for recycling schemes. Copper, zinc, and lead are also recycled in the UK.

Many countries also rethread tires and reuse them. For example, every year in the UK, between 25 and 30 million scrap tyres are generated. Approximately 21% of these tyres are retreaded and reused. The old tread is ground off the tyre and replaced with a new tread. However, about half of all used tyres are dumped in landfill sites throughout the country; other tyres may be incinerated.

11.3.1.3 Incineration with Energy Recovery

Scrubbing the (flue) gas released during incineration of pollutants

The incineration of wastes is the combustion of wastes.

Modern incinerators used in economically developed countries such as in the EU and the USA are designed to avoid or minimize the short-comings of the older incinerators.

Modern incinerators generally burn MSW and the heat produced is used to boil water to produce steam which in turn drives turbines for the production of electricity. This, in the parlance of modern energy

Table 11.2 Generation and Recovery of Materials in MSW, 2008 in the US. (From http://www.epa.gov/wastes/nonhaz/municipal/pubs/msw2008rpt.pdf, Anonymous, 2009)

Material	Weight generated	Weight recovered	Recovery as %age of generation
Paper and glass			
Paper and board	77.42	42.94	55.50%
Glass	12.15	2.81	23.10%
Metals			
Steel	15.68	5.29	33.70%
Aluminum	3.41	0.72	21.10%
Other nonferrous metals	1.76	1.21	68.80%
Total metals	**20.85**	**7.22**	**34.60%**
Plastics	30.05	2.12	7.10%
Rubber and leather	7.41	1.06	14.30%
Textiles	12.31	1.89	15.30%
Wood	16.39	1.58	9.60%
Other materials	4.50	1.15	25.60%
Total materials in products	**181.14**	**60.77**	**33.50%**
Other wastes			
Food	31.79	0.80	2.50%
Yard trimmings	32.90	21.30	64.70%
Miscellaneous inorganic wastes	3.78	Negligible	Negligible
Total other wastes	**68.47**	**22.10**	**32.30%**
Total municipal solid wastes	**249.61**	**82.67**	**33.20%**

generation is Waste-to-Energy (WtE) or Energy-from-Waste (EfW) activity.

When MSW and other materials are ordinarily burnt, the composition of combustion gas flue gas (i.e., the gas coming off the chimney or pipe leading the gases into the atmosphere) will depend on the composition of the material being burnt. In general, the flue gas from burning MSW will be contain 66% nitrogen, CO_2, some O_2 water vapor, as well as materials regarded as pollutants: Particulate matter, carbon monoxide, oxides of nitrogen and sulfur, as well as hydrochloric acid, which make solutions of the flue gas acidic. Other items which may be found in flue gas are heavy metals, dioxins, and furans (Anonymous 2010h).

Modern incinerators have facilities which ameliorate the problems of the older ones by passing the flue gas through scrubbers which remove the pollutants. There are many different types of scrubbing arrangements depending on the components of flue gases. In some, particles are removed by passing the gas through special filters where the particles, including those of heavy metals such as mercury, are captured by electrostatic attraction; the sulfur dioxide present in the flue gas is then passed through $CaCO_3$, which neutralizes the sulfurous acid yielding $CaSO_4$. Nitrogen oxides are treated either by modifications to the combustion process to prevent their formation, or by high temperature or catalytic reaction with ammonia or urea; in both cases, the objective is to produce nitrogen gas.

In the first stage of one scrubber, the Van Roll Flue Gas Scrubber, scrub water is sprayed into the hot flue gases. As it vaporizes, it cools the gas stream to about 70°C. The scrub water washes coarse particulate matter out of the flue gases and absorbs most of the HCl. This stage also handles the absorption of the HCl and the removal of mercury. In the second stage, the gases pass through a packed bed stage in countercurrent to the scrub water, which cools them to about 60°C. The remaining pollutants, such as HCl and HF, are absorbed with good efficiency here, while heavy metal vapors condense to aerosols. If necessary, sodium hydroxide may be added at this stage to further remove any sulfur dioxide. In the third stage, aerosols and submicron dust particles generated in the cooling and absorption stage are collected. The flue gases pass a manifold of ring jets which are part of a multiple venturi configuration that subdivides the gas stream in order to boost separation efficiency. Controlled water supply makes it possible to hold a constant pressure drop and thus a constant removal efficiency over a wide range of loads. The ring jet also offers high efficiency in collecting gaseous contaminants such as sulfur dioxide. After passing through another demister, the flue gases – now free of all pollutants in gas, particulate, and aerosol forms – are routed to the stack or to a downstream process stage, for example, the carbon entrainment process. The scrubber stages can be combined in any way to achieve an optimal solution for individual requirements (Anonymous 2009b, 2010h).

Controversy over the health effects of incineration

Inspite of the scrubbing of the flue gases, it was still claimed in a report by the UK Society for Ecological Medicine *The Health Effects of Waste Incinerators* (Thompson and Anthony 2008) that "incinerator emissions are a major source of fine particulates, of toxic metals and of more than 200 organic chemicals, including known carcinogens, mutagens, and hormone disrupters. Emissions also contain other unidentified compounds whose potential for harm is as yet unknown, as was once the case with dioxins. Since the nature of waste is continually changing, so is the chemical nature of the incinerator emissions and therefore the potential for adverse health effects." The report concluded thus:

> Recent research, including that relating to fine and ultrafine particulates, the costs of incineration, together with research investigating non-standard emissions from incinerators, has demonstrated that the hazards of incineration are greater than previously realised. The accumulated evidence on the health risks of incinerators is simply too strong to ignore and their use cannot be justified now that better, cheaper and far less hazardous methods of waste disposal have become available. We therefore conclude that *no more incinerators should be approved.* (My italics).

However another report on *The Impact on Health of Emissions to Air from Municipal Waste Incinerators*, published by the British Government's Health Protection Agency a year later, disputes these findings and concludes that modern, well managed incinerators make only a small contribution to local concentrations of air pollutants and any such small additions, if they have any effects on health at all, "are likely to be very small and not detectable."

Modern and Emerging (and Safer?) Methods of the Thermal Treating of MSW

Some of the newer and emerging technologies which may obviate the negative aspects of incineration are: Plasma gasification, pyrolysis, supercritical water oxidation, and Sonophotochemical Oxidation. They are sometimes called Advanced Thermal Technologies or Alternative Conversion Technologies.

1. Plasma arc gasification (also called plasma pyrolysis)

   Very hot plasma is formed by ionized gas in a strong electrical arc with the power ranging from 2 to 20 MW and temperatures ranging from 2,000°C to 6,000°C. An example in nature is lightning, capable of producing temperatures exceeding 6,980°C. A gasifier vessel utilizes plasma torches operating at 5,540°C which is about the surface temperature of the Sun. In such high temperature, all waste constituents, including metals, toxic materials, silicon, etc. are totally melted forming nontoxic dross. Plastic, biological and chemical compounds, toxic gases yield complete dissociation into simpler gases mainly H_2 and CO_2. The resulting gas mixture is called synthesis gas or syngas and is itself a fuel. When municipal solid waste is subjected to this intense heat within the vessel, the waste's molecular bonds break down into elemental components. The process results in elemental destruction of waste and hazardous materials. Gasification is a method for extracting energy from many different types of organic materials; the minimum temperature for it to occur is 1,500°C. Simpler gases, mainly H_2 can be used as ecological fuel to generate heat energy and electrical energy decreasing significantly the cost of plasma formation and waste utilization. Regained metals from dissociation process can safely return to metallurgic industry, and slag can be used as an additive to road and construction materials. The utilization of municipal waste using this method does not cause the emission of foul odors and does not produce a harmful ash, which is something that normally takes place in an incinerating plant.

   Some of the advantages of plasma technology are as follows:

   (a) Plasma method can be used on any type of waste (hazardous, toxic, or lethal) because of the very high temperatures which disassociate molecular bonds.

   (b) Plasma waste utilization method takes place in a close system, without releasing ashes, waste remnants, dusts, and toxic gases into environment. Regained metals return to metallurgic industry and created slag is used as an additive to road construction materials. Nontoxic gases, which are created, are stored in special containers (gas cylinders) and used as fuel and energy creators.

(c) The ratio of the reduction in volume after waste is treated by plasma is about 300:1, whereas with conventional incineration, it is only about 5:1, because a large quantity of ash is produced.

(d) Plasma technology allows converting large quantities of municipal waste in the range of 10–500 tons a day.

(e) This method of waste reduction is the only method available to stabilize electronic waste, which does not undergo biodegradation.

(f) The costs of using plasma technology are significantly low (as low as $40/ton or less) when the value of the products made because of it are included; the costs of using conventional incineration are in the range of $100/ton.

(g) Contaminates in slag and gases created during plasma utilization with elements such as mercury, cadmium, sulfur, SO_2, HCl, dioxins, selenium, chromium, lead, barium, arsenic, radioactive elements are strictly controlled by usage of special water or dry scrubbers and filters. Using this method, elements released are considerably minimized below environmental standards. The remainder of the pollutants sink into glassy slag and can be treated further in a closed system, which is a major distinction to conventional incineration.

(h) The ashes that are formed as a result of conventional incineration can be further burned down using plasma technology to make them harmless.

(i) Contemporary plasma converters are computer controlled, safe, quiet, and can be stationary or mobile; incinerators are always stationary.

(j) Plasma waste utilization will improve public health and safely achieve total and irreversible destruction of hazardous and toxic compounds, lethal viruses, bacteria, and prions that are dangerous to human health (Kowalski and Kopinski 2010).

In addition syngas, which is produced with this technology has the following advantages:

(k) Syngas has also to be scrubbed as is the case with incineration flue gas; however, the volume of syngas is much less than that of flue gas.

(l) Electric power may be generated in engines and gas turbines, which are much cheaper and more efficient than the steam cycle used in incineration

(m) No product of incineration has the versatility of syngas, which can not only produce electricity but also other chemicals, including other fuels.

2. Pyrolysis

Pyrolysis come from two Greek words (pyros = fire; lysis = breakdown) i.e., breakdown by heat. Pyrolysis is the thermal decomposition of large molecules when heated in the absence of oxygen at temperatures over 500°C. One of the products of this process is pyrolysis oil which has the potential to be a low-cost transport fuel with low greenhouse gas emissions when derived from waste. Pyrolysis typically occurs under pressure and at operating temperatures above 430°C. In practice, it is not possible to achieve a completely oxygen-free atmosphere; on account of this, in any pyrolysis system, a small amount of oxidation occurs.

Pyrolysis is a special case of thermolysis, and is most commonly used for organic materials, then being one of the processes involved in charring. The pyrolysis of wood, which starts at 200–300°C, and in general, pyrolysis of organic substances produces gas and liquid products and leaves a solid residue richer in carbon content.

Pyrolysis is used in the chemical industry, for example, to produce charcoal, activated carbon, methanol, and other chemicals from wood, to convert ethylene dichloride into vinyl chloride to make PVC, to produce coke from coal, to convert biomass into syngas, to turn waste into safely disposable substances, and for transforming medium-weight hydrocarbons from oil into lighter ones like gasoline. These specialized uses of pyrolysis are called by various names, such as dry distillation, destructive distillation, or cracking.

Pyrolysis also plays an important role in several cooking procedures, such as baking, frying, grilling, and caramelizing.

Pyrolysis is usually the first chemical reaction that occurs in the burning of many solid organic fuels, like wood, cloth, and paper, and also of some kinds of plastic. In a wood fire, the visible flames are not due to combustion of the wood itself, but rather of the gases released by its pyrolysis; whereas the flame-less burning of embers is the combustion of the solid residue (charcoal) left behind by it.

3. Supercritical water decomposition (hydrothermal monophasic oxidation)

Supercritical water oxidation (SCWO) is the destruction technology for organic compounds and toxic wastes using the unique properties of water in supercritical condition that is high temperature and pressure (above 374°C and 22 MPa). In supercritical water, organic materials, such as chlorinated organic compounds, are quickly oxidized and decomposed with oxidants. Carbon in the organic compounds is converted to carbon dioxide, hydrogen to water, and chlorine atoms to chloride ion.

A supercritical (SC) fluid is defined as a substance that is at conditions of temperature and pressure that are above its vapor-liquid critical point. At supercritical conditions, a fluid does not behave entirely as a liquid or as a gas, but somewhere in between. The properties of supercritical fluids combine the solvating powers of liquids with the diffusivities of gases. The critical point for water is at 3740 C, (70SoF) and 218 atm, (22 MPa, 3,191 psi). The changes in the properties of water once supercriticality has been reached are remarkable. The familiar, polar liquid with its high dielectric constant of 78.5 changes to an almost nonpolar fluid with a value of less than five, approaching that of ambient hexane at 1.8. The density of SCW is found to decrease to around 0.15 g/ml, depending upon conditions. SCW possesses properties which enable it to become miscible with organic molecules and with gases.

Gases including oxygen and organic compounds are completely soluble in supercritical water and become a single phase. Such single phase contact under high density and high temperature allows rapid and almost complete oxidation reaction. Quite high destruction efficiencies for various compounds have been demonstrated.

SCWO is a high temperature and pressure technology that uses the properties of supercritical water in the destruction of organic compounds and toxic wastes. Under SC conditions, the oxidation reactions occur in a homogeneous phase where carbon is converted to carbon dioxide, hydrogen to water,

nitrogen-containing substances to nitrogen, and sulfur-containing substances to sulfuric acid. An important factor in the context of this application of SCWO is that the reactions are exothermic and the process can become thermally self-sustaining if the appropriate concentration of oxidizable substances is present. SCW is known to be highly effective at rapidly oxidizing organic matter, for example, aqueous waste streams. Its application to the complete destruction of hazardous and toxic wastes has been extensively studied (Hamley et al. 2001).

4. Combinative sonochemical oxidations of pollutants in water

Sonic and ultra-sonic sound waves in combination with oxidative methods are receiving growing attention as ways of destroying pollutants in water. Some processes have combined sonochemical methods with UV or chemical oxidants such as hydrogen peroxide, H_2O_2, ozone O_3 with some degree of success. It appears to be a method which may find application in special circumstances (see Adewuyi 2005).

MSW can be directly combusted in waste-to-energy facilities to generate electricity. Because no new fuel sources are used other than the waste that would otherwise be sent to landfills, MSW is often considered a renewable power source (see Table 11.3). Although MSW consists mainly of renewable resources such as food, paper, and wood products, it also includes nonrenewable materials derived from fossil fuels, such as tires and plastics.

At the power plant, MSW is unloaded from collection trucks and shredded or processed to ease handling. Recyclable materials are separated out, and the remaining waste is fed into a combustion chamber to be burned. The heat released from burning the MSW is used to produce steam, which turns a steam turbine to generate electricity. Over one-fifth of the US municipal solid waste incinerators use refuse derived fuel (RDF); the United States has about 89 operational MSW-fired power generation plants, generating approximately 2,500 MW, or about 0.3% of total national power generation. However, because construction costs of new plants have increased, economic factors have limited new construction.

The combustion of MSW reduces MSW waste streams, reducing the creation of new landfills. MSW combustion creates a solid waste called ash,

Table 11.3 Generation, materials recovery, composting, combustion with energy recovery, and discards of MSW, 1960–2008 (in millions of tons) in the US (From USEPA: Municipal solid waste generation, recycling, and disposal in the United States: Facts and Figures for 2008 http://www.epa.gov/osw/nonhaz/municipal/pubs/msw2008rpt.pdf, Anonymous 2009a)

Activity	1960	1970	1980	1990	2000	2003	2005	2007	2008
Generation	88.1	121.1	151.6	205.2	239.1	242.2	249.7	254.6	249.6
Recovery for recycling	5.6	8.0	14.5	29.0	52.9	55.6	58.6	62.5	60.8
Recovery for composting[a]	Negligible	Negligible	Negligible	4.2	16.5	19.1	20.6	21.7	22.1
Total materials recovery	5.6	8.0	14.5	33.2	69.4	74.7	79.2	84.2	82.9
Combustion with energy recovery[b]	0.0	0.4	2.7	29.7	33.7	33.1	31.6	32.0	31.6
Discards to landfill, other disposal[c]	82.5	112.7	134.4	142.3	136.0	134.4	138.9	138.4	135.1

[a] Composting of yard trimmings, food scraps, and other MSW organic material. Does not include backyard composting
[b] Includes combustion of MSW in mass burn or refuse-derived fuel form, and combustion with energy recovery of source separated materials in MSW (e.g., wood pallets, tire-derived fuel)
[c] Discards after recovery minus combustion with energy recovery. Discards include combustion without energy recovery

which may contain any of the elements that were originally present in the waste. MSW power plants reduce the need for landfill capacity because disposal of MSW ash requires less land area than does unprocessed MSW. However, because ash and other residues from MSW operations may contain toxic materials, the power plant wastes must be tested regularly to assure that the wastes are safely disposed of, so as to prevent toxic substances from migrating into ground-water supplies.

Under current regulations, MSW ash must be sampled and analyzed regularly to determine whether it is hazardous or not. Hazardous ash must be managed and disposed of as hazardous waste. Depending on state and local restrictions, nonhazardous ash may be disposed of in a MSW landfill or recycled for use in roads, parking lots, or daily covering for sanitary landfills.

A variety of pollution control technologies significantly reduce the gases emitted into the air, including scrubbers, devices that use a liquid spray to neutralize acid gases and filters, that remove tiny ash particles.

5. Disposal through composting, landfilling or combustion without energy generation

11.3.1.4 Composting

Composting is the aerobic decomposition of organic materials by microorganisms under controlled conditions. During composting, the microorganisms consume O_2 while subsisting on organic matter (Fig. 11.3). Active composting generates considerable heat, large quantities of CO_2, and water vapor. The CO_2 and water

losses can amount to half the weight of the initial materials, thereby reducing the volume and mass of the final product.

There are three stages in the composting process: Preprocessing, processing and post-processing.

Preprocessing

These are steps which enable the optimization of the composting process and include the following:

Sorting the feedstock material and removing materials that are difficult or impossible to compost such as woody stems.

Reducing the particle size of the feedstock. In large scale composting, machinery exist which chop the material into desirable particle sizes. Particle size reduction increases the surface area to volume ratio of the feedstock materials and this facilitates decomposition by increasing the area exposed to microorganisms. A balance must be drawn between the size which will increase the surface and that may increase compaction of the material and hence limit the free flow of air within the pile (Fig. 11.5).

Optimizing composting conditions. To enhance composting, the materials need to be adjusted for the optimal conditions of moisture content, carbon-to-nitrogen (C:N) ratio, and acidity/alkalinity (pH) (see Table 11.4).

Mixing. Mixing entails either blending certain ingredients with feedstock materials or combining different types of feedstock materials together; for example, bulking agents (such as wood chips) are often added to feedstock materials that have a fine particle size (such as grass).

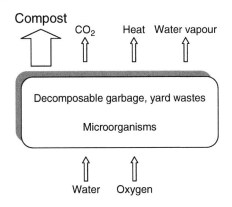

Fig. 11.5 Schematic diagram depicting the composting process

Table 11.4 Optimum conditions for composting (Modified from Palmisano and Barlaz 1998)

Condition	Reasonable range	Preferred range
Carbon:nitrogen (C:N) ratio	20:1–40:1	25:1–30:1
Moisture content	40–65%	50–60%
Oxygen concentration	Greater 5%	Much greater than 5%
Particle size (diameter, in.)	1/8–1/2	Variable
pH	5.5–9.0	6.5–8.0
Temperature (°F)	110–150	130–140

Compost Processing Conditions

Processing methods are chosen to enhance the rapid decomposition of the materials and to minimize negative effects, such as odor release and leachate runoff. Maximum conditions are provided for the decomposition of the composting feed stock.

(a) *Oxygen and aeration*

A minimum oxygen concentration of 5% within the pore spaces of the compost is necessary for aerobic composting. Oxygen levels within the windrows or piles may be replenished by turning the materials over; in large scale composting, this may be done by a front-end loader, or by means of special compost turner.

(b) *C:N ratio*

Carbon (C), nitrogen (N), phosphorous (P), and potassium (K) are the primary nutrients required by the microorganisms in composting. An appropriate C:N ratio usually ensures that the other required nutrients are present in adequate amounts. Raw materials are blended to provide a C:N ratio of 25:1 to 30:1. For C:N ratios below 20:1, the avail-

able carbon is fully utilized without stabilizing all of the nitrogen which can lead to the production of excess ammonia and unpleasant odors. For C:N ratios above 40:1, not enough N is available for the growth of microorganisms and the composting process slows dramatically.

(c) *Moisture*

Composting materials should be maintained within a range of 40–65% moisture. Water displaces much of the air in the pore spaces of the composting materials when the moisture content is above 65%. This limits air movement and leads to anaerobic conditions. Moisture content generally decreases as composting proceeds; therefore, water may need to be added to the compost. As a rule of thumb, the materials are too wet if water can be squeezed out of a handful and too dry if the handful does not feel moist to the touch.

(d) *Particle size*

The rate of aerobic decomposition increases with smaller particle size. Smaller particles, however, may reduce the effectiveness of oxygen movement within the pile. Optimum composting conditions are usually obtained with particle sizes ranging from 1/8 to 2 in. average diameter.

(e) *Temperature*

Composting will essentially take place within two temperature ranges: Mesophilic (10–35°C) and Thermophilic (over 45°C). The thermophilic temperatures are desirable because they destroy more pathogens, weed seeds, and fly larvae in the composting materials.

(f) *Time*

The length of time required to transform raw materials into compost depends upon the factors listed above. In general, the active decomposition period may be between 3 and 4 months.

Compost Post Processing or the Curing Stage

When the compost material no longer releases heat, it has entered the curing stage; curing stage begins. The curing stage of compost usually lasts for about a month. Curing occurs at mesophilic temperatures. Curing piles undergo slow decomposition; care must be taken during this period so that these piles do not become anaerobic. The C:N ratio of finished compost should not be greater than 20:1. C:N ratios that are too low can result in phytotoxins being emitted when composts are used. Compost becomes dry and crumbly in texture.

Fig. 11.6 Microbial groups dominant at each temperature regime during composting (Modified from Palmisano and Barlaz 1998)

Stages in the production of compost: A, mesophilic stage; B, thermophilic stage; C, cooling stage; D, maturation stage

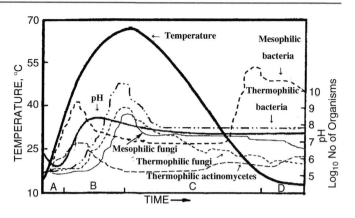

When the curing stage is complete, the compost is considered "stabilized" or "mature". Any further microbial decomposition will occur very slowly.

Microbiology of Composting

Composting proceeds through a succession of non-actinomycete mesophilic bacteria, followed by actinomycetes, then fungi; the simplest substrates are metabolized quickly, while substrates which are more complex or difficult to metabolize remain. In the composting of grass straw, the sugars from cell wall polysaccharides are degraded initially, while the less available cellulose is decomposed in the latter stages of composting.

In the first stage of composting (see "A" in Fig. 11.6), the temperature rises from ambient into the thermophilic range. Mesophilic bacterial populations, (mainly non-actinomycete) multiply very rapidly utilizing simple and readily available substrates.

The activity of the mesophilic organisms (non-actinomycete bacteria followed by mesophilic fungi) generates heat which is retained within the compost pile. As the temperature rises to about 45°C, mesophilic populations die off and thermophilic bacterial populations begin to flourish (B in Fig. 11.6). Thermophilic non-actinomycete bacteria, thermophilic fungi, and thermophilic actinomycete in that order become dominant.

In the third stage (see "C" in Fig. 11.6), the temperature drops as the activity by thermophilic declines because they have used up available food in the compost.

In the fourth and final stage (see "D" in Fig. 11.6), the curing stage, the temperature once again returns to the ambient and mesophilic bacteria increase, some-times even beyond their initial original numbers. The mesophilic fungal and actinomycete populations gradually attack the more complex materials such as cellulose and lignin. Overall, microbial activity drops progressively to very low rates,

Some of the mesophilic bacteria encountered in compost are *Alcaligenes faecalis, Bacillus brevis, B. circulans* complex. Among the thermophilic bacteria are: *Bacillus subtilis, B. polymyxa, B. pumilus, B. sphaericus, and B. licheniformis, B. stearothermophilus, B. acidocaldarius, and B. schleglii, B. circulans complex types i and ii, and B. subtilis.*

Mesophilic actinomycetes include *Actinobifida chromogena, Microbispora bispora, Micropolyspora faeni, Nocardia* sp., and *Streptomyces rectus*, while some thermophilic actinomycetes are *Thermomonospora* sp., and *Thermoactinomyces vulgaris*

Mesophilic fungi include: *Absidia corymbifer, A. ramose, Mortierella turficola Mucor miehe, M. pusillus, and Rhizomucor* sp., while thermophilic fungi are *Aspergillus fumigatus, Chaetomium thermophile, Humicola lanuginosa, Mucor pusillus, Thermoascus aurantiacus, and Torula thermophila.* Of these, only *Chaetomium thermophile* has been found to be cellulolytic.

11.3.1.5 Landfills Used in the Disposal of MSW

An MSW landfill is not simply a hole in the ground where refuse is dumped. Rather it is a carefully engineered structure built into the ground (where a natural valley exists) or on top of the ground (if the terrain is flat) in which trash is isolated from the surrounding environment, especially groundwater, but also from rain and the surrounding air. This isolation is accomplished with a daily covering of compacted soil over the day's trash and a bottom liner of clay or specially designed thick plastic material. Some authors designate landfalls using clay as sanitary landfill and those using a synthetic (plastic) liner as Municipal solid waste (MSW) landfill. Many factors must be considered and taken into account before the construction of a landfill occurs. After construction, many steps must be taken for it to continue to function well (Anonymous 2010i).

A number of other types of landfills exist. For example, special landfills can be used to reclaim land from the oceans, rivers, or swamps; the materials dumped therein are usually building materials. Some special landfills are also used for disposing of industrial wastes. Hazardous waste landfills are waste disposal units constructed to specific design criteria and which receive hazardous wastes. These landfills are generally constructed to securely hold materials which are regarded as injurious to human health such as radioactive wastes. Inert waste landfills are units which receive wastes that are chemically and physically stable and do not undergo decomposition, including bricks, concrete sand etc. Dumps are simple un-engineered depositories which store various materials and have no facility for leachate and water contamination control. They are not common, are often used in rural areas. Some authorities do not allow their use.

In this section, the discussion will be on landfills used in the disposal of domestic wastes or municipal solid wastes (MSW).

Factors to be Considered in Planning an MSW Landfill

1. *Approval by the appropriate government authority*
   The world over, there are regulations that govern where a landfill can be placed and how it can operate. Permission must therefore be sought from the appropriate authority. In the United States, taking care of trash and building landfills are local government responsibilities and are built by city or other local government authorities. However, a few privately established and operated landfills do exist. Before a city, other local government authority, or private entrepreneurs can build a landfill, an environmental impact assessment (EIA) or study must be carried out to determine the availability of the other factors.

2. *The nature of the underlying soil and bedrock*
   The nature of the underlying soil and bedrock must be determined. The rocks should be as impervious and watertight as possible to prevent any leakage from reaching groundwater. It must not be cracked so as to be sure that leachates from the landfill do not enter the underground water. The site should not be near mines or quarries because these structures frequently contact the groundwater supply. It must finally be possible to sink wells at various points around the site to monitor the groundwater or to capture any escaping wastes.

3. *There must be sufficient land for the landfill and the activities associated with it*
   Although the actual landfill is at the center stage of the landfill establishment, the essential support services and activities must also be present. These include runoff access roads, collection ponds, leachate collection ponds, drop-off stations, areas for borrowing soil, and 50- to 100-foot buffer areas etc. These ancillary activities take up three to four times more space than the actual landfill and must be taken into account when a landfill is being planned.

4. *The flow of water over the area must be studied*
   It is necessary to study the flow of water so as to ensure that excess water from the landfill does not drain into to neighboring property or vice versa. Similarly, the landfill should not be close to rivers, streams, or wetlands so that any potential leakage from the landfill does not enter the groundwater or watershed.

5. *The ground water should be as low as possible*
   The level of the underground water table should be as low as possible so that leachates from the landfill do not easily enter the ground water.

6. *Avoidance of wild life or historical locations*
   The landfill should not be located near areas important to wild life such as nesting areas of local or migrating birds or fisheries. Similarly, it should not be located near any sites containing any historical or archaeological artifacts.

Parts of a Landfill

Figure 11.7 is a vertical section of a generalized landfill reproduced with the permission of Waste Management Inc. of Houston, Texas, USA; the description of parts of a landfill which follows below are based on the legend of the generalized landfill and is used with the permission of Waste Management Inc.

The parts of a landfill may be divided into: The protective cover (A); the moisture barrier cap (B); working

landfill (C); leachate collection system (D); ground water protection base (E); monitoring liner system (F); secondary protection base (G). The details of the various parts of a landfill are in Parts A to G of Fig. 11.5. The aspects of landfills and their ancillary activities dealing with ground water, surface water, and methane gas monitoring are discussed in Fig. 11.8 (Anonymous 2010i).

(A) **Protective Cover**

1. *Grass and flower cover*

   As portions of the landfill are completed, they are planted over with native grass and shrubs, no trees whose roots will penetrate deep are included. The vegetation is aesthetically pleasing and helps prevent erosion. It will resemble an open space with rolling hills of wild flowers.

2. *Top soil*

   This provides the root system with support and nutrients for the vegetation growing on the erstwhile landfill.

3. *Soil, sand, and gravel*

   This protects the liner system from direct rain and borrowing animals and provides additional moisture retention to help support the plant on the top soil.

(B) **Moisture Barrier Cap**

4. *Filtering system*

   A fabric composed of felt-like plastic acts as a filtration system. The geotext prevents the overlying soil and small particles from clogging the underlying drainage system. High density polyethylene geo net is a heavy plastic with large mesh-like openings. The geo net allows liquid to flow away from the landfill and helps prevent the infiltration of rain water.

5. *Plastic shield*

   High density polyethylene liner shields the landfill from liquid penetration.

6. *Clay shield*

   When the landfill reaches its permitted height, a minimum of 18 in. layer of re-compacted clay is placed over the garbage. This liner is another system which provides an excellent

Fig. 11.7 Vertical section of a generalized landfill (From http://tontitown.wm.com/documents/tontitown_landfill_anatomy.pdf. With the kind permission of Waste Management Inc., Texas, Houston)

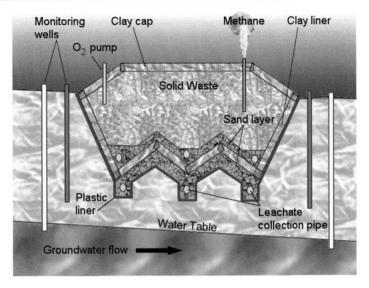

Fig. 11.8 A composite figure of a landfill showing a cell and accompanying landfill activities (From Anonymous 2010j) *Note:* (a) There are pipes for sampling the ground water around the landfill for contamination from the landfill. (b) The bottom of the landfill is slanted to facilitate the drainage of the leachates to the pipes located at the bottom of the landfill. (c) The methane (landfill gas) is being flared, although many modern landfills have facilities for collecting the gas and using it for generating electricity, for heating homes. (d) The ground water is slanting, and some way from the bottom of the landfill

barrier against rainwater infiltration; it also keeps out gas (and hence odors) and vermin.

(C) Working Landfill

7. *Daily cover*

    At the end of each working day, garbage is covered with a 6–12 in. layer of soil. Daily cover reduces odors, blowing litter, and helps deter scavengers.

8. *Garbage*

    In compliance with the requirement of most authorities, as garbage arrives it is compacted in layers within a small area known as a *cell* to reduce the volume consumed within the landfill. It helps to control odors, stop the refuse from scattering and keeps off rats and other vermin. Perhaps, the most precious commodity and overriding problem in a landfill is air space. The amount of space is directly related to the capacity and usable life of the landfill. To increase the space, trash is compacted into cells that contain only 1 day's trash (see Fig. 11.6). In one landfill, a cell is approximately 50 ft long by 50 ft wide by 14 ft high (15.25 m × 15.25 m × 4.26 m). The amount of trash within the cell is 2,500 ton and is compressed at 1,500 lb per cubic yard using heavy equipment (tractors, bulldozers, rollers, and graders) that go over the mound of trash several times. Once the cell is made, it is covered with 6 in of soil and compacted further. Cells are arranged in rows and layers of adjoining cells known as *lifts.* As a further way of conserving space, bulky materials, such as carpets, mattresses, foam, and yard waste, are excluded from most landfill.

(D) Leachate Collection System

Leachates consist of water which enter the landfill during rainfall or formed as a result of biological activity within the landfill. As the water percolates through the trash, it picks up contaminants (organic and inorganic chemicals, metals, biological waste products of decomposition) and is typically acidic. The function of the leachate collection system is to control the flow of leachates so that it can be properly removed from the landfill and treated.

9. *Sand and gravel*

    A 12-inch layer of sand and/or gravel provides protection for a layer of thick plastic mesh (called a geonet) which collects leachate and allows it to drain by gravity to the leachate collecting pipe system.

10. *Filtering (geotextile) system*

A blanket of nonwoven geotextile fabric, composed of felt-like plastic is placed below the gravel. This porous material allows liquid to flow downward while preventing fine particles from clogging and blocking the drainage and collection systems. Underneath, a geo net made of felt-like plastic diverts made of mesh-like plastic diverts the leachate toward the underlying collection pipes and a low-lying sump.

11. *Collection system*

Six-inch perforated PVC perforated plastic pipes surrounded by a bed of porous rock/and sand directs the leachate to the lowest part of each disposal cell. The liquid is conveyed via the leachate collection system to a sump located at the lowest point of each disposal cell. When a small amount of the liquid has built up, a pump automatically removes the liquid from the landfill to specially designed storage tanks or pond, where they are tested for acceptable levels of various chemicals (biological and chemical oxygen demands, organic chemicals, pH, calcium, magnesium, iron, sulfate, and chloride) and allowed to settle. After testing, the leachate must be treated like any other sewage/wastewater; the treatment may occur on-site or off-site at approved wastewater treatment plants.

(E) **Ground Water Protection Base**

12. *The HDPE plastic (geomembrane) shield*

The primary bottom shield consists of two components. First is a man-made synthetic thick high density polyethylene liner, the geomembrane, which is used in MSW landfills and consists of a thick (30–100 mm) type of durable, puncture-resistant synthetic plastic, polyethylene, polyvinylchloride, high-density polyethylene (HDPE). It is impermeable, and highly resistant, to chemicals which might be present in the liquids leaving the landfill (i.e., leachates) as well as gases emanating therefrom. This plastic is also called the bottom liner and is key to the achievement of the purpose of the landfill which is to keep the refuse or any items emanating from it from entering the environment. The plastic liner may also be surrounded on either side by a fab-

ric mat (geotextile mat) that will help to keep the plastic liner from tearing or puncturing from the nearby rock and gravel layers.

The second part of the ground water base is the clay shield described in 13.

13. *Clay shield*

The shield consists of a minimum of 24 in. of compacted clay. It is the second component of the primary liner system and provides added environmental protection, preventing leachates and gas from entering the environment from the landfill.

(F) **Monitoring Liner System**

14. *Protection against clogging of the monitoring layer*

A blanket of geotextile fabric composed of felt-like plastic fibers is laid below the primary ground water protection base. Underneath, a geonet of mesh-like plastic is used to permit any liquids movement. These materials, collectively, prevent the fine clay particles from clogging the monitoring layer below.

15. *Collection system*

Perforated plastic pipes surrounded by a bed of porous rock and/or sand at the lowest point of the liner system. This feature is used to measure the performance of the liner system and groundwater protection base. Regular monitoring is performed to ensure the integrity of the landfill components.

(G) **Secondary Protection Base**

16. *HPDE plastic shield*

The secondary bottom liner consists of two components. First is a man-made synthetic 60 mil high density polyethylene liner, impermeable to liquids. The second is the clay shield described in 17 below.

17. *Clay shield*

The second component of the secondary bottom liner is a minimum of 24 in. of re-compacted clay. The clay and the synthetic liner located above it provide an added measure of environmental protection.

Operation of a Landfill

Landfills are used differently in different localities, including different countries. In some, MSW is dumped into landfills without much possessing, and no attempt is made to recover recyclable items. In

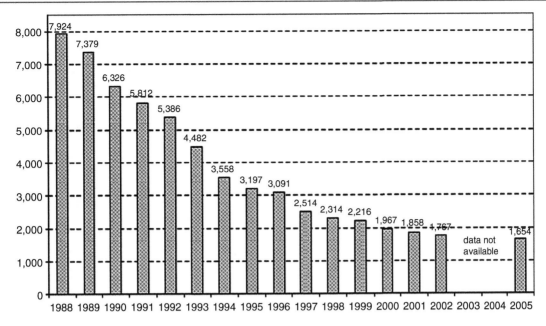

Fig. 11.9 Number of landfills in the US. 1988–2005 (From Municipal Solid Waste in the United States: *2005 Facts And Figures.* U.S. EPA www.epa.gov/osw. Accessed at on Monday, June 11, 2007, Anonymous 2006)

others, the items taken to a landfill would have passed through other activities (see Fig. 11.7). In general, in the USA, the items taken to a landfill are the recalcitrant ones which are left over; the MSW has been subjected to other treatments. Consequently, the number of landfills in the USA has been declining (see Fig. 11.9). In a similar vein, in recent years, some countries, such as Germany, Austria, and Switzerland, have banned the disposal of untreated waste in landfills. In these countries, only the ashes from incineration or the stabilized output of mechanical biological treatment plants may still be deposited.

Problems of Landfills

Major concerns in a landfill include the following, which routine operations are designed to take care of.

(a) *Leachate collection and treatment and/or removal*

Leachate is liquid generated from moisture brought in with the waste, from rainfall which percolates into the landfill, and from the waste decomposition processes. It contains dissolved and insoluble chemicals. A network of perforated pipes within the landfill collects the leachate, which then is pumped to a treatment plant, where it may be treated on site for reuse (e.g., to control dust) or may be piped to a treatment plant for safe disposal.

(b) *Checking for leachate leakage*

Many landfills have wells located around them at appropriate depths and are sampled on a daily basis to determine if leachates from the landfills are escaping into underground waters.

(c) *Control of odors, flies, and vermin*

At the end of each day's operation, bulldozers spread and compact the MSW and further compacts at least 6 in. of earth over all of the waste. The daily cover prevents the emergence of flies and other insects, and controls odors, blowing litter, and the infiltration of rainfall. The compacting also limits hiding places for rats and other vermin.

(d) *Control of gases released from the landfill*

As a result of anaerobic decomposition in the landfills (see below), gases, methane, and carbon dioxide, are released. A series of perforated pipes connected to vacuum blowers collects and removes the gas to be burned off or used to generate electricity in a gas turbine.

Gas monitoring wells are placed all around the site, and sampled frequently, to make certain that there is no migration of the gas beyond the site boundaries as gas leakages have been known to cause explosions around landfills.

(e) *Drainage and Erosion Control*

The landfill must be constructed with graded decks to ensure proper drainage and prevent flooding.

(f) *Closure and Post-Closure Care Requirements for Municipal Solid Waste Landfills*

The EPA requires that when a landfill is closed, it must have a final cover which:

1. Has a permeability less than or equal to the permeability of any bottom liner system or natural subsoils present, using an infiltration material that contains a minimum 18-inches of earthen material

2. Minimizes erosion of the final cover by the use of an erosion layer that contains a of minimum 6-inches of earthen material that is capable of sustaining native plant growth

After closure, the EPA has the following post-closure requirements which must be observed for 30 years, although this period may be shortened:

1. Maintaining the integrity and effectiveness of any final cover, including making repairs to the cover as necessary to correct the effects of settlement, subsidence, erosion, or other events, and preventing run-on and run-off from eroding or otherwise damaging the final cover

2. Maintaining and operating the leachate collection system

3. Monitoring for possible the ground water contamination

4. Maintaining and operating the gas monitoring system

11.4 Anaerobic Breakdown of Organic Matter in Landfills (and Aquatic Sediments)

Anaerobic activity by microorganisms is usually indicated by offensive odors due to H_2S, CH_4, amines, and skatoles.

Methanogenesis or biomethanation is the formation of methane, also known as marsh gas by microorganisms. Methanogens, microorganisms capable of producing methane, are found only among Archaea (see Chap. 2). Recently, it has been demonstrated that leaf tissues of living plants emit methane, although the mechanism

by which such methane production occurs is, as yet, unknown.

Methanogens belong to the Domain Archaea that produce methane as a metabolic byproduct in anoxic (anaerobic) conditions. Methanogens are the most common and widely dispersed of the Archaea being found in anoxic sediments and swamps, lakes, marshes, paddy fields, landfills, hydrothermal vents, and sewage works as well as in the rumen of cattle, sheep and camels, the cecae of horses and rabbits, the large intestine of dogs and humans, and in the hindgut of insects such as termites and cockroaches. In marine sediments, biomethanation is generally confined to where sulfates are depleted, below the top layers. Ecologically, methanogens play the vital role in anaerobic environments of removing excess hydrogen and fermentation products that have been produced by other forms of anaerobic respiration (see Fig. 11.10).

Methanogens typically thrive in environments in which all other electron acceptors (such as oxygen, nitrate, sulfate, and trivalent iron) have been depleted. Most methanogens grow on CO_2 and H_2 as their sole energy source. There are a few exceptions which only metabolize acetate, or reduce methanol with H_2, or use methylamine and methanol (see Table 11.5).

For the majority that reduce CO_2 to CH_4, there are a few key coenzymes they need; coenzyme bound C_1-intermediates Methanofuran (MFR), tetrahydromethanopterin (H_4MPT), and coenzyme M (H-S-CoM). Other key coenzymes worth noting are F_{420} and N-7-mercaptoheptanoyl-O-phospho-L-threonine (H-S-HTP). Coenzyme F_{420} acts analogously to a quinone in electron transfer sequences by accepting the H^+ ions from the electron donor and supplying them to the electron acceptor. The other coenzyme, H-S-HTP, does the same task as F_{420} only in the last step of methanogenesis from CO_2 and H_2 (Fig. 11.11).

11.4.1 Some Properties of Methanogens

Seventeen genera of methanogenic Archaea exist (see Table 11.6). They have a few peculiarities (Anonymous 2010k).

(a) *Shapes*

Two shapes, rods and coccoid, seem to dominate the methanogens. Some examples of rod shaped cells include *Methanobacterium* spp. and *Methanopyrus*

Fig. 11.10 Breakdown of organic matter in the anaerobic regimes of landfill and aquatic sediments

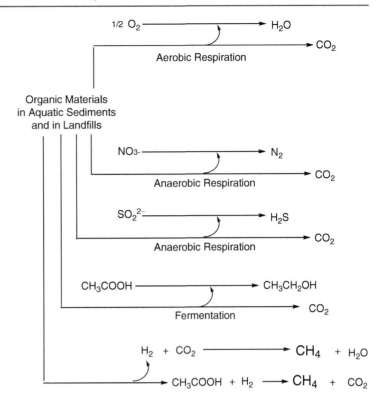

Table 11.5 Some substrates utilized by methanogens (Modified from Rehm and Reed 2008)

$CO_2 + H_2$	Carbon dioxide + hydrogen gas
$HCOO^-$	Formate
$CO + H_2$	Carbon monoxide + Hydrogen gas
CH_3OH	Methanol
CH_3NH^{3+}	Methylamine
$(CH_3)2NH^{2+}$	Dimethylamine
$(CH_3)3NH^+$	Trimethylamine
CH_3SH	Mehtylmercaptan
$(CH_3)2S$	Dimethylsulphide
CH_3COO^-	Acetate

kandleri. Examples of the coccoid methanogens include species from *Methanococcus* and *Methanosphaera* to name a few. *Methanoculleus* and *Methanogenium* are coccoid as well but are irregularly shaped, possibly due to S-layers not being so strongly bonded like other wall structures. Methanogens are not just limited to these shapes, but include a plate shaped genus *Methanoplanus*, *Methanospirillum* that are long thin spirals, and *Methanosarcina* that are cluster of round cells.

(b) *Lack of mureins in their cell walls*

Methanogens lack murein typical of bacteria (eubacteria), but some contain pseudomurein, which can only be distinguished from its bacteria counterpart through chemical analysis. Those methanogens that do not possess pseudomurein have at least one paracrystalline array (S-layer). An S-layer is made up of proteins that fit together in an array like jigsaw pieces that do not covalently bind to one another, in contrast to a cell wall that is one giant covalent bond. Also, some methanogens have S-layer proteins that are glycosylated, which could increase stability, while others do not (e.g., *Methanococcus* spp).

(c) *Ecological distribution in Extreme environments*

Methanogens are found in extreme environments: from the hot vents in the ocean floor to the polar ice to hot springs.

11.4.2 Landfill Gas

Landfill gas is the gas released from landfills. The gas produced by landfills contains about 50% each of methane (CH_4) and carbon dioxide (CO_2). Methane is a

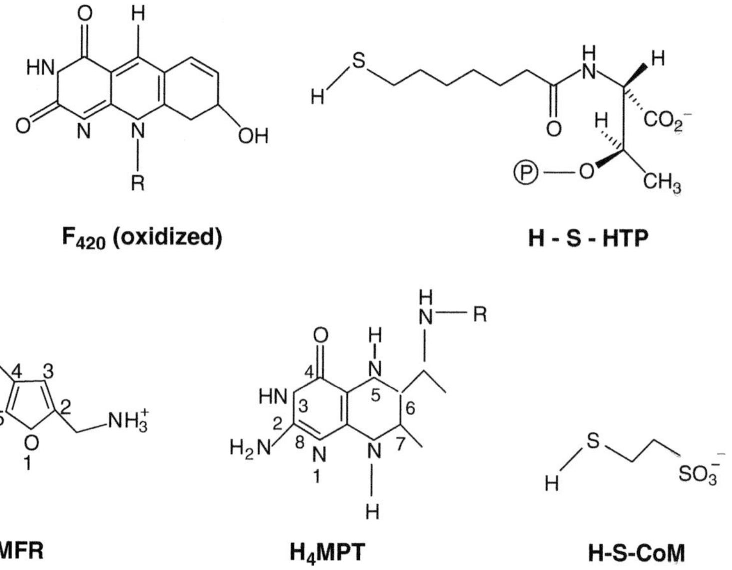

Fig. 11.11 Some
coenzymes peculiar
to methanogens (From
DiMarco et al. 1990. With
permission)

F_{420} (oxidized)   H - S - HTP

MFR H_4MPT H-S-CoM

Table 11.6 Methanogenic genera among the archeae (Modified from http://www.earthlife.net/prokaryotes/euryarchaeota.
html, and http://en.wikipedia.org/wiki/Methanogen, Anonymous 2010i)

Methanogenic archaea	Number of species	Cell morphology	Example
Methanobacterium	19	Long rods	*Methanobacterium bryantii*
Methanobrevibacter	7	Short rods	*Methanobrevibacter ruminantium*
Methanosphaera	2	Cocci	*Methanosphaera stadtmanae*
Methanothermus	2	Rods	*Methanothermus fervidus*
Methanococcus	11	Irregular cocci	*Methanococcus deltae*
Methanomicrobium	2	Short rods	*Methanomicrobium mobile*
Methanogenium	11	Irregular cocci	*Methanogenium frigidum*
Methanospirillium	1	Sprilla	*Methanospirillium hungatei*
Methanoplanus	3	Plate-like	*Methanoplanus limicola*

greenhouse gas and has 25 times more heat trapping potential than carbon dioxide (CO_2). However, methane is only present in the atmosphere at concentrations around 1,800 parts per billion (as opposed to CO2's 385 ppm).

There are many factors that determine how much landfill gas, including the valuable methane, a given landfill will produce. The most important factors for predicting the quality of the landfill methane that will be produced over the lifetime of a landfill are:

1. Waste composition

The more organic waste present in a landfill, the more landfill gas (e.g., carbon dioxide, methane,

nitrogen, and hydrogen sulfide) is produced by the bacteria during decomposition. The more chemicals disposed of in the landfill, the more likely non-methane organic compounds (NMOCs) will be produced either through volatilization or chemical reactions.

2. Age of refuse

Generally, more recently buried waste (i.e., waste buried less than 10 years) produces more landfill gas through bacterial decomposition, volatilization, and chemical reactions than does others.

3. Moisture content

The presence of moisture in a landfill increases gas production because it encourages bacterial decomposition. Moisture may also promote chemical reactions that produce gases.

4. Temperature

As the landfill's temperature rises, bacterial activity increases, resulting in increased gas production. Increased temperature may also increase rates of volatilization and chemical reactions.

11.4.2.1 Landfill Methane Capture Technology

The basic process behind capturing the methane that is emitted by landfills is to place a cap on it (see the operation of a landfill), using a variety of different materials based in part on the waste contents of the landfill, to block the direct emissions of methane into the atmosphere. A common landfill gas capture system is made up of an arrangement of vertical wells and horizontal collectors, usually placed after the landfill has been capped, that is used to direct the flow of the gas This common type of collection is known as a "passive gas collection system" and the collection wells can be installed during the initial construction of the landfill or after the landfill is permanently closed.

In the other method, the "active gas collection system", a series of pumps move the gas to collection wells and through a series of low-pressure chambers to help direct and control the flow of the gas. The active gas collection method is more expensive than the passive version, but the ability to control the flow rate of gas, coupled with the ability to have multiple collection wells, helps to make the active gas collection system an economically viable option for many landfills. There are nearly 500 landfills in the United States that are capturing methane and either burning it for electricity generation or flaring it, which converts the methane into carbon dioxide, which has a lower global warming potential than methane.

11.5 Options for Municipal Solid Wastes Management

Several options exist for the disposal of wastes and what is adopted in any community, municipal, state or country, depends much on the economic status of the

unit. As seen in Fig. 11.8, MSW can be sorted into seven categories: Yard trimmings, food scraps, paper and paper board, metals, glass, plastics and wood. The first two, yard trimmings and food scraps, may be composted, while the other five may recycled and used for manufacturing. The compost is used as soil amendment or organic manure in agriculture, while any materials not decomposable is either sent to the landfill or incinerated.

The non-decomposable and/or nonrecyclable remnants of the recycled materials among paper, metals, glass, plastics, and wood are sent to the landfill or for incineration, depending on their nature.

The breakdown of organic materials in the landfill leads to methane release which can be used to produce electricity. The leachate from the landfill must be carefully managed to prevent it from getting into ground water.

All seven categories of MSW may however be incinerated; sometimes, the heat generated from the incineration may be used to generate electricity. The ash generated from the incineration is sent to the landfill (Fig. 11.12).

The disposal of wastes (solid or sewage) costs money and the community must not only have the means to pay for the service, but must also be willing to pay for it.

When the various options are arranged in descending order of desirability, the order would be:

1. Source reduction
2. Recycling
3. Incineration with energy recovery
4. Disposal through:
   (a) Incineration
   (b) Landfilling
   (c) Incineration without energy recovery

The more affluent a society is, the more likely it is that the options it will select will be nearer to the top of the list. Many developing countries, for example, do not have the resources to establish the type of well-engineered landfills described above, or to invest in the infrastructure to generate electricity from combusted MSW. They end up simply combusting the wastes, or just dumping them in dumpsites to be collected and buried or incinerated at a latter date (see Bassey et al. 2006).

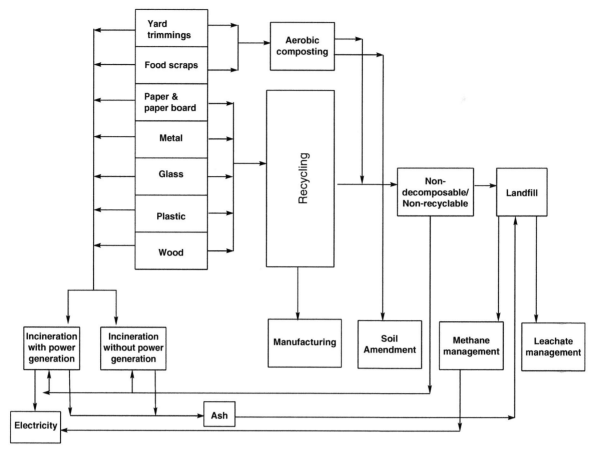

Fig. 11.12 Options for managing municipal solid wastes

References

Adewuyi, Y. G. (2005). Sonochemistry in environmental remediation. 1.Combinative and hybrid sonophotochemical oxidation processes for the treatment of pollutants in water.
Environmental Science and Technology, 39, 3408–3420.

Anonymous (2004, 13–16 April). *Integrated waste management.* Papers at the Integrated Waste Management & Life
Cycle Assessment Workshop and Conference, Prague. http://
viso.ei.jrc.it/iwmlca/aboutiwm.html. Accessed 4 Sept 2010.

Anonymous. (2005). *Waste generated and treated in Europe (Data
1995–2003).* Luxembourg: Office for Official Publications of
the European Communities. ISBN 92-894-9996-6.

Anonymous (2006) Municipal solid waste in the united states:
2005 facts and figures. U.S. Environmental Protection
Agency. Accessed at http://www.epa.gov/osw on Monday
June 11 2007.

Anonymous (2009a). Municipal solid waste in the United States:
2008 facts and figures. U.S. Environmental Protection
Agency. Accessed at http://www.epa.gov/wastes/nonhaz/
municipal/pubs/msw2008rpt.pdf on 5 Oct 2010.

Anonymous (2009b). The impact on health of emissions to air
from municipal waste incinerators http://www.hpa.org.uk/

web/HPAwebFile/HPAweb_C/1251473372218. Accessed 30
Aug 2010.

Anonymous (2010a). A brief history of environmentalism.
www.public.iastate.edu/~sws/enviro%20and%20society.

Anonymous (2010b). Timeline of history of environmentalism.
http://en.wikipedia.org/wiki/Timeline_of_history_of_
environmentalism. Accessed 3 Sept 2010.

Anonymous (2010c). United Nations environment programme:
Environment for development. http://www.unep.org/
Documents.Multilingual/Default.asp?documentid=97.
Accessed 5 Sept 2010.

Anonymous (2010d). The European environmental agency.
http://www.eea.europa.eu/. Accessed 4 May 2009.

Anonymous (2010e). Ministry of the environment: Government
of Japan. http://www.env.go.jp/en/. Accessed 26 Aug 2010.

Anonymous (2010f). Welcome to Ghana EPA. http://www.epa.
gov.gh/. Accessed 6 Sept 2010.

Anonymous (2010g). Egyptian environmental affairs agency.
http://www.eeaa.gov.eg/. Accessed 8 April 2010.

Anonymous (2010h). Flue gas scrubber. http://aeevonroll.thom-
asnet.com/Asset/fluegas.pdf. Accessed 1 Sept 2010.

Anonymous (2010i). "Anatomy of a Landfill". Waste management Inc. http://tontitown.wm.com/documents/tontitown_
landfill_anatomy.pdf. Accessed 29 Aug 2010.

Anonymous (2010j). Ecology labs. http://cas.bellarmine.edu/tietjen/ environmental/waterpollution.htm. Accessed 4 April 2010.

Anonymous (2010k). Euryarchaeota, extremely halophilic archeae. http://www.earthlife.net/prokaryotes/euryarchaeota. html and methanogens http://en.wikipedia.org/wiki/ Methanogen). Accessed 11 Aug 2010.

Bassey, B. E., Benka-Coker, M. O., & Aluyi, H. S. A. (2006). Characterization and management of solid medical wastes in the Federal Capital Territory, Abuja Nigeria. *African Health Sciences, 6*, 58–63.

Carson, R. (1962). *Silent spring*. Boston: Houghton Mifflin Harcourt.

DiMarco, A. A., Bobik, T. A., & Wolfe, R. S. (1990). Unusual coenzymes of methanogenesis. *Annual Review of Biochemistry, 59*, 355–394.

Hamley, P., Lester, E., Thompson A., Cloke, M., Poliakoff, M. (2001). *The removal of carbon from fly ash using supercritical water oxidation*. International Ash Utilization Symposium, Center for Applied Energy Research, University of Kentucky, Paper no: 85. www.flyash.com.

Kowalski, M., & Kopinski, S. (2010). Plasma waste disposal. http://plasmawastedisposal.com. Accessed 2 Sept 2010.

Palmisano, A. C., & Barlaz, M. A. (1998). *Microbiology of solid wastes*. Boca Raton: CRC Press.

Rehm, H.-J., & Reed, G. (2008). Methanogens. *Encyclopedia of Chemical Technology, 7*, 249–264. Wiley.

Thompson, J., & Anthony, H. (Moderators) (2008). *The health effects of waste incinerators* (2nd ed.), 4th Report of the British Society for Ecological Medicine. http://www.ecomed. org.uk/content/IncineratorReport_v3.pdf.

Index

A

Actinomycetes, 63, 118, 183, 219, 231, 292
Activated charcoal, 231, 269, 271, 288
Albedoes, 142–144
Algae, 16–19, 29, 34, 48, 50, 61, 65, 66, 68, 71, 72, 78,
 82–85, 89, 90, 104–106, 118–122, 129, 130,
 133–135, 137, 139, 141–145, 152–156, 182, 183,
 209–210, 215, 217, 218, 221, 231, 250–251, 255,
 256, 261–265
Allochthonous, 17
American Petroleum Institute (API), 175–177
Ammonox, 147
API. *See* American Petroleum Institute
Atmospheric, 5, 6, 9, 70, 111–112, 144, 147
Authochthonous, 17

B

Bacteriophages, 22, 84, 86–98, 101, 102, 104, 153,
 170, 171
Benthic, 17, 118, 120, 125, 127, 137,
 156, 264
Bergey's manual, 57
Biochemical oxygen demand (BOD), 156, 157, 228, 244,
 250–254, 258–260, 262, 267
Biotoxins, 193–200, 211
Blooms, 83, 129, 135, 142, 144–145, 154–156
BOD. *See* Biochemical oxygen demand

C

Caulobacter, 16, 61, 100
Cesspools, 254, 267–268
Chemical oxygen demand (COD), 228, 244,
 253–254, 296
Chemo-autrotrophic, 52, 70, 73, 74, 118, 124
Cholera, 64, 99, 158, 189–194, 201, 211, 222
Classification, 9–10, 47–48, 50–57, 59, 81, 82, 86, 95, 96, 137,
 172, 205, 210–212, 242
Climate, 5, 6, 62, 83, 112, 114, 128, 130, 141–147, 210, 271,
 276, 279–281, 285
COD. *See* Chemical oxygen demand
Cold seeps, 65, 124, 125
Compatible, 141, 142
Composts, 275, 283, 284, 290–292, 301, 302
Crude oil, 74, 154, 173–178, 182, 183, 186, 280
Cryptosporidiosis, 194–195, 202

Cyanobacteria, 18, 28–29, 52, 61, 62, 66, 68–70, 72, 74, 117,
 129, 131–133, 139–141, 144–146, 152, 154–156, 183,
 190, 192, 193, 213
Cyanotoxins, 192
Cytometry, 27–29, 97

D

Dimethyl sulphide (DMS), 72, 135, 142–144, 299
Domains, 48–50, 56, 57, 75, 78, 298

E

Electrodialysis, 269–272
Enteroviruses, 87, 92, 191, 198, 199, 202, 225
Estuaries, 9, 18, 61, 153, 154, 209, 210, 271
Eutrophic, 18
Eutrophication, 18, 118–121, 151–186

F

Faecal, 292
Flavobacterium, 118, 120, 183, 256, 264
Floc, 152, 219–220, 255–257
Fungi, 47, 48, 52, 73, 74, 78, 80–82, 84, 90–92, 95, 105,
 118–120, 133, 134, 137, 156, 183, 208, 256, 261, 292

G

Gasification, 287–288
Geomembrane, 296
Ground water, 9, 10, 71, 75, 114–116, 268, 292, 293, 295, 296

H

Hybridization, 38, 51, 55, 56, 169
Hydrothermal, 58, 60, 65, 72, 76, 77, 124–126, 133,
 137–138, 298

I

Ice, 4–6, 9, 10, 19, 78, 115, 144, 299
Imhoff, 254, 264, 267, 268
Incineration, 284–290, 297, 301
Indicator, 53, 97, 104, 105, 156–173, 211, 243, 251
Ingoldian, 119
Ion exchange, 217, 230, 232–233, 269, 271–272

N. Okafor, *Environmental Microbiology of Aquatic and Waste Systems*,
DOI 10.1007/978-94-007-1460-1, © Springer Science+Business Media B.V. 2011